Springer Undergraduate Mather

Springer

London
Berlin
Heidelberg
New York
Barcelona
Hong Kong
Milan
Paris
Singapore
Tokyo

Advisory Board

Other books in this series

Mícheál Ó Searcóid

Elements of Abstract Analysis

 Springer

Mícheál Ó Searcóid, PhD
Department of Mathematics, University College Dublin, Belfield, Dublin 4, Ireland

Cover illustration elements reproduced by kind permission of:
Aptech Systems, Inc., Publishers of the GAUSS Mathematical and Statistical System, 23804 S.E. Kent-Kangley Road, Maple Valley, WA 98038, USA. Tel: (206) 432 - 7855 Fax (206) 432 - 7832 email: info@aptech.com URL: www.aptech.com
American Statistical Association: Chance Vol 8 No 1, 1995 article by KS and KW Heiner 'Tree Rings of the Northern Shawangunks' page 32 fig 2
Springer-Verlag: Mathematica in Education and Research Vol 4 Issue 3 1995 article by Roman E Maeder, Beatrice Amrhein and Oliver Gloor 'Illustrated Mathematics: Visualization of Mathematical Objects' page 9 fig 11, originally published as a CD ROM 'Illustrated Mathematics' by TELOS: ISBN 0-387-14222-3, German edition by Birkhauser: ISBN 3-7643-5100-4.
Mathematica in Education and Research Vol 4 Issue 3 1995 article by Richard J Gaylord and Kazume Nishidate 'Traffic Engineering with Cellular Automata' page 35 fig 2. Mathematica in Education and Research Vol 5 Issue 2 1996 article by Michael Trott 'The Implicitization of a Trefoil Knot' page 14.
Mathematica in Education and Research Vol 5 Issue 2 1996 article by Lee de Cola 'Coins, Trees, Bars and Bells: Simulation of the Binomial Process' page 19 fig 3. Mathematica in Education and Research Vol 5 Issue 2 1996 article by Richard Gaylord and Kazume Nishidate 'Contagious Spreading' page 33 fig 1. Mathematica in Education and Research Vol 5 Issue 2 1996 article by Joe Buhler and Stan Wagon 'Secrets of the Madelung Constant' page 50 fig 1.

British Library Cataloguing in Publication Data
Ó Searcóid, Mícheál
    Elements of abstract analysis. - (Springer undergraduate
    mathematics series)
    1. Mathematical analysis
    I. Title
    515
ISBN 185233424X

Library of Congress Cataloging-in-Publication Data
Ó Searcóid, Mícheál, 1948-
    Elements of abstract analysis / Mícheál Ó Searcóid.
        p.   cm. -- (Springer undergraduate mathematics series, ISSN 1615-2085)
    Includes bibliographical references and index.
    ISBN 1-85233-424-X (alk. paper)
    1. Functional analysis. I. Title. II. Series.
QA320.O74   2001
515'.7--dc21   2001042665

Springer Undergraduate Mathematics Series ISSN 1615-2085
ISBN 1-85233-424-X Springer-Verlag London Berlin Heidelberg
a member of BertelsmannSpringer Science+Business Media GmbH
http://www.springer.co.uk

Typesetting: Camera ready by author
Printed and bound at the Athenæum Press Ltd., Gateshead, Tyne & Wear
12/3830-543210 Printed on acid-free paper SPIN 10791475

Hail to thee, lady! and the grace of heaven,
Before, behind thee, and on every hand,
Enwheel thee round!   *Othello, II,i.*

Tiomnaím an leabar seo
faoi ċoimirce na Maiġoine Muire
i noil-ċuiṁne ar mo ṁáċair
a o'imiġ ar ṡlí na fírinne
12.2.2000

Críost roṁat, Críost i oo ḃiaiḃ;
Críost ar ḃeis agus Críost ar ċlé;
Críost i oo ċimpeall le saol na saol.

Preface

In nature's infinite book of secrecy
A little I can read.   Antony and Cleopatra, I,ii.

This is a book about a few elementary concepts of analysis and the mathematical structures which enfold them. It is more concerned with the interplay amongst these concepts than with their many applications.

The book is self-contained; in the first chapter, after acknowledging the fundamental rôle of mathematical logic, we present seven axioms of Set Theory; everything else is developed from these axioms. It would therefore be true, if misleading, to say that the reader requires no prior knowledge of mathematics. In reality, the reader we have in mind has that level of sophistication achieved in about three years of undergraduate study of mathematics and is already well acquainted with most of the structures discussed—rings, linear spaces, metric spaces, and so on—and with many of the principal analytical concepts—convergence, connectedness, continuity, compactness and completeness. Indeed, it is only after gaining familiarity with these concepts and their applications that it is possible to appreciate their place within a broad framework of set-based mathematics and to consolidate an understanding of them in such a framework. To aid in these pursuits, we present our reader with things familiar and things new side by side in most parts of the book—and we sometimes adopt an unusual perspective. That this is not an analysis textbook is clear from its many omissions. It is rather an essay on the foundations of modern analysis, and it is hoped that mastery of its contents will prove a sound basis for pleasurable study of more advanced areas of Functional Analysis and Operator Theory. But our primary aim has been to stimulate in the reader delight at the very beautiful arrangement of abstract mathematics which has been conceived and nurtured by great mathematicians of the last hundred years or so.

Mathematics is a discipline which abounds with insights to be savoured and wondered at; and the perspective we have tried to maintain throughout the book is one of *Mathematics as Art*. Like music, drama and literature, it is to be enjoyed in all its humours and profundity. But the mathematics being presented is not only art; it is also a firm basis for science—for beginning to understand the miracles of creation. It is an awesome thought that, through the achievements of modern mathematics—with its identification of important concepts and its highly developed language and concise notation—quite ordinary mathematicians can see further and delve deeper than the great Newton could ever have imagined. One important motif, which weaves its way like a golden thread from start to finish of the book, is the Axiom of Choice in a variety of guises. The quotations are all from plays of William Shakespeare.

In mathematics, as in nature, small variations from general patterns are a source of interest and vitality—and their recognition brings new levels of complexity to the understanding of those patterns. Intuition depends on a knowledge of patterns and is blind to unnoticed variations. So, although well informed intuition in mathematics is often correct, it can be spectacularly erroneous. It must always be confirmed by proof. The student is therefore advised to read slowly through proofs, filling gaps where necessary, and to develop a habit of checking in particular those statements which the author says are obviously true. Definitions should be read with care, as different authors may use the same name for different objects or different names for the same object. For clarity, we never use the word *contain* to refer to set inclusion: a set *contains* its members, but *includes* its subsets. More often than not, use of a *slanted* typeface indicates that the term in question has not yet been defined or is to remain undefined in the book. Most of the notation we use is either standard or has been used by other authors; but the reader will find a few innovations.

My thanks are due to the SUMS advisors for their helpful comments and to the staff of Springer-Verlag in London—especially to Karen Borthwick who has always dealt promptly and efficiently with my enquiries. I am deeply indebted to those colleagues who have read and commented on parts of the typescript—to Thomas Unger, Thomas J. Laffey and Oren Kolman; to Mary Hanley who noticed errors in early drafts; to Robin E. Harte who made various contributions to an early draft; to Karim Zahidi whose reading of most of the chapters has led to many fruitful discussions and corrections; and to Rod Gow who has examined much of the material with meticulous care and has eliminated several errors. I shall be similarly grateful to readers who advise me, through my website or otherwise, of errors which yet remain. I wish also to thank my wife Máire and my family Ciar, Eoghan, Oisín, Aoife and Colm for their great patience with me while I was engaged in the writing and rewriting of this book.

Mícheál Ó Searcóid
University College Dublin
30 May 2001
`http://mathsa.ucd.ie/~mos`

Contents

Let us from point to point this story know,
To make the even truth in pleasure flow.   *All's Well that Ends Well, V,iii.*

List of Symbols

1
Sets

The discovery of the AXIOMATIC METHOD and its use in the development of geometry as a deductive science—an achievement which stands without parallel in the history of mathematics—belong to antiquity. But that discovery is the foundation stone of modern mathematics; and the development of geometry, expounded in the *Elements* of Euclid of Alexandria some three hundred years before the birth of Christ, is the primary model for the development of mathematics along axiomatic lines.

We employ the axiomatic method when we state clearly the assumptions we are making, describe precisely the rules of deduction we shall permit, and then proceed by means of those rules to make deductions from our assumptions. The totality of assumptions we make is called an AXIOM SYSTEM.

The success of the axiomatic method applied to any particular theory depends on the *consistency* of the axiom system—we should like an assurance that contradictory results cannot be deduced from it—and, to a lesser extent, on its *completeness*, in that we might hope for an assurance that the truth or falsehood of all *propositions* in the theory can, at least theoretically, be decided by it. It also depends on the precision of definitions.

The concept of *proof* is at the heart of mathematics. From time to time in its history, gaps have been discovered in arguments, hidden assumptions exposed and, more seriously, the validity of procedures disputed. Mathematicians have responded by applying the axiomatic method rigorously to the foundations of mathematics. The result of their endeavours is that a great part of modern mathematics is axiomatized, being presented as a discipline which deals with *sets* exhibiting certain types of structure determined by precise axioms. Under-

1

lying all of this mathematics is the Theory of Sets which is itself an axiomatic theory of mathematical logic.

Because Set Theory is fundamental to modern mathematics, some knowledge of it is essential for all serious students of mathematics. In this chapter, we shall present its axioms in a less formal way than mathematical logic would demand. We shall show how *relations* and *functions* can be presented within the theory. We shall introduce some notation, give some formal definitions and establish some conventions; and we shall record some basic results which will subsequently be used freely.

We shall see that Set Theory contains some perhaps unexpected innate difficulties and that mathematicians may be faced with choices which have far reaching implications for the universe described by their mathematics; in particular, we shall introduce the Axiom of Choice which, as we shall show throughout the subsequent chapters, plays a vital rôle in the development of modern analysis.

1.1 Set Theory

Set Theory might attempt to formalize the intuitive idea that a set is a *class* of objects which are its members. It soon becomes clear, however, that some restriction is necessary. Even if we confine our research to classes consisting of members which satisfy some well defined property which is formalizable in a sense determined by the underlying logical theory, we cannot avoid the difficulty demonstrated by Bertrand Russell's famous question: does the class of all those sets which are not members of themselves have itself as a member or not? RUSSELL'S PARADOX is that if this class were to be regarded as a set, then it would be a member of itself if and only if it were not a member of itself. The reader who has not encountered this before might like to pause here to check that this observation is correct.

Russell's Paradox exposes a contradiction at the very foundation of intuitive set theory which we cannot afford to ignore. Evidently, if Set Theory—and the mathematics depending on it—is to be free of contradiction, it is necessary to circumscribe very carefully the notion of a set. One way of doing this is to regard some accumulations of objects as being too 'big' (in a sense which can be properly formulated) to be called sets. Following this line, we postulate the existence as sets of only those objects without which we could not reasonably expect to present any foundation for modern mathematics. From now on the term SET, and its synonym COLLECTION, will be used only in the restricted sense determined by the theory we are going to develop. When we wish to refer to an accumulation of objects determined by some well defined and formalizable property we shall call it a CLASS; such a class may or may not be a set. A MEMBER of a set will variously be called an ELEMENT or a POINT.

Logical Foundation

There are several different ways in which the Theory of Sets can be axiomatized. We shall describe informally that of Zermelo, Fraenkel and Skolem, usually denoted by ZF. It is based on an axiomatic theory of mathematical logic in which a few self-evident truths, called LOGICAL AXIOMS, are represented by finite sequences of symbols; one infers from these, in a finite number of deductive steps, using a finite number of simple rules of deduction, other finite sequences of symbols which are called THEOREMS. The whole process is independent of the intended meaning of the symbols. This underlying axiomatic theory is known as a PREDICATE CALCULUS; it can be shown to be both consistent and complete.

ZF grafts onto the Predicate Calculus a single primitive symbol $\in$ and postulates a number of new axioms in which $\in$ appears; they are called PROPER AXIOMS to distinguish them from the logical axioms. The basic assumptions which mathematicians have thought it reasonable to take for granted are mirrored by these proper axioms—six in number in our presentation below; they are given in an order which suits the mathematical development rather than the demands of the logical theory. These axioms are joined later in the chapter by the Axiom of Choice to produce an augmented set theory denoted by ZFC. In the logical theory, these axioms are simply finite sequences of symbols which are used together with the logical axioms, in the way already described, in order to make deductions and prove theorems. The validity of arguments is independent of the interpretation of the symbols.

Deductive steps in our arguments, usually introduced by words like *therefore, so, whence*, and so on, correspond to one or more applications of the rules of deduction which are stated precisely within the logical theory. Some standard techniques of proof, such as PROOF BY CONTRADICTION, are validated within the Predicate Calculus; PROOF BY INDUCTION is not, and we must develop it within Set Theory.

ZFC does not fulfil all our hopes; Gödel has shown that ZF cannot provide a proof of its own consistency. So we must tacitly assume that no contradictory result can emerge in ZF; it is then provable that none occurs in ZFC either. The cautious reader might take consolation in the fact that proofs of consistency of ZF are available in other systems.

Neither is ZFC complete. Although its scope is wide enough to formalize the questions which are of most interest to us in modern mathematics, there are amongst those questions some which it is incapable of answering. Gödel has shown that, in any formal theory sufficiently complicated to give us arithmetic, there will always be propositions which are *independent* of the axiom system—propositions which can be neither proved nor disproved within it. In short, no such theory encompasses the whole of mathematics. But the richness of the theory which does ensue from this foundation is beyond dispute.

Notation

Even in intuitive set theory, members of sets may themselves be sets, so that a particular object may be regarded as a set or as a member, depending on the context. In ZFC, and within the mathematics that emerges from it, every object of the theory is a set, even if it is being regarded only as a member of a set. Indeed, all objects of interest, including *functions*, individual *natural numbers*, *real numbers* and so on, are presented as sets. It is therefore unnecessary to provide different axioms for objects which might otherwise be perceived as being of different types. Moreover, it is tacitly assumed when basing set theory on the Predicate Calculus that at least one such object exists.

We shall employ the following logical symbols: the UNIVERSAL QUANTIFIER $\forall$ and the EXISTENTIAL QUANTIFIER $\exists$ which we shall interpret as meaning *for all* and *there exists* respectively; $\exists!$ which will mean *there exists exactly one*; and the connectives $\Rightarrow$ and $\Leftrightarrow$, which we interpret as meaning *implies* and *if and only if* respectively.

We shall use symbols from various alphabets to denote sets and the symbol $\in$, already mentioned, to denote membership. Statements of the same type will often be coalesced. Such abbreviation is always conjunctive; thus $a, b \in x$ means ($a \in x$ and $b \in x$) and $a \in b \in c$ means ($a \in b$ and $b \in c$). Where it appears helpful to do so, we shall use lower case, upper case, script and Fraktur typefaces to indicate a hierarchy of membership, as in $a \in A \in \mathcal{A} \in \mathfrak{A}$. We shall assume that the notion of equality has been incorporated into the Predicate Calculus and use the symbol $=$ to indicate that sets are identical. Negations will be provided for in the usual mathematical way: $x \neq y$ means that x and y are DISTINCT SETS; $x \notin y$ means that the set x is not a member of the set y. Parentheses will be used in the usual way to ward off ambiguity; some standard notation requires their use as delimiters, but they will sometimes be omitted if such omission increases legibility without introducing ambiguity. Other symbols will be used as indicated in the text.

In our discussion of Set Theory, we shall use notation of the type $\phi(x)$ or $\phi(x, y)$ to indicate a LOGICAL CONDITION on x, or on x and y (where the order may matter), as appropriate. Such conditions are properties which objects x or x and y of a logical theory may or may not have; a simple example of a condition on sets is $x \in y$. A condition $\phi(x, y)$ will be called a FUNCTIONAL CONDITION if and only if, for each object x of the theory, there is at most one object y for which $\phi(x, y)$ holds. Despite the informality attending its presentation here, the term *condition* is not vague; it cannot be applied to every possible property that we might conceive, but only to those properties which are precisely formalized in the logical theory. A condition $\phi(x)$ on sets determines a class, namely the class of all sets which satisfy it; the point which has been made by Russell's Paradox is that the class determined by the condition $x \notin x$ is not a set.

Choice

The question of choice will arise later in our exposition, but its most elementary case should be dispensed with now. Well defined objects of a logical theory can be denoted by particular symbols; but is the same true for objects which are not well defined? Specifically, if there exists a set which satisfies a particular condition, is it valid to assign to some such object a previously unused symbol? If one particular object satisfying the condition can be properly defined, as, say, the only such object, then we certainly can. But the question involves a little subtlety because the phrase *there exists* in mathematical logic, and in mathematics generally, is intrinsically non-specific: *there exists an object which satisfies ...* is the negation of *for every object, that object does not satisfy ....* It is possible that there exist objects which satisfy a given condition, no one of which is exactly specified—perhaps because such specification is not possible.

Despite this consideration, assignment of a symbol within a proof is justified by the fact that, within the Predicate Calculus, any proposition which can be proved using such an assignment can be proved without it. The implication of this for Set Theory is that, given a condition $\phi(x)$ which is known to be satisfied by some set, it is always valid within a proof to say *let z be a set which satisfies the condition* or *choose a set z for which $\phi(z)$*, provided the symbol z is not used ambiguously. This is true even if it is impossible to construct such a z or to distinguish one from other sets with the same property. In fact, it is common practice in mathematics to collapse the two, technically different, statements *there exists a set x such that $\phi(x)$* (where x is variable) and *let z be a set for which $\phi(z)$* (where z is fixed), writing perhaps *there exists z such that $\phi(z)$*, and to proceed to use z as a label for a fixed but undetermined chosen set with the specified property. We shall follow this practice.

It happens often that we want to adopt labelling which applies to an infinite number of objects which, though they cannot be determined, can each separately be validly labelled in the way we have described. Neither the Predicate Calculus nor ZF Set Theory provides the justification for doing so. This undesirable situation will be remedied by postulating the Axiom of Choice.

The Axioms

Axiom I AXIOM OF EXTENSIONALITY
For all sets a and b, if a and b have precisely the same members, then $a = b$.

The Axiom of Extensionality tells us that each set is fully and uniquely determined by its elements; the term *extensional* is a technical term which refers to the property described by the axiom itself. Sets, however they are defined, are identical if they have the same members; in particular, the order in which the members are specified is irrelevant, as also is the frequency of their specification. For example, the set whose members are a and b is the same as the set whose members are b, a, a and a. Axiom I suggests that we might safely adopt the familiar practice of declaring a set by listing its members in between braces; so the set referred to above can be presented as $\{a, b\}$ or as $\{b, a\}$; if $a = b$ then it will more usually be presented as $\{a\}$ or as $\{b\}$.

Notation sometimes hides pitfalls. If x is a set whose members are specified as, say, a, b and c, then we use the notation $\{a, b, c\}$ for the set x. But consider the situation in which we are given sets a, b and c; it is easy to write down $\{a\}$ or $\{a, b\}$ or $\{a, b, c\}$ and these look like sets. But do such sets exist? Indeed they do, but we are not yet able to prove it; the proof for any particular case requires repeated use of 1.1.5 and Axiom IV, each given below.

Axiom II AXIOM OF POWER SETS
For each set a, there exists a set whose members are precisely those sets x which satisfy the condition that every member of x is a member of a. It is unique by Axiom I and is called the POWER SET of a; it will be denoted by $\mathcal{P}(a)$.

Definition 1.1.1
Suppose a and b are sets and $a \in \mathcal{P}(b)$. Then a will be called a SUBSET of b and b will be called a SUPERSET of a; we shall write $a \subseteq b$ and say that b INCLUDES a and that a IS INCLUDED IN b. If also $a \neq b$, then we shall write $a \subset b$ and say that a is a PROPER SUBSET of b and that b is a PROPER SUPERSET of a. Thus every set is a subset of itself and no set is a proper subset of itself. But we do not yet know that any set has a proper subset.

Axiom III AXIOM OF REPLACEMENT
If $\phi(x, y)$ is a functional condition on sets, then, for each set a, there exists a set b whose members are precisely those sets y for which there exists $x \in a$ for which $\phi(x, y)$ holds.

Axiom III involves a non-specific condition and is therefore more properly called an AXIOM SCHEMA; it represents a whole raft of different axioms got by varying the condition. This axiom has its origin in the intuitive idea of a function. A functional condition $\phi(x, y)$ associates with each set x either nothing at all or one specific set, namely the unique y such that $\phi(x, y)$ holds. This looks rather like the action of a *function* whose *domain* consists of those x for which such a y exists. What the axiom states is that, if we restrict this 'function' to a 'domain' a which is known to be a set, then the 'range' is also a set. This cannot be proved from the other axioms of ZFC, but it is not an unreasonable assertion, since the 'range' cannot be bigger in size than the 'domain', in the intuitive sense of pairing.

It is more proper in ZF to precede Axiom III by two other propositions which can be proved using it and are presented below as the Subset Principle and the Pairing Principle. But the Axiom of Replacement is stronger than these two principles put together and we shall need its greater power later on.

Theorem 1.1.2 SUBSET PRINCIPLE

If $\phi(x)$ is a condition on sets, then, for each set a, there exists a unique set whose members are precisely those members z of a for which $\phi(z)$ holds.

Proof

Suppose a is a set and $\psi(x, z)$ is the condition ($x = z$ and $\phi(z)$). This is clearly a functional condition. By Axiom III, there exists a set b whose members are those sets z which satisfy $\psi(x, z)$ for some $x \in a$; these are precisely the members z of a which satisfy $\phi(z)$. Moreover, b is unique by Axiom I. □

Russell's Paradox forced us to abandon the intuitive idea that a set is a class whose members are those objects which satisfy some well defined property. This intuitive idea has re-emerged in the Subset Principle. But there are two important restrictions. The first is that the members must belong to some previously known set and the second is that the well defined property must be a formal condition of the underlying logical theory. The second restriction we shall pass over lightly, since all the properties we are likely to consider can be properly formalized; but we cannot afford to disregard the first restriction.

It is tempting to believe that, given a set a, every subset of a is determined by some condition. But the Subset Principle leaves open the question of whether or not there exist subsets which are not so determined. We shall see later that the Axiom of Choice postulates that such subsets do exist.

The method of denoting sets by listing their members is not always practical or possible and the Subset Principle suggests that a more flexible way of denoting many sets is by those properties which identify their members. Axiom I ensures that the set consisting of all members x of a which satisfy the condition $\phi(x)$ is well defined by that property; it will normally be written $\{x \in a \mid \phi(x)\}$. Sometimes we shall be less specific and write $\{x \mid \phi(x)\}$ for the set of all sets x which satisfy $\phi(x)$, but, mindful of Russell's Paradox, we shall do this only where it can be established that there exists a set a to which every such x belongs; where we adopt this loose convention, the reader is encouraged to make the mental check that some such set a does indeed exist. Sometimes several conditions will be indicated; they might be separated by an appropriate phrase or by punctuation; in all cases, the intended meaning should be clear. If a is a set and $\phi(x)$ is a condition on sets, then we shall adopt the notation $\{x \subseteq a \mid \phi(x)\}$ as an abbreviation for $\{x \in \mathcal{P}(a) \mid \phi(x)\}$; it requires both Axiom II and the Subset Principle to establish that this is a set.

Definition 1.1.3

Suppose a and b are sets. We define the SET DIFFERENCE $a \backslash b$ to be the set $\{x \in a \mid x \notin b\}$; this is a set by the Subset Principle. If $b \subseteq a$, the set difference $a \backslash b$ is also called the SET COMPLEMENT of b in a; it may sometimes be denoted by b^c when the intended superset is clear from the context.

Theorem 1.1.4

There exists exactly one set which has no members; it is a subset of every set. There is no set which has every set as a member.

Proof

We have already tacitly assumed (1.1) that there is at least one set. Suppose a is a set. Then the set $a \backslash a$ clearly has no members; Axiom I tells us that this empty set is unique, and the definition of subsets ensures that it is a subset of every set. Towards the second assertion, suppose z is a set and let $y = \{x \in z \mid x \notin x\}$, which is a set by the Subset Principle. If we suppose that $y \in z$, then we get the contradiction $(y \in y \Leftrightarrow y \notin y)$; so $y \notin z$. □

The fact that there is no set of all sets implies that, given any set a, there exists a set x which satisfies the condition $x \notin a$; it is therefore always valid within a proof to let z be a set for which $z \notin a$,

The unique set with no members will be called the EMPTY SET and will be denoted by $\varnothing$. The only subset of $\varnothing$ is $\varnothing$, so that $\{\varnothing\}$ is a set by Axiom II; it is $\mathcal{P}(\varnothing)$. Moreover the only subsets of this set are $\varnothing$ and itself, so that $\{\varnothing, \{\varnothing\}\}$ is the set $\mathcal{P}\mathcal{P}(\varnothing)$ by Axiom II. These two sets are used below to give us the Pairing Principle.

Theorem 1.1.5 PAIRING PRINCIPLE

Suppose a and b are sets. Then there is a unique set whose members are just a and b.

Proof

Let $\phi(x, y)$ be the condition $((x = \varnothing$ and $y = a)$ or $(x = \{\varnothing\}$ and $y = b))$; this is a functional condition because $\{\varnothing\} \neq \varnothing$ by Axiom I. By applying Axiom III to the set $\mathcal{PP}(\varnothing)$, we infer that there exists a set whose members are precisely those sets y for which $y = a$ or $y = b$. It is unique by Axiom I. $\qquad\square$

Example 1.1.6

Suppose a is a set. Apply the Pairing Principle with $b = a$. The result is that there exists a set whose only member is a. Such a set is called a SINGLETON SET. The fact that every set a spawns a set $\{a\}$ allows us to identify sets by those sets to which they belong rather than by their members. Specifically, if a and b are sets and if, for every set x, we have $a \in x \Rightarrow b \in x$, then it follows, because one such x is $\{a\}$, that $b \in \{a\}$ and hence, by Axiom I, that $b = a$.

Example 1.1.7

Suppose a is a set. Two applications of the Pairing Principle establish that $\{a, \{a\}\}$ is a set.

Axiom IV AXIOM OF UNIONS

For each set a, there exists a set whose members are precisely the members of the members of a. This set is unique by Axiom I; it will be denoted by $\bigcup a$ and called the UNION of the members of a.

Definition 1.1.8

Suppose a is a set. We define the INTERSECTION of the members of a, denoted by $\bigcap a$, to be $\{x \in \bigcup a \mid \forall z \in a, \, x \in z\}$, which is a set by Axiom IV and the Subset Principle.

The TRIVIAL UNION $\bigcup \varnothing$ is clearly $\varnothing$ and the TRIVIAL INTERSECTION $\bigcap \varnothing$ is then also $\varnothing$. Moreover, $\bigcup\{\varnothing\} = \bigcap\{\varnothing\} = \varnothing$. (It is more usual in ZF to refrain from defining $\bigcap \varnothing$ and to define $\bigcap a$, where $a \neq \varnothing$, as $\{x \in y \mid \forall z \in a, \, x \in z\}$ where y is a member of a, first showing that the particular choice of $y \in a$ is irrelevant. In some systems, $\bigcap a$ is defined by the same formula, except that y is some universal set; in such a system, $\bigcap \varnothing = y$.)

Notation 1.1.9

If $x = \{a, b\}$, then we shall habitually denote the union $\bigcup x$ by $a \cup b$ or $b \cup a$ and the intersection $\bigcap x$ by $a \cap b$ or $b \cap a$. This type of notation will be extended to three, four or more sets where appropriate.

Definition 1.1.10

Sets a and b will be said to be DISJOINT SETS if and only if $a \cap b$ is empty. A collection x of sets will be called a DISJOINTED SET and its members will be called MUTUALLY DISJOINT SETS if and only if, for each $a, b \in x$, we have $a \cap b = \varnothing$. A collection of mutually disjoint subsets of a set z whose union is z is called a PARTITION of z.

Example 1.1.11

Suppose a, b and c are distinct sets. Three applications of the Pairing Principle show that $\{\{a\}, \{b, c\}\}$ is a set; and $\{a, b, c\}$ is a set by Axiom IV. Then $\{a\}$ and $\{b, c\}$ are subsets of $\{a, b, c\}$ and $\{\{a\}, \{b, c\}\}$ is a partition of $\{a, b, c\}$.

Definition 1.1.12

Suppose a is a set. We shall say that a is CLOSED UNDER UNION if and only if, for every $x, y \in a$, we have $x \cup y \in a$; we shall say that a is CLOSED UNDER ARBITRARY UNIONS if and only if, for every subset c of a, we have $\bigcup c \in a$. Variations of this terminology will also be used. For example, if, for every non-empty subset c of a, we have $\bigcup c \in a$, then we say that a is CLOSED UNDER NON-TRIVIAL UNIONS. Similar definitions hold for intersection.

Theorem 1.1.13 DE MORGAN'S LAWS

Suppose a and b are sets and b is not empty. Then $a \backslash \bigcap b = \bigcup \{a \backslash y \mid y \in b\}$ and $a \backslash \bigcup b = \bigcap \{a \backslash y \mid y \in b\}$.

Proof

For each $y \in b$, $a \backslash y$ is a member of $\mathcal{P}(a)$ by the Subset Principle, so that $\{a \backslash y \mid y \in b\}$ is indeed a set. Suppose that $x \in a \backslash \bigcap b$. Then $x \in a$ and there exists $z \in b$ such that $x \notin z$. So $x \in a \backslash z$, whence $x \in \bigcup \{a \backslash y \mid y \in b\}$. It follows that $a \backslash \bigcap b \subseteq \bigcup \{a \backslash y \mid y \in b\}$. Conversely, suppose $x \in \bigcup \{a \backslash y \mid y \in b\}$; then there exists $z \in b$ such that $x \in a \backslash z$. So $x \in a$ and $x \notin z$, whence also $x \notin \bigcap b$. Therefore $x \in a \backslash \bigcap b$. It then follows that $\bigcup \{a \backslash y \mid y \in b\} \subseteq a \backslash \bigcap b$. So $a \backslash \bigcap b = \bigcup \{a \backslash y \mid y \in b\}$ by Axiom I. This proves the first assertion; proof of the second is left as an exercise (Q 1.1.2). □

Axiom V AXIOM OF INFINITY

There exists a set which has $\varnothing$ as a member and has $x \cup \{x\}$ as a member whenever x is a member.

The Axiom of Infinity is the mathematician's passport to an infinite universe. Although it is possible to define *finite sets* and *infinite sets* without this axiom, the definitions are deficient because it cannot be proved that any infinite set exists. The axiom is designed to ensure that all *natural numbers* can be gathered together in a set; but it also ensures, perhaps surprisingly, that *real numbers* can be presented as sets and that they too form a set. It thus opens the way for the development of mathematical analysis.

Axiom VI AXIOM OF FOUNDATION

Every non-empty set has a member disjoint from itself.

Axioms II to V are all concerned with making new sets out of old ones. Unlike those axioms, Axiom VI does not allow any more objects to be called sets, but rather restricts the application of the term *set*. An immediate consequence of Axiom VI is that no set can be a member of itself (1.1.14); it also follows that looped memberships of the type $a \in b \in c \in d \in a$ never occur in ZF (see Q 1.1.2). This axiom is not absolutely necessary for mathematics; its main purpose is to tidy up Set Theory itself. In this book, we shall use it only once outside this section (after 12.1.12) and could easily have arranged matters slightly differently to avoid its use there. The attitude to the Axiom of Foundation expressed in [3] is that, if at some stage the axiom impedes significant mathematical research, then it may be discarded.

Theorem 1.1.14

Suppose a is a set. Then $a \notin a$.

Proof

By 1.1.6, $\{a\}$ is a set. By Axiom VI, $\{a\}$ has a member disjoint from itself. So $a \cap \{a\} = \varnothing$, and therefore $a \notin a$. $\qquad\square$

EXERCISES

Q 1.1.1 Suppose a, b, c and d are sets. Show that there exists a set $\{a, b, c, d\}$ whose members are precisely a, b, c and d.

Q 1.1.2 Suppose a and b are sets and $b \neq \varnothing$. Prove $a \backslash \bigcup b = \bigcap \{a \backslash y \mid y \in b\}$.

1.2 Relations and Functions

Sets are unordered; this is stipulated by Axiom I. But mathematics needs ordering, and the *ordered pair* is the means of introducing it into set-based mathematics. A little ingenuity is used in defining an ordered pair as a set: specifically, if a and b are sets, then three applications of the Pairing Principle establish that $\{\{a\}, \{a, b\}\}$ is also a set, and it is this set which is called the ordered pair (a, b). Sets of ordered pairs are called *relations* and relations with a certain uniqueness property are called *functions*.

Ordered Pairs

Definition 1.2.1

Suppose a, b, c, and d are sets. We define the ORDERED PAIR (a, b) to be the set $\{\{a\}, \{a, b\}\}$. We define the ORDERED TRIPLE (a, b, c) to be $((a, b), c)$ and the ORDERED QUADRUPLE (a, b, c, d) to be $((a, b, c), d)$.

Theorem 1.2.2

Suppose a, b, c and d are sets. Then $(a, b) = (c, d) \Leftrightarrow (a = c$ and $b = d)$.

Proof

The backward implication holds by Axiom I. So suppose that $(a, b) = (c, d)$, that is $\{\{a\}, \{a, b\}\} = \{\{c\}, \{c, d\}\}$. By Axiom I, we have either $\{a\} = \{c\}$ or $\{a\} = \{c, d\}$ and $a = c$ in either case. So $\{a\} = \{c\}$ and, since $(a, b) = (c, d)$, Axiom I yields $\{a, b\} = \{c, d\}$ and hence also $b = d$.   □

Definition 1.2.3

Suppose a and b are sets. We define the CARTESIAN PRODUCT $a \times b$ to be $\{(x, y) \mid x \in a, y \in b\}$. Each (x, y) with $x \in a$ and $y \in b$ is a member of the set $\mathcal{PP}(a \cup b)$, so that $a \times b$ is a set by the Subset Principle.

Relations

Definition 1.2.4

A set is called a RELATION if and only if all its members are ordered pairs. Suggestive notation, such as $\sim$, will often be used to denote a relation, and we shall usually write $x \sim y$ instead of $(x, y) \in \sim$.

Example 1.2.5

Suppose a and b are sets. Then $a \times b$ and $b \times a$ are relations. Except in trivial cases, they are not the same. But relations need not be Cartesian products; $\{(a,a),(b,b)\}$ is a relation which is not so unless $a = b$.

Example 1.2.6

Suppose a, b, c, d and e are sets and $r = \{(a,b),(b,c),(d,e),(e,a)\}$. A simple calculation reveals that $\bigcup r = \{\{a\}, \{a,b\}, \{b\}, \{b,c\}, \{d\}, \{d,e\}, \{e\}, \{e,a\}\}$ and hence that $\bigcup\bigcup r = \{a,b,c,d,e\}$. Generally, suppose r is a relation; then the members of the set $\bigcup r$ are all sets $\{x\}$ and all sets $\{x,y\}$ where $(x,y) \in r$, and $\bigcup\bigcup r$ is the set whose members are the members of such sets. In other words $\bigcup\bigcup r$ is the set whose members are all x and y such that $(x,y) \in r$.

Definition 1.2.7

Suppose r is a relation. The FIELD of r, denoted by field(r), is defined to be $\bigcup\bigcup r$; this is a set by Axiom IV. It has two important subsets, namely the DOMAIN $\{x \in \text{field}(r) \mid \exists y : (x,y) \in r\}$ of r, denoted by dom(r), and the RANGE $\{y \in \text{field}(r) \mid \exists x : (x,y) \in r\}$ of r, denoted by ran(r). It is easy to check that field$(r) = \text{dom}(r) \cup \text{ran}(r)$ and that $r \subseteq \text{dom}(r) \times \text{ran}(r)$. If A is any superset of field(r), then r may be described as a relation ON A. The relation r is said to be ONE-TO-ONE or INJECTIVE if and only if, for each y in ran(r), there is precisely one x in dom(r) such that $(x,y) \in r$.

Definition 1.2.8

Suppose A is a set and r is a relation on A. We say that r is

- REFLEXIVE ON A if and only if $(a,a) \in r$ for each $a \in A$;
- ANTI-REFLEXIVE ON A if and only if, for each $a \in A$, $(a,a) \notin r$;
- SYMMETRIC if and only if, for each $(a,b) \in r$, we have also $(b,a) \in r$;
- ANTI-SYMMETRIC if and only if, for all $(a,b) \in r$, we have $(b,a) \notin r$;
- TRANSITIVE if and only if, for all $a,b,c \in A$, we have $(a,c) \in r$ whenever both $(a,b) \in r$ and $(b,c) \in r$.

Equivalence Relations

Definition 1.2.9

Suppose A is a set. A relation on A which is reflexive, symmetric and transitive is called an EQUIVALENCE RELATION on A. If $\sim$ is an equivalence relation on A, then A can be split into EQUIVALENCE CLASSES, two elements $a, b \in A$

belonging to the same class if and only if $a \sim b$. Note that each equivalence class $\{b \in A \mid b \sim a\}$ is a set, despite the fact that we refer to it as a class. The collection of all such equivalence classes is a partition of A. It may be denoted by $A/\sim$, and is called the QUOTIENT of A by $\sim$.

Example 1.2.10

If x is a set, then $\{(a,a) \mid a \in x\}$ is an equivalence relation on x; we call it EQUALITY and, confident that there will be no confusion, denote it by $=$. Let $A = \{a, b, c, d\}$; then $A/=$ is $\{\{a\}, \{b\}, \{c\}, \{d\}\}$.

Functions

Here we establish the terms and notation we shall use concerning *functions*.

Definition 1.2.11

A set f is called a FUNCTION if and only if f is a relation and, for each $x \in \text{dom}(f)$, there exists a unique set y such that $(x, y) \in f$. The unique y for which $(x, y) \in f$ is called the IMAGE of x under f or the VALUE of f at x, and is usually denoted by $f(x)$; we say that f MAPS x to y. A function f is called a CONSTANT FUNCTION if and only if its range is a singleton set. If f is an injective function, then $\{(y, x) \mid (x, y) \in f\}$ is a function; it is denoted by f^{-1} and called the INVERSE of f; this inverse is clearly injective also and has inverse f. If f is a function and X and Y are supersets of $\text{dom}(f)$ and $\text{ran}(f)$ respectively, then we say that (f, X, Y), is a MAPPING OUT OF X into Y; we call Y the CO-DOMAIN of f and denote it by $\text{codom}(f)$; in the case where $X = \text{dom}(f)$, we call (f, X, Y) a MAPPING FROM X into Y and say that f MAPS X into Y. Then (f, X, Y) is said to be SURJECTIVE or ONTO Y if and only if $Y = \text{ran}(f)$; it is said to be BIJECTIVE and called a BIJECTION if and only if f is injective and $X = \text{dom}(f)$ and $Y = \text{ran}(f)$. A bijection from a set X onto itself is called a PERMUTATION of X.

- Adoption of the notation $f(x)$ for the value of f at each $x \in \text{dom}(f)$ involves labelling for a possibly *infinite* collection of objects. But this is quite different from the labelling we alluded to in our discussion on Choice (1.1). Here there is no choice involved; each value is uniquely determined and the labelling is justified in the Predicate Calculus.

- The parentheses used in denoting the value of a function are habitually dispensed with in certain cases, especially where this practice improves clarity without introducing ambiguity; if the domain of a function f is a set of ordered pairs, for example, we write $f(a, b)$ rather than $f((a, b))$.

- Sometimes a function f is expressed by notation like $x \mapsto f(x)$, particularly where $f(x)$ is determined by some formula; thus $x \mapsto (x, x)$ denotes the function $\{(x, (x, x)) \mid x \in X\}$ where the domain X is either stated or is understood from the context.

- For each function f, we have $f = \{(x, f(x)) \mid x \in \mathrm{dom}(f)\}$. Nonetheless, it is more usual in analysis to call this set the GRAPH of f.

- The notation $f: X \rightarrowtail Y$ is used to indicate that (f, X, Y) is a mapping out of X into Y, and $f: X \to Y$ that (f, X, Y) is a mapping from X into Y. We shall habitually employ the terms *function* and *mapping* synonymously and replace either by the term *map*; thus we say simply that f is a function out of or from X into Y and make no reference to the ordered triple (f, X, Y).

- Suppose X and Y are sets. If $f: X \to Y$, then $f \subseteq X \times Y$. By Axiom II and the Subset Principle, $\{f \mid f: X \to Y\}$ is a set. We denote it by Y^X.

- Suppose $f: X \rightarrowtail Y$. If $A \subseteq X$, then the set $\{f(a) \mid a \in A\}$ is often denoted by $f(A)$ and is called the IMAGE of A under f; in particular, $f(X)$ is the range of f. If $B \subseteq Y$, then the set $\{a \in X \mid f(a) \in B\}$ is often denoted by $f^{-1}(B)$ and is called the INVERSE IMAGE of B under f (this does not imply the existence of a function f^{-1}). If $\mathcal{C} \subseteq \mathcal{P}(X)$, the subset $\{f(A) \mid A \in \mathcal{C}\}$ of $\mathcal{P}(Y)$ may be denoted by $f(\mathcal{C})$; similarly, if $\mathcal{D} \subseteq \mathcal{P}(Y)$, the subset $\{f^{-1}(B) \mid B \in \mathcal{D}\}$ of $\mathcal{P}(X)$ may be denoted by $f^{-1}(\mathcal{D})$. In general, this type of notation will be used only if there is no possibility of confusion; in particular, there may exist $a \in X$ for which also $a \subseteq X$ or $a \subseteq \mathcal{P}(X)$; in such cases $f(a)$ will always denote the value of f at a and the extended notation will not be used.

Example 1.2.12

The empty set is a function; its domain and range are both empty; indeed, $\varnothing: \varnothing \to \varnothing$ is bijective. Suppose X is a non-empty set. Then $\varnothing^X$, $X^\varnothing$ and $\varnothing^\varnothing$ have been defined. To see what they are, notice firstly that there is exactly one relation, namely $\varnothing$, whose domain or range is $\varnothing$. This empty relation is a function from $\varnothing$ into X; it is a function from $\varnothing$ into $\varnothing$; but it is not a function from X into $\varnothing$. So $\varnothing^X = \varnothing$, whereas $X^\varnothing = \{\varnothing\}$ and $\varnothing^\varnothing = \{\varnothing\}$.

Example 1.2.13

Suppose X is a set. The mapping from X to X which maps each member of X to itself is called the IDENTITY FUNCTION on X. It will be denoted by ι or by ι_X. It is bijective. Suppose Y is a superset of X. The mapping inc: $X \to Y$ which maps each member of X to itself is called the INCLUSION FUNCTION. The functions ι and inc are identical, but they may have different co-domains.

Example 1.2.14

Suppose X is a set and $\sim$ is an equivalence relation on X. The function
$\pi\colon X \to X/\!\!\sim$ which maps each $x \in X$ to its equivalence class is called the
QUOTIENT MAP associated with $\sim$. We may denote the equivalence class of
$x \in X$ by $x/\!\!\sim$. Then $\pi = \{(x, x/\!\!\sim) \mid x \in X\}$. This quotient map is surjective
onto $X/\!\!\sim$ but is not, except in trivial cases, injective.

Example 1.2.15

Suppose that X is a set and that $\mathcal{F}$ is a collection of functions with domain
X. For each $x \in X$, the set $\{(f, f(x)) \mid f \in \mathcal{F}\}$ is a function. We shall generally
denote this function by $\hat{x}$, its domain $\mathcal{F}$ being understood from the context; it
is called the POINT EVALUATION FUNCTION at x (with respect to $\mathcal{F}$). Notice
that, for each $x \in X$ and $f \in \mathcal{F}$, we have $\hat{x}(f) = f(x)$.

Compositions, Restrictions and Extensions

Definition 1.2.16

Suppose f and g are functions and $\operatorname{ran}(f) \subseteq \operatorname{dom}(g)$. We define the COMPOSI-
TION of g after f to be the function $\{(x, g(f(x))) \mid x \in \operatorname{dom}(f)\}$. This composition is denoted by $g \circ f$.

Example 1.2.17

Suppose f and g are functions and $\operatorname{ran}(f) \subseteq \operatorname{dom}(g)$. If f and g are injective,
then clearly $g \circ f$ is injective. If $\operatorname{ran}(f) = \operatorname{dom}(g)$ then $g \circ f\colon \operatorname{dom}(f) \to \operatorname{ran}(g)$
is surjective. If X and Y are sets and $h\colon X \to Y$ is bijective, then $h \circ h^{-1} = \iota_Y$
and $h^{-1} \circ h = \iota_X$.

Definition 1.2.18

Suppose f is a function and $g \subseteq f$. Then it is clearly true that g itself is a
function; we call g the RESTRICTION of f to the set $\operatorname{dom}(g)$ and we call f an
EXTENSION of g to the set $\operatorname{dom}(f)$. For $A \subseteq \operatorname{dom}(f)$, we write $f|_A$ to denote
the restriction $\{(a, b) \in f \mid a \in A\}$ of f to A. If a co-domain has been specified
for f, then this restriction will be considered to have the same co-domain as
f unless we state otherwise; we shall use the term SURJECTIVE RESTRICTION
if the domain is unchanged and the co-domain is the range of f. If $f(A) \subseteq A$,
then A is said to be INVARIANT under f and the map $f|_A$ with co-domain A is
called the COMPRESSION of f to A.

One-to-One Correspondence

The process of pairing all the members of one set with all the members of another in a one-to-one fashion is possible if and only if there exists a bijective function between the two sets. In this case, the sets are said to be in *one-to-one correspondence* with each other. Although we might like to use this property to define an equivalence relation amongst sets, we cannot do so because there is no set of all sets on which to make such a definition. But 1.2.20 shows that one-to-one correspondence behaves exactly like an equivalence relation.

Definition 1.2.19
Suppose A and B are sets. We say that A is in ONE-TO-ONE CORRESPONDENCE with B or that A is EQUINUMEROUS with B, and write $A \approx B$, if and only if there exists a bijective function from A to B.

Theorem 1.2.20
Suppose A, B and C are sets. Then

- $A \approx A$;
- $A \approx B \Leftrightarrow B \approx A$;
- $(A \approx B$ and $B \approx C) \Rightarrow A \approx C$.

Proof
The function ι_A proves the first assertion. $f\colon A \to B$ is bijective if and only if $f^{-1}\colon B \to A$ is, proving the second. For the third, suppose $f\colon A \to B$ and $g\colon B \to C$ are bijective. Then $g \circ f\colon A \to C$ is bijective (1.2.17). $\qquad\square$

Families

We make use of functions in order to *index* sets. In this process, it is the range of the function which assumes importance; the domain might be changed and the same effect achieved. Such functions are called *families*, and we may make only token distinction between the family and its range. One reason for indexing is that repetition of members is prohibited in sets by Axiom I, and indexing allows us to simulate such repetitions, the same element appearing many times with different indices. For example, if $X = \{x\}$ and $I = \{a, b, c\}$, then $\{(a, x), (b, x), (c, x)\}$ is a function from I to X, and so a family. It represents as well as is possible a set with three *copies* of x as its members.

Definition 1.2.21

Suppose I and X are sets. A function from I to X is called also a FAMILY in X. The set I will then be called the INDEXING SET of the family, the members of I will be called INDICES and the image of $i \in I$, normally written using i as a suffix (as, for example, x_i rather than $x(i)$), will be called a TERM of the family. The range $\{x_i \in X \mid i \in I\}$ of an injective family $x \colon I \to X$ is called an INDEXED SUBSET of X; if the family is bijective, then the range, X itself, is called an INDEXED SET and we say that X is INDEXED by I. If the members of X are sets whose members are of some particular interest, then we might refer to the family as a FAMILY OF SETS; if those members are functions, we might refer to it as a FAMILY OF FUNCTIONS, and so on. The family $x = \{(i, x_i) \in I \times X \mid i \in I\}$ will usually be denoted by $(x_i)_{i \in I}$, or more simply by (x_i). Where an indexing set is a set of ordered pairs, it is usual to leave out parentheses; we write, for example, $x_{j,k}$ rather than $x_{(j,k)}$. If $(X_i)_{i \in I}$ is a family, then $\bigcup(X_i)$ and $\bigcap(X_i)$ will be used as abbreviations for the sets $\bigcup\{X_i \mid i \in I\}$ and $\bigcap\{X_i \mid i \in I\}$ respectively.

Example 1.2.22

For any set X, the identity function ι_X is a bijective family. So X can be indexed by itself.

Products

The Cartesian product does not lend itself well to generalization when an arbitrary collection of sets is involved. Indexing is the standard device used to simulate such a generalization. Firstly, we define the product of an arbitrary family; the indexing of a set (possibly using the set itself as the indexing set) is the creation of a particular type of family, and the product of this family is the desired simulation.

Definition 1.2.23

Suppose $(X_i)_{i \in I}$ is a family of sets. Then $\bigcup(X_i)$ is a set by Axiom IV, whence $(\bigcup(X_i))^I$ is a set (1.2.11) and $\{x \in (\bigcup(X_i))^I \mid \forall j \in I,\ x_j \in X_j\}$ is a set by the Subset Principle. We define the CARTESIAN PRODUCT of $(X_i)_{i \in I}$, denoted by $\prod_{i \in I} X_i$, or more simply by $\prod_i X_i$, to be $\{x \in (\bigcup(X_i))^I \mid \forall j \in I,\ x_j \in X_j\}$. The members of this product are the families $(x_i)_{i \in I}$ which exhibit the property that $x_j \in X_j$ for each $j \in I$. For each $j \in I$, x_j is called the jth CO-ORDINATE of (x_i) and the map $\pi_j \colon \prod_i X_i \to X_j$ given by $(x_i) \mapsto x_j$ is called the NATURAL PROJECTION of the product onto the CO-ORDINATE SET X_j. If the co-ordinate sets are all equal to one set A, then the product $\prod_i A$ is simply A^I.

If any one of the co-ordinate sets defined above is empty then so, clearly, is the product. It is tempting to believe that the converse is also true: that, if no co-ordinate set is empty, then the product is not empty. This, however, cannot be proved in ZF and is an assertion equivalent to the Axiom of Choice (1.5.2).

Example 1.2.24

We may loosely refer to a product $\prod_i X_i$ as the product of the sets X_i rather than of the family, the product being regarded as the generalization of the Cartesian product of two sets. There is, however, a difference, as we illustrate here. Consider the collection $\mathcal{C} = \{A, B\}$ of distinct sets A and B. We index $\mathcal{C}$ by itself. Then $\prod_{i \in \mathcal{C}} \mathcal{C} = \{\{(A, x), (B, y)\} \mid x \in A,\ y \in B\}$. This is not the same set as $A \times B = \{(x, y) \mid x \in A, y \in B\}$, but the difference has little mathematical significance because the bijective map $(x, y) \mapsto \{(A, x), (B, y)\}$ enables us to transfer results proved for one type of product to the other.

Example 1.2.25

Suppose $(X_i)_{i \in I}$ is a family with non-empty product. Then the natural projections are surjective and they have natural bijective restrictions onto the co-ordinate sets. Suppose $z \in \prod_i X_i$ and $j \in I$; let $X_{j,z} = \prod_i S_i$, where $S_i = \{z_i\}$ if $i \neq j$ and $S_j = X_j$. Then $X_{j,z}$ is a subset of $\prod_i X_i$ which may be regarded as a copy of X_j in the product $\prod_i X_i$. It is clear that the restriction $\pi_{j,z}$ of π_j to $X_{j,z}$ is injective; and, for each $a \in X_j$, $y = \{(j, a)\} \cup z \setminus \{(j, z_j)\} \in X_{j,z}$ and $\pi_{j,z}(y) = a$, so that $\pi_{j,z}$ (and hence also π_j) is surjective. Note that surjectivity depends on the non-emptiness of the product; as we stated earlier, this non-emptiness cannot in general be proved in ZF.

Example 1.2.26

Suppose X and Y are sets. Then Y^X is the product $\prod_{x \in X} Y$ of indexed copies of the one set Y. Notice also that, for each $x \in X$ and $f \in Y^X$, we have $\pi_x(f) = f_x = f(x) = \hat{x}(f)$, where $\hat{x}$ denotes the point evaluation function (1.2.15). So the natural projections are precisely the point evaluation functions.

EXERCISES

Q 1.2.1 Suppose X and Y are sets and $f \colon X \to Y$. Suppose $\mathcal{V} \subseteq \mathcal{P}(Y)$ and show that $f^{-1}(\bigcup \mathcal{V}) = \bigcup f^{-1}(\mathcal{V})$ and that $f^{-1}(\bigcap \mathcal{V}) = \bigcap f^{-1}(\mathcal{V})$. Now suppose that $\mathcal{U} \subseteq \mathcal{P}(X)$ and show that $f(\bigcup \mathcal{U}) = \bigcup f(\mathcal{U})$ and that $f(\bigcap \mathcal{U}) \subseteq \bigcap f(\mathcal{U})$, but that this last inclusion may be proper; find a condition on f under which it is not.

Q 1.2.2 Suppose that X and Y are sets, $A \subseteq X$ and $f: X \to Y$. Show that $A \subseteq f^{-1}(f(A))$ with equality if f is injective.

Q 1.2.3 Suppose X and Y are sets, $B \subseteq Y$ and $f: X \to Y$. Show that $f(f^{-1}(B)) = B \cap f(X)$. Deduce that $f(f^{-1}(B)) = B$ if f is surjective.

Q 1.2.4 Suppose $f: Y \to Z$ and $g: X \to Y$ are bijective functions. Show that $f \circ g: X \to Z$ is bijective and that $(f \circ g)^{-1} = g^{-1} \circ f^{-1}$.

Q 1.2.5 Suppose a and b are sets. Show that $a \cap b^a = \varnothing$ and $a \cap \mathcal{P}(b^a) = \varnothing$.

Q 1.2.6 Suppose r is a relation. Show that the INVERSE $\{(a, b) \mid (b, a) \in r\}$ of r is also a relation. Then show that a relation is a function if and only if it is the inverse of an injective relation.

1.3 Ordered Sets

Order Relations

Definition 1.3.1
Suppose S is a set. A relation on S which is both anti-reflexive and transitive is called a PARTIAL ORDER RELATION on S. If $<$ is such a relation on S, then the ordered pair $(S, <)$ is called a PARTIALLY ORDERED SET and we say that S is partially ordered by $<$. In such a case, $x < y$ may be written $y > x$; and $x \leq y$ and $y \geq x$ are abbreviations for $(x < y$ or $x = y)$. Negations follow standard usage: for example, $x \nleq y$ is the negation of $x \leq y$, which is $(x \nless y$ and $x \neq y)$.

Example 1.3.2
Suppose $\mathcal{C}$ is a set. The members of $\mathcal{C}$ are sets which may or may not be subsets one of another and the relation $\{(A, B) \in \mathcal{C} \times \mathcal{C} \mid A \subset B\}$, denoted also by $\subset$, is a partial order relation on $\mathcal{C}$. We express this fact by saying that $(\mathcal{C}, \subset)$ is a partially ordered set or that $\mathcal{C}$ is PARTIALLY ORDERED BY INCLUSION (meaning proper inclusion). The relation $\subset$ may, of course, be empty.

Theorem 1.3.3
Every partial order relation is anti-symmetric.

Proof
Suppose $(S, <)$ is a partially ordered set. Suppose $a, b \in S$ and $a < b$ and $b < a$. Then, since the order relation is transitive, we have $a < a$, which contradicts its anti-reflexivity. So $<$ is anti-symmetric.   □

This book is concerned with sets endowed with various forms of structure, partial ordering being an example. An ordered pair such as $(S, <)$ is used in the definition in order to ensure that the object is a set; and it is always possible to retrieve S as the first member of this pair. But, not wishing to be pedantic, we shall refer to S itself as being a partially ordered set (or *group*, or *vector space*, or otherwise as appropriate). We talk of the members of this partially ordered set, meaning the members of S rather than those of $(S, <)$ (which are $\{S\}$ and $\{S, <\}$). We shall often duplicate notation where we expect no confusion, using the same symbol, perhaps $<$, for more than one relation in the same context.

Definition 1.3.4

Suppose $(S, <)$ is a partially ordered set, $s \in S$ and $A \subseteq S$.

- We define the LOWER SEGMENT of s in S to be $\grave{s} = \{a \in S \mid a < s\}$. We define the UPPER SEGMENT of s in S to be $\acute{s} = \{a \in S \mid s < a\}$.

- s is called a MAXIMAL ELEMENT of S if and only if $\acute{s} = \varnothing$; s is called a MINIMAL ELEMENT of S if and only if $\grave{s} = \varnothing$.

- s is called an UPPER BOUND for A in S if and only if $A \backslash \{s\} \subseteq \grave{s}$; it is called a LOWER BOUND for A in S if and only if $A \backslash \{s\} \subseteq \acute{s}$. A is said to be BOUNDED ABOVE or BOUNDED BELOW in S according as such an upper or lower bound, respectively, exists. We say that A is ORDER-BOUNDED in S if and only if A is both bounded above and bounded below in S.

- If the set of upper bounds for A in S has a minimal element, this is called a LEAST UPPER BOUND for A in S; if there is precisely one such least upper bound it is called the SUPREMUM of A in S and is denoted by $\sup A$. If $\sup A$ exists and is a member of A, it is called the GREATEST MEMBER or MAXIMUM ELEMENT of A and is denoted by $\max A$.

- If the set of lower bounds for A in S has a maximal element, this is called a GREATEST LOWER BOUND for A in S and, if there is just one such element, then it is called the INFIMUM of A in S and is denoted by $\inf A$. If $\inf A$ exists and is a member of A, it is called the LEAST MEMBER or MINIMUM ELEMENT of A and is denoted by $\min A$.

- If $\acute{s}$ has a minimum element, then this is called the IMMEDIATE SUCCESSOR of s in S and will be denoted by s_+. If $\grave{s}$ has a maximum element, then this is called the IMMEDIATE PREDECESSOR of s in S and is denoted it by s_-.

Example 1.3.5

Let $\mathcal{C}$ be a set partially ordered by inclusion (1.3.2). Then $A \subseteq \bigcup \mathcal{C}$ for all $A \in \mathcal{C}$, so that $\bigcup \mathcal{C}$ is an upper bound—indeed, the supremum—for $\mathcal{C}$ in any superset $\mathcal{B}$ of $\mathcal{C}$ for which $\bigcup \mathcal{C} \in \mathcal{B}$. In particular, if $\bigcup \mathcal{C} \in \mathcal{C}$, then $\bigcup \mathcal{C} = \max \mathcal{C}$.

Example 1.3.6

Suppose S is a set. $\mathcal{P}(S)$ is partially ordered by inclusion; provided $S \neq \varnothing$, the order relation is not empty. $\mathcal{P}(S)$ has maximum element S and minimum element $\varnothing$.

Example 1.3.7

Suppose X and S are sets and S is endowed with a partial ordering $<$. Then S^X is endowed with a standard partial ordering given by stating that, for $f, g \in S^X$, $f < g$ if and only if $f(x) < g(x)$ for all $x \in X$.

Definition 1.3.8

Suppose f and g are functions, $X \subseteq \operatorname{dom}(f) \cap \operatorname{dom}(g)$, and $f(X) \cup g(X)$ is endowed with a partial ordering $<$. Then we say that f is DOMINATED by g on X if and only if $f(x) \leq g(x)$ for all $x \in X$.

Notation 1.3.9

Mathematical notation often suppresses information, assuming that the context makes it clear. When using $\sup A$ or $\inf A$, for example, we assume that it is clear from the context to which superset S of A the supremum or infimum belongs; their meaning may indeed depend on which superset is understood. Whenever there is a likelihood of confusion caused by using this or other notation, we shall either dispel it explicitly or augment the notation to make the meaning clear; accordingly, we might adopt the convention of using a subscript or superscript, writing, for example, $\sup_S A$ or $\inf_S A$ instead of $\sup A$ or $\inf A$. In such cases, we shall assume that the reader will take our meaning. Some notation, such as $\dot{s}$, s_+ and s_-, does not lend itself well to such augmentation; where the context does not dispel all ambiguity, we shall use other notation.

Definition 1.3.10

Suppose $(S, <)$ is a partially ordered set. S is said to be

- TOTALLY ORDERED by $<$ if and only if, for every pair of distinct members $x, y \in S$, either $x < y$ or $y < x$;
- DENSELY ORDERED by $<$ if and only if S is totally ordered by $<$ and, for each $x, y \in S$ with $x < y$, there exists $z \in S$ such that $x < z < y$;
- COMPLETELY ORDERED by $<$ if and only if S is totally ordered by $<$ and each non-empty subset of S which has an upper bound in S has a supremum in S;

- WELL ORDERED by $<$ if and only if each non-empty subset of S has a minimum element.

$(S, <)$, or simply S, is called a TOTALLY ORDERED, DENSELY ORDERED, COMPLETELY ORDERED or WELL ORDERED set as may be appropriate. It is an easy calculation that every well ordered set is totally ordered and, indeed, completely ordered (Q 1.3.3). We shall call $<$ a total ordering on S only if S is totally ordered by $<$, with a similar definition and proviso for each of the other three types of ordering.

Example 1.3.11

If $(W, <)$ is a well ordered set with no maximum element, then $x \mapsto x_+$ with domain W is an injective function. Specifically, it is $\{(x, x_+) \mid x \in W\}$.

Definition 1.3.12

Suppose S is a set. Then $(S, \subset)$, or simply S, is called a NEST or a NESTED SET if and only if $\subset$ is a total ordering on S; if $S = \varnothing$, this is called the TRIVIAL NEST or the EMPTY NEST. If a collection $\mathcal{C}$ has the property that, for every nest $\mathcal{N}$ which is included in $\mathcal{C}$, we have $\bigcup \mathcal{N} \in \mathcal{C}$, then we say that $\mathcal{C}$ is CLOSED UNDER NESTED UNIONS.

Notation 1.3.13

We shall habitually use the same symbol for an INHERITED order relation as was used for the parent order relation. So, if $(S, <)$ is a partially ordered set and Y is a subset of S, then the relation $\{(a, b) \in < \mid a, b \in Y\}$ on Y will also be denoted by $<$. Y is partially, totally, or well ordered by $<$ if S is partially, totally, or well ordered, respectively, by $<$; but Y may fail to be completely or densely ordered when S is. We shall use terms like ORDERED SUBSET, WELL ORDERED SUBSET and so on only when we assume that the order relation is inherited from a superset identified by the context.

Example 1.3.14

Suppose $(S, <)$ is a totally ordered set. Then $<$ induces an ordering on $S \times S$ as follows: for $a, b, c, d \in S$, set $(a, b) < (c, d)$ if and only if $\max\{a, b\} < \max\{c, d\}$ or $(\max\{a, b\} = \max\{c, d\}$ and $\min\{a, b\} < \min\{c, d\})$ or $a = d < b = c$. A short calculation establishes that this is a total ordering on $S \times S$. Suppose that $(S, <)$ is well ordered and that U is a non-empty subset of $S \times S$. Then define $u = \min\{\max\{a, b\} \mid (a, b) \in U\}$ and $v = \min\{s \in S \mid (u, s) \in U$ or $(s, u) \in U\}$. Then $v \le u$ and $(v, u) \le (u, v)$. If $(v, u) \in U$, we have $(v, u) = \min U$; otherwise, $(u, v) \in U$ and $(u, v) = \min U$. So $S \times S$ is well ordered by $<$.

Similarity

Some bijective maps between sets endowed with a total ordering preserve that ordering. Such maps are called *similarity functions*. Although there is no set of all totally ordered sets, similarity behaves like an equivalence relation amongst totally ordered sets.

Definition 1.3.15

Suppose $(X, <)$ and $(Y, <)$ are totally ordered sets and $f: X \to Y$. Then f is said to be

- INCREASING if and only if, for all $a, b \in X$, $a < b \Rightarrow f(a) \le f(b)$;
- DECREASING if and only if, for all $a, b \in X$, $a < b \Rightarrow f(b) \le f(a)$.

If f is also injective, the terms STRICTLY INCREASING, and STRICTLY DECREASING may be used.

Definition 1.3.16

Suppose $(A, <)$ and $(B, <)$ are totally ordered sets. We say that $(A, <)$ is SIMILAR to $(B, <)$ if and only if there exists a strictly increasing function from A onto B; we call such a map a SIMILARITY FUNCTION.

When speaking of similar ordered sets, we usually say simply that sets A and B (on which orderings are defined) are similar, and write $A \simeq B$, where the particular orderings are assumed to be clear from the context. However, it is possible for sets to be similar with respect to one pair of total orderings and not similar with respect to another; see, for example, Q 4.1.1.

It is left as an exercise (Q 1.3.5) to prove that inverses of similarity maps are also similarity maps. The following proposition that similarity behaves like an equivalence relation then follows from 1.2.20.

Theorem 1.3.17

Suppose A, B and C are totally ordered sets. Then

- $A \simeq A$;
- $A \simeq B \Leftrightarrow B \simeq A$;
- $(A \simeq B$ and $B \simeq C) \Rightarrow A \simeq C$.

Theorem 1.3.18

Suppose $(A, <)$ and $(B, <)$ are similar well ordered sets. Then there is precisely one similarity function from A onto B.

Proof

Suppose g and h are distinct strictly increasing maps from A to B. Then $\{a \in A \mid h(a) \neq g(a)\}$ is a non-empty subset of A; it therefore has a least member, m, with respect to $<$. Either $h(m) < g(m)$ or $g(m) < h(m)$, because B is totally ordered. Suppose $h(m) < g(m)$; then, for each $a \in A$, we have $(a < m \Rightarrow g(a) = h(a) < h(m))$ and $(m \leq a \Rightarrow h(m) < g(m) \leq g(a))$, from which we infer that $h(m) \notin g(A)$ and hence that g is not surjective. The assumption $g(m) < h(m)$ similarly implies that h is not surjective. We conclude that not both g and h are similarity maps from A onto B. Since one similarity map certainly exists, it is unique.   □

Example 1.3.19

Suppose X is a set indexed by a set I and I is endowed with an ordering $<$. Since the family is bijective, this ordering imposes an ordering on X itself. Specifically, suppose $x: I \to X$ is bijective; then $\{(x_j, x_k) \in X \times X \mid j < k\}$ is an order relation on X; and, if $(I, <)$ is totally ordered or well ordered, then so is X with this ordering—indeed, X is similar to I. Imposed orderings of this type may or may not coincide with any natural or predefined ordering on X.

Theorem 1.3.20

Suppose $(W, <)$ is a well ordered set and $\omega \in W$. Then no ordered subset of $\grave{\omega}$ is similar to W.

Proof

Suppose $f: W \to \grave{\omega}$. Then $f(\omega) < \omega$, whence $A = \{\mu \in W \mid f(\mu) < \mu\}$ is non-empty. Since W is well ordered, A has a least member, $\alpha \leq \omega$. Then $f(\alpha) < \alpha$; and $f(f(\alpha)) \geq f(\alpha)$ by minimality of α; so f is not strictly increasing.   □

Theorem 1.3.21

Suppose $(A, <)$ and $(C, <)$ are well ordered sets and $A \simeq C$. Suppose $a \in A$. Then there exists a unique $c \in C$ such that $\grave{a} \simeq \grave{c}$. If $s: A \to C$ denotes the similarity map, then $c = s(a)$.

Proof

Let $s: A \to C$ be the similarity map from A onto C. Set $c = s(a)$. Then s^{-1} is a similarity map and we have $y \in \grave{c} \Leftrightarrow y < s(a) \Leftrightarrow s^{-1}(y) \in \grave{a} \Leftrightarrow y \in s(\grave{a})$ for each $y \in C$, which yields $s(\grave{a}) = \grave{c}$. Then $s|_{\grave{a}}: \grave{a} \to \grave{c}$ is a similarity map; so $\grave{a} \simeq \grave{c}$. Uniqueness of c follows from 1.3.20 and 1.3.17.   □

Example 1.3.22

It follows from 1.3.20 that a well ordered set is not similar to any lower segment of itself. This might seem intuitive, but its counterpart for upper segments is not true. Suppose W is non-empty and well ordered; let $\alpha = \min W$. Then, if every member of $W \backslash \{\alpha\}$ has an immediate predecessor and every member of W has an immediate successor, $\beta \mapsto \beta_+$ is a similarity function from W to $\acute{\alpha}$. The Axiom of Infinity is needed to show that such a set W exists, but the astute reader will expect the set of *natural numbers* to have the properties ascribed to W. Observe that 1.3.20 itself is not in general true for totally ordered sets.

Induction and Recursion

An important elementary result concerning well ordered sets is the *Principle of Induction*, which we present below. The reader will be familiar with a variation of this principle which is used when we are dealing with the set of *natural numbers*; that form becomes possible because the natural numbers have a property not shared by every well ordered set, namely that every natural number except the smallest has an immediate predecessor. We shall present the variation later on; here, we concern ourselves with arbitrary well ordered sets.

Theorem 1.3.23 PRINCIPLE OF INDUCTION

Suppose $(W, <)$ is a well ordered set and $S \subseteq W$. If, for each $w \in W$, we have $\grave{w} \subseteq S \Rightarrow w \in S$, then $S = W$.

Proof

Suppose $W \backslash S \neq \varnothing$. Then, because W is well ordered, there exists a minimum member w of $W \backslash S$. So $\grave{w} \subseteq S$. The hypothesis then implies that $w \in S$, which is a contradiction. So $W \backslash S = \varnothing$, and, since $S \subseteq W$, we have $S = W$.   □

Some functions are defined by specifying their values; others by declaring that the values are to be determined by some rule. However, it is often the case that we want to define a function on a well ordered domain (such as the *natural numbers*, yet to be defined), by specifying each value indirectly in terms of values already defined at previous members of the domain. Such a process is called RECURSIVE or INDUCTIVE. But it is not immediately clear that it determines a function; indeed, attempts to write down such a function as a set of ordered pairs might be fraught with difficulty.

Recursive definition is used so widely and freely in mathematics that its justification cannot be regarded as a luxury. The Recursion Theorem is the justification. It states that, if a given rule determines a specific value at each member α of a well ordered domain W in terms of pre-determined values at each member of the segment $\grave{\alpha}$, then the rule does indeed determine a function defined on W. The proof is a little tricky. It requires more than a straightforward application of the Principle of Induction, as we see in the presentation below. In this presentation, the recursive rule is represented by the function r and the recursively defined function by f. Because each value of what will eventually be the function f may depend on some or all of the previously defined values of f, the domain of r contains the restrictions of (what will eventually be) f to lower segments of W; in effect, r tells us how each such restriction of f in turn is to be extended to the next remaining member of W until f has been defined on the whole of W. The condition imposed on dom(r) in 1.3.24 is no more than is necessary; in practice, r will often be defined on a set of functions out of W into X which automatically satisfies the condition.

Theorem 1.3.24 RECURSION THEOREM

Suppose that $(W, <)$ is a well ordered set and that X is a non-empty set. Suppose that r is a function into X whose domain satisfies the hypothesis that $\left\{ g \in X^{\grave{\alpha}} \mid \alpha \in W;\ \beta \in \grave{\alpha} \Rightarrow (g|_{\grave{\beta}}, g(\beta)) \in r \right\} \subseteq \mathrm{dom}(r)$. Then there exists a unique function $f \colon W \to X$ such that $f(\alpha) = r(f|_{\grave{\alpha}})$ for all $\alpha \in W$.

Proof

As there is no set of all sets, there exists a set z with $z \notin W$. Let $W' = W \cup \{z\}$ and extend the well ordering of W to W' by setting $\beta < z$ for all $\beta \in W$. Let

$$S = \left\{ \alpha \in W' \mid \exists! g \in X^{\grave{\alpha}} \text{ such that } \beta \in \grave{\alpha} \Rightarrow (g|_{\grave{\beta}}, g(\beta)) \in r \right\}.$$

For each $\eta \in S$, let f_η be the unique member of $X^{\grave{\eta}}$ alluded to in the definition of S. If $\mu \in W'$ and there exists $g \in X^{\grave{\mu}}$ such that $\beta \in \grave{\mu} \Rightarrow (g|_{\grave{\beta}}, g(\beta)) \in r$, then, for each $\eta \in \grave{\mu}$ and for all $\beta \in \grave{\eta}$, we have $g|_{\grave{\eta}}(\beta) = g(\beta) = r(g|_{\grave{\beta}}) = r(g|_{\grave{\eta}}|_{\grave{\beta}})$ and, if $\eta \in S$, it follows from the uniqueness of f_η that $g|_{\grave{\eta}} = f_\eta$ and hence that $g(\eta) = r(f_\eta)$.

We now use induction to show that $S = W'$. Suppose $\mu \in W'$ and $\grave{\mu} \subseteq S$. Then $\grave{\mu} \subseteq S \cap W$; so the hypothesis on dom(r) ensures that $f_\gamma \in \mathrm{dom}(r)$ for all $\gamma \in \grave{\mu}$; also, the calculation above yields $f_\gamma = \{(\alpha, r(f_\alpha)) \mid \alpha \in \grave{\gamma}\}$. So $g = \{(\alpha, r(f_\alpha)) \mid \alpha \in \grave{\mu}\}$ is a well defined function in $X^{\grave{\mu}}$ and, for all $\beta \in \grave{\mu}$, we have $g|_{\grave{\beta}} = f_\beta$, yielding $g(\beta) = r(f_\beta) = r(g|_{\grave{\beta}})$. Since $\grave{\mu} \subseteq S$, the calculation performed in the paragraph preceding this one ensures that g is the only function with these properties and hence that $\mu \in S$. By the Principle of Induction, $S = W'$. In particular, $z \in S$, and since $\grave{z} = W$, the result is proved. □

If r of 1.3.24 is not a function but merely a relation, then the proof given above breaks down. Specifically, uniqueness has to be abandoned; then the simultaneous labelling of the f_η, on which the inductive part of the proof depends, is tantamount to the assertion that there exists a function $\eta \mapsto f_\eta$ with domain S in which each f_η is chosen rather than specified. This assertion cannot be proved in ZF.

However, it is quite often the case in mathematics that so-called recursive or inductive definition does indeed involve the arbitrary selection of each value from several possible candidates satisfying a given rule, rather than unique determination; moreover, mathematicians will sometimes make such choices even when determination is possible, simply to make proofs shorter and easier to read. This is permitted by the axioms of ZF only if there is a way in which choice can be eliminated from the process.

The Axiom of Choice will allow us to use recursion where choice is a necessary feature. In such cases, the term *definition*, though often used, is not strictly accurate, because the process does not specify a function; it merely ensures the existence of some undetermined function.

EXERCISES

Q 1.3.1 Suppose $(S, <)$ is a totally ordered set and $A \subseteq S$. Show that S can have no more than one maximal and no more than one minimal element; and that A can have no more than one least upper bound in S and no more than one greatest lower bound in S.

Q 1.3.2 There is an apparent lack of symmetry in the definition of a completely ordered set. It is dispelled by the following: suppose $(S, <)$ is a completely ordered set; show that every non-empty subset of S which has a lower bound in S has an infimum in S.

Q 1.3.3 Show that every well ordered set is completely ordered.

Q 1.3.4 Suppose $\sim$ is a symmetric transitive relation. What is wrong with the following argument which purports to prove that $\sim$ is reflexive: by symmetry, $a \sim b$ implies $b \sim a$; then, by transitivity, $a \sim b$ and $b \sim a$ together imply $a \sim a$.

Q 1.3.5 Show that the inverse of a similarity map is also a similarity map.

Q 1.3.6 Suppose $(U, <)$ and $(V, <)$ are similar well ordered sets and $s: U \to V$ is the similarity mapping from U onto V. Suppose $x \in U$. Show that, if x has an immediate successor in U, then so has $s(x)$ in V and $s(x_+) = s(x)_+$; show also that, if x has an immediate predecessor in U, then so has $s(x)$ has in V and $s(x_-) = s(x)_-$.

1.4 Ordinals

Perhaps the strangest of all sets which can be shown to exist from the axioms of ZF are the *ordinals*. Corresponding to each set z, the Pairing Principle and Axiom IV give us a set $z \cup \{z\}$ of which z is both a member and a subset. Consider now the empty set $\varnothing$, which we shall denote also by 0; then consider the set $0 \cup \{0\} = \{0\}$ which we shall denote by 1; and then the set $1 \cup \{1\} = \{0, 1\}$ which we shall denote by 2; and then the set $2 \cup \{2\} = \{0, 1, 2\}$ which we shall denote by 3; and then the set $3 \cup \{3\} = \{0, 1, 2, 3\}$ which we shall denote by 4; and so on, as far as we wish to go using the familiar symbols. Observe the curious relationships: $0 \in 1 \in 2 \in 3 \in 4 \in \ldots$ and $0 \subset 1 \subset 2 \subset 3 \subset 4 \subset \ldots$. These features inform the definition of an ordinal, and the sets 0, 1, 2, 3 and 4 are our first examples of ordinals. Note that each of the sets 0, 1, 2, 3 and 4 is well ordered by membership in the sense that each of them is well ordered by a relation of the type $\{(a, b) \mid a \in b\}$, which we shall denote also by $\in$.

Definition 1.4.1

A set α is called an ORDINAL if and only if α is well ordered by membership and $x \in \alpha \Rightarrow x \subseteq \alpha$.

Well Ordering amongst the Ordinals

We should like to be able to say that $\in$ well orders the ordinals. But the ordinals, as we shall see, do not form a set. Nonetheless, all the properties of a well ordering are satisfied by $\in$ amongst the ordinals, and every set of ordinals is well ordered by $\in$, as the following sequence of results shows.

Theorem 1.4.2

Suppose α, β and γ are sets, α is an ordinal, $\beta \in \alpha$ and $\gamma \in \beta$. Then $\gamma \in \alpha$.

Proof

Since every element of an ordinal is a subset of that ordinal, we have $\beta \subseteq \alpha$ and hence $\gamma \in \alpha$ as required. $\square$

Theorem 1.4.3

Every member of an ordinal is also an ordinal; and no ordinal is a member of itself.

Proof

Suppose α is an ordinal and $\beta \in \alpha$. Then $\beta \subseteq \alpha$, so that β is also well ordered by membership. We must show that every member of β is a subset of β. Suppose therefore that $\gamma \in \beta$ and $\delta \in \gamma$. Then 1.4.2 ensures firstly that $\gamma \in \alpha$ and then that $\delta \in \alpha$. So each of the three sets δ, γ and β is a member of α; since α is well ordered by $\in$ and $\delta \in \gamma \in \beta$, we have $\delta \in \beta$. Since δ is an arbitrary element of γ, we have $\gamma \subseteq \beta$, and since γ is an arbitrary member of β, it follows that β is an ordinal. This proves the first assertion. We do not need to invoke the Axiom of Foundation to establish the second; indeed, if α is an ordinal and $\alpha \in \alpha$, then the anti-reflexivity of the order relation $\in$ on the set α gives the contradiction $\alpha \notin \alpha$; so we must have $\alpha \notin \alpha$. $\square$

Theorem 1.4.4

Suppose α is an ordinal, x is a proper subset of α and each member of x is a subset of x. Then $x \in \alpha$. In particular, x is an ordinal.

Proof

Since x is a proper subset of α, the set $\alpha \backslash x$ is not empty. Since α is well ordered by $\in$, there is a least member of $\alpha \backslash x$; call it β. Now, for each $\gamma \in \beta$, we have $\gamma \in \alpha$, by 1.4.2, and the minimality with respect to $\in$ of β in $\alpha \backslash x$ ensures that $\gamma \in x$. It follows that $\beta \subseteq x$. We show that $x = \beta$. Suppose $x \neq \beta$; then there exists $\delta \in x \backslash \beta$. By hypothesis, every member of x is a subset of x; so we have $\delta \subseteq x$. Moreover, since $x \subset \alpha$, we have $\delta \in \alpha$. But $\beta \in \alpha$ and α is totally ordered, so we must have $\delta \in \beta$ or $\delta = \beta$ or $\beta \in \delta$. The first is ruled out by the definition of δ; the second is ruled out because $\delta \in x$ and $\beta \notin x$; and the third is ruled out because $\delta \subseteq x$ and $\beta \notin x$. Thus we have a contradiction and must conclude that $x = \beta$ and hence that $x \in \alpha$. $\square$

Corollary 1.4.5

Suppose α and β are ordinals. Then $\beta \in \alpha$ if and only if $\beta \subset \alpha$. In particular, if $\alpha \neq 0$, then $0 \in \alpha$.

Theorem 1.4.6

Suppose α and β are ordinals. Then exactly one of $\alpha \in \beta$ and $\alpha = \beta$ and $\beta \in \alpha$ holds.

Proof

Firstly, 1.4.2 ensures that we cannot have both $\alpha \in \beta$ and $\beta \in \alpha$, for then we should have $\alpha \in \alpha$, contradicting 1.4.3. Similarly, we cannot have either of them if $\alpha = \beta$. So at most one of the three occurs; we must show that one actually does occur. Every member of $\alpha \cap \beta$ is a subset of $\alpha \cap \beta$ because each of α and β is an ordinal. Now, if $\alpha \cap \beta$ were a proper subset of both α and β, then two invocations of 1.4.4 would reveal $\alpha \cap \beta$ to be an ordinal which is a member of itself, contradicting 1.4.3. So $\alpha \cap \beta$ is either α or β, whence $\alpha \subset \beta$ or $\alpha = \beta$ or $\beta \subset \alpha$. Invoking 1.4.4 yields the result.   □

Theorem 1.4.7

Suppose that $\phi(x)$ is a condition on sets and that there exists an ordinal which satisfies this condition. Then there exists an ordinal β such that $\phi(\beta)$ holds and, for every $\gamma \in \beta$, $\phi(\gamma)$ does not.

Proof

Suppose α is an ordinal for which $\phi(\alpha)$ holds. If there is no member of α which also satisfies the condition, then we set $\beta = \alpha$ and the proposition is proved. So we may suppose that $S = \{x \in \alpha \mid \phi(x)\}$ is non-empty. But α, being an ordinal, is well ordered by $\in$, so there exists $\beta = \min S$, the minimality here being with respect to $\in$; and β is an ordinal by 1.4.3. Now suppose that z is any set for which $\phi(z)$ holds. If we had $z \in \beta$, then 1.4.2 would give $z \in \alpha$ and we should get $z \in S$, contradicting minimality of β. So $z \notin \beta$ and the proposition is proved.   □

Theorem 1.4.8

Suppose x is a set of ordinals. Then x is well ordered by $\in$ and $\bigcup x$ is the least ordinal (with respect to $\in$) which is not a member of a member of x.

Proof

That every set of ordinals is totally ordered by $\in$ has been proved in 1.4.6 and 1.4.2, and it then follows from 1.4.7 that such a set is well ordered by $\in$. So x is well ordered by $\in$ and, since each member of $\bigcup x$ is an ordinal by 1.4.3, $\bigcup x$ is also well ordered by $\in$. Moreover, if $\gamma \in \bigcup x$, then $\gamma \in \alpha$ for some $\alpha \in x$, and, since α is an ordinal, it follows that $\gamma \subseteq \alpha$ and hence that $\gamma \subseteq \bigcup x$. So $\bigcup x$ is an ordinal. Moreover, if $\bigcup x$ were a member of a member of x, then we should have $\bigcup x \in \bigcup x$, contradicting 1.4.3. Finally, suppose β is any ordinal which is not a member of any member of x. Then $\beta \notin \bigcup x$. So, since β and $\bigcup x$ are ordinals, either $\beta = \bigcup x$ or $\bigcup x \in \beta$ by 1.4.6.   □

Theorem 1.4.9

Suppose α is an ordinal. Then $\alpha \cup \{\alpha\}$ is also an ordinal.

Proof

The well ordering of $\alpha \cup \{\alpha\}$ with respect to $\in$ follows immediately from that of α; also, if $\beta \in \alpha \cup \{\alpha\}$ then $\beta \in \alpha$ or $\beta = \alpha$, whence $\beta \subseteq \alpha$, since α is an ordinal; therefore $\beta \subseteq \alpha \cup \{\alpha\}$. So $\alpha \cup \{\alpha\}$ is an ordinal. $\square$

Theorem 1.4.10

There is no set which has every ordinal as a member.

Proof

Suppose z is a set; let $x = \{\gamma \in z \mid \gamma \text{ is an ordinal}\}$ and $\alpha = \bigcup x \cup \{\bigcup x\}$. Then $\bigcup x$ is an ordinal by 1.4.8 and α is an ordinal by 1.4.9. Also, since $\bigcup x \notin \bigcup x$ by 1.4.3, α is not a subset of $\bigcup x$; in particular, $\alpha \notin x$; therefore $\alpha \notin z$. $\square$

Notation 1.4.11

The ordinals do not form a set by 1.4.10. Nonetheless, since the foregoing results show that $\in$ behaves like a well ordering amongst the ordinals, we shall henceforth use the language and notation which we have set down for well ordered sets as indicated below. It is clear that this involves no ambiguity.

- It was shown in 1.4.5 that, if α and β are ordinals, then $\beta \in \alpha$ if and only if $\beta \subset \alpha$. We shall generally prefer to write $\beta < \alpha$ to indicate this ordering from now on, and we shall freely use all the standard terms like *less than* to compare ordinals.

- If α is any ordinal it is easy to check that the ordinal $\alpha \cup \{\alpha\}$ of 1.4.9 is the least ordinal strictly greater than α. We shall therefore denote the ordinal $\alpha \cup \{\alpha\}$ by α_+ and call it the IMMEDIATE SUCCESSOR of α amongst the ordinals. Also α will be called the IMMEDIATE PREDECESSOR of $\alpha \cup \{\alpha\}$ amongst the ordinals. Since every ordinal has an immediate successor, there is no maximum ordinal.

- If an ordinal α has an immediate predecessor amongst the ordinals, this will be denoted by α_- . Then $\alpha = \alpha_{-+} = \alpha_- \cup \{\alpha_-\}$.

- If α is an ordinal, then the lower segment $\grave{\alpha}$ is the set of all ordinals less than α; so $\grave{\alpha} = \alpha$. There is no upper segment for α, because the ordinals do not form a set. These two facts render the concept of segments redundant in this context.

- In view of 1.4.8, whenever x is a set of ordinals, we call the ordinal $\bigcup x$ the SUPREMUM of x amongst the ordinals and denote it by $\sup x$. Moreover (Q 1.4.1), if α is an ordinal, then either α has an immediate predecessor amongst the ordinals and $\sup \alpha = \alpha_- = \max \alpha$, or α has no immediate predecessor and $\sup \alpha = \alpha$.

Similarity of Well Ordered Sets to Ordinals

Ordinals and their well orderings seem at first sight to be very special. But they are the prototypes for all well ordered sets, in that every well ordered set is similar to a unique ordinal. This fact is the cornerstone for a comprehensive theory of counting.

Lemma 1.4.12

Suppose $(A, <)$ is a well ordered set. Then there is at most one ordinal α such that $(A, <)$ is similar to $(\alpha, \in)$.

Proof

Suppose α and β are ordinals and both $(\alpha, \in)$ and $(\beta, \in)$ are similar to $(A, <)$. Then, by 1.3.17, $(\beta, \in)$ and $(\alpha, \in)$ are similar. Since each ordinal is a segment of every larger ordinal, 1.3.20 yields $\alpha \notin \beta$ and $\beta \notin \alpha$; so $\beta = \alpha$ by 1.4.6. □

Lemma 1.4.13

Suppose $\mathcal{C}$ is a set. Then there exists an ordinal which is not equinumerous with any member of $\mathcal{C}$.

Proof

Every ordering on any member of $\mathcal{C}$ is a subset of $\bigcup \mathcal{C} \times \bigcup \mathcal{C}$ and it follows that $\mathcal{D} = \{(A, <) \mid A \in \mathcal{C}, < \text{ is a well ordering on } A\}$ is a set. By 1.4.12, each member of $\mathcal{D}$ is similar to $(\alpha, \in)$ for at most one ordinal α. So, by Axiom III, the class of such ordinals is a set, which we label W. But the ordinals do not form a set, by 1.4.10. So there exists an ordinal γ which is not in W. Suppose B is any set which is equinumerous with γ and let $f: \gamma \to B$ be a bijective function. Then $\mathcal{O} = \{(f(\delta), f(\epsilon)) \mid \delta \in \epsilon \in \gamma\}$ is a well ordering on B, and $(B, \mathcal{O})$ is similar to $(\gamma, \in)$; therefore $B \notin \mathcal{C}$. So γ has the required property. □

Theorem 1.4.14

Every well ordered set is similar to exactly one ordinal.

Proof

Suppose $(W, <)$ is a well ordered set. Invoking 1.4.13, let μ be an ordinal which is not equinumerous with any proper subset of W. For every injective function g from a proper subset of W into μ, let $r(g) = \min(\mu \backslash \mathrm{ran}(g))$. It is immediately observed that $\mathrm{dom}(r)$ satisfies the condition of the Recursion Theorem (1.3.24). So there exists a function $f: W \to \mu$ such that $f(w) = r(f|_{\dot{w}})$ for each $w \in W$. If $\alpha \in \mathrm{ran}(f)$, then $\alpha = f(v) = \min(\mu \backslash f(\dot{v}))$ for some $v \in W$, so that, for each

$\beta \in \alpha$, $\beta \in f(\hat{v})$, yielding $\alpha \subseteq \mathrm{ran}(f)$; and, using 1.4.4, $\mathrm{ran}(f)$ is an ordinal. Also, if $a, b \in W$ and $a < b$, then $f(a) = \min(\mu \backslash f(\hat{a})) < \min(\mu \backslash f(\hat{b})) = f(b)$. So $\mathrm{ran}(f)$ is similar to W; that it is the only ordinal which is similar to W follows from 1.4.12.   □

Corollary 1.4.15

Suppose x is a set of ordinals and κ is the unique ordinal which is similar to x. Then $\sup \kappa \leq \sup x$.

Proof

Let $s: x \to \kappa$ be the similarity map from x onto κ. Let $S = \{\alpha \in x \mid s(\alpha) \leq \alpha\}$. We claim that $S = x$. Suppose otherwise and let $\delta = \min(x \backslash S)$. It follows that $\delta < s(\delta) < \kappa$. Since s is surjective, there exists $\rho \in x$ such that $s(\rho) = \delta$. Then $s(\rho) < s(\delta)$ and, because s is increasing, $\rho < \delta$. Therefore $\rho \in S$ and so $s(\rho) \leq \rho$, which gives the contradiction $\delta \leq \rho$. So $S = x$ as claimed. Then $\sup \kappa = \bigcup s(x) \subseteq \bigcup x = \sup x$.   □

Cardinals

It has not been ruled out that there may be distinct ordinals which are in one-to-one correspondence with each other; we shall establish their existence later on (2.1.8) using the Axiom of Infinity. But certainly none of the ordinals 0, 1, 2, 3 or 4 is equinumerous with any smaller ordinal. They are our first examples of *cardinals*.

Definition 1.4.16

An ordinal is called a CARDINAL if and only if it is not equinumerous with any smaller ordinal.

Theorem 1.4.17

Every well ordered set is equinumerous with exactly one cardinal.

Proof

Let S be a well ordered set. By 1.4.14, there exists an ordinal which is similar to, and therefore equinumerous with, S. By 1.4.7, there exists a least ordinal β which is equinumerous with S. Then, using 1.2.20, β is a cardinal and is, moreover, the only cardinal which is equinumerous with S.   □

Theorem 1.4.18

Suppose C is a set. Then there exists a cardinal which is not in one-to-one correspondence with any member of C.

Proof

By 1.4.13, there exists an ordinal which is not equinumerous with any member of C; so, by 1.4.7, there exists a least such ordinal. By 1.2.20, this ordinal is not equinumerous with any smaller ordinal, so is a cardinal.   □

Corollary 1.4.19

The cardinals do not form a set and, for each cardinal α, there exists a least cardinal strictly greater than α.

Proof

Suppose S is a set. By 1.4.18, there exists a cardinal which is not equinumerous with any member of S, and is therefore not a member of S. Since S is arbitrary, the first assertion is proven. The second follows easily from 1.4.7.   □

EXERCISES

Q 1.4.1 Suppose α is an ordinal. Show that $\bigcup \alpha = \alpha_-$ if α has an immediate predecessor amongst the ordinals and that $\bigcup \alpha = \alpha$ otherwise.

Q 1.4.2 Show that a set α which is well ordered by $\in$ is an ordinal if and only if $\bigcup \alpha \subseteq \alpha$.

1.5 The Axiom of Choice

Although ZF lays the foundation for Set Theory, it cannot answer all its questions. We have noted already that the proposition that every non-trivial product of non-empty sets is non-empty cannot be proved or disproved within ZF. Such a proposition is said to be INDEPENDENT of the axioms of ZF. There are many known examples of propositions of ZF which are independent of the axioms of ZF; some are equivalent to each other, some are stronger or weaker than others and some neither imply nor are implied by others.

Propositions which are independent of ZF present mathematicians with three choices. Such a proposition can be incorporated as an axiom to create a theory which is then richer than ZF; or the negation of the proposition can be incorporated to create a somewhat different richer theory. Gödel's work has shown that neither of these decisions will produce a complete theory, in the sense that however widely we extend the axiom system in a consistent manner, independent propositions are bound to occur. The third, and often most prudent, choice is to defer making any such decision to extend the axiom system until there is ample informal evidence within mathematics itself to justify it.

The pages of this book contain a number of neat theorems of modern analysis which cannot be proved within ZF but which have wide applications in many areas of research; they also contain some propositions of an elementary nature which cannot be proved in ZF. But all of these follow from the axioms of ZF together with the Axiom of Choice; this axiom we present below with a number of equivalent statements which will be used in the text.

Definition 1.5.1

Suppose a is a set. A function f defined on $a\backslash\{\varnothing\}$ is called a CHOICE FUNCTION for a if and only if $f(x) \in x$ for all $x \in a\backslash\{\varnothing\}$. We say that a has a choice function if and only if such a function exists.

Theorem 1.5.2

The following propositions are equivalent in ZF:

(i) (ZERMELO'S AXIOM OF CHOICE) Every set has a choice function.

(ii) (SINGLETON INTERSECTION THEOREM) For each collection $\mathcal{C}$ of mutually disjoint non-empty sets, there exists a subset of $\bigcup\mathcal{C}$ which has singleton intersection with each member of $\mathcal{C}$.

(iii) (PRODUCT THEOREM) Every product of a non-empty family of non-empty sets is non-empty.

(iv) (INCLUDED FUNCTION THEOREM) Every relation includes a function with the same domain.

(v) (RECURSIVE CHOICE THEOREM) If $(W, <)$ is a well ordered set, X is a non-empty set and r is a relation with range included in X whose domain satisfies $\left\{f \in X^{\alpha} \ \middle| \ \alpha \in W; \ \beta \in \grave{\alpha} \Rightarrow (f|_{\grave{\beta}}, f(\beta)) \in r\right\} \subseteq \mathrm{dom}(r)$, then there exists a function $f\colon W \to X$ such that $(f|_{\grave{\alpha}}, f(\alpha)) \in r$ for all $\alpha \in W$.

(vi) (CARDINALITY THEOREM) Every set is in one-to-one correspondence with a unique cardinal.

(vii) (CANTOR'S WELL ORDERING PRINCIPLE) Every set can be well ordered.

Proof

(i) $\Rightarrow$ (ii):   Suppose $\mathcal{C}$ is a set whose members are non-empty and mutually disjoint. If f is a choice function for $\mathcal{C}$, then $\mathrm{ran}(f)$ has singleton intersection with each member of $\mathcal{C}$.

(ii) $\Rightarrow$ (iii):   Suppose $(X_i)_{i \in I}$ is a non-empty family of non-empty sets. Then the members of the collection $\mathcal{C} = \{\{i\} \times X_i \mid i \in I\}$ are mutually disjoint and non-empty. Every subset of $\bigcup \mathcal{C}$ which has singleton intersection with each member of $\mathcal{C}$ is, by definition, a member of $\prod_i X_i$.

(iii) $\Rightarrow$ (iv): Suppose r is a relation. If r is empty then the result is trivial, so we suppose otherwise. For each $x \in \mathrm{dom}(r)$, let $Y_x = \{y \in \mathrm{ran}(r) \mid (x,y) \in r\}$. Certainly each Y_x is non-empty; and each member of the product $\prod_{x \in \mathrm{dom}(r)} Y_x$ is a function which is included in r and which has domain $\mathrm{dom}(r)$.

(iv) $\Rightarrow$ (v):   Suppose $(W, <)$ is a well ordered set, X is a non-empty set and r is a relation with range included in X which satisfies the specified condition. If there exists a function $r' \subseteq r$ with $\mathrm{dom}(r') = \mathrm{dom}(r)$, then r' clearly satisfies a similar condition and, by the Recursion Theorem (1.3.24), there exists $f : W \to X$ such that $f(\alpha) = r'(f|_{\grave{\alpha}})$, whence also $(f|_{\grave{\alpha}}, f(\alpha)) \in r$, for all $\alpha \in W$.

(v) $\Rightarrow$ (vi):   Let X be a set. By 1.4.13, let ω be an ordinal which is not equinumerous with any member of $\mathcal{P}(X)$. Let $r = \{(f, x) \mid f : \omega \rightarrowtail X, \, x \in X \backslash \mathrm{ran}(f)\}$. By hypothesis, there is no injective function from ω into X. So, if (v) holds, $\{f \in X^\alpha \mid \alpha \in \omega; \, \beta \in \alpha \Rightarrow (f|_\beta, f(\beta)) \in r\} \backslash \mathrm{dom}(r) \neq \varnothing$, yielding some $\alpha \in \omega$ and $g \in X^\alpha$ such that $(g|_\beta, g(\beta)) \in r$ for all $\beta < \alpha$, whence g is injective, with $g \notin \mathrm{dom}(r)$, whence g is surjective; the conclusion is that X is equinumerous with the ordinal α and therefore with a unique cardinal.

(vi) $\Rightarrow$ (vii): If X is a set and $f : X \to \omega$ is a bijective function onto an ordinal ω, then $\{(a,b) \in X \times X \mid f(a) < f(b)\}$ is a well ordering on X.

(vii) $\Rightarrow$ (i):   Suppose $\mathcal{C}$ is a set. If $\bigcup \mathcal{C}$ admits a well ordering then, with respect to such an order, the map $S \mapsto \min S$ with domain $\mathcal{C} \backslash \{\varnothing\}$ is a choice function for $\mathcal{C}$. $\square$

The Product Theorem has a claim to being intuitively obvious. It is easy to prove in ZF if the indexing set is *finite* or if revealing information is available about the co-ordinate sets; if, for example, they are all equipped with a well ordering. With appropriate definitions and a little algebraic structure, it is easy to show that the number of elements in the product of a finite number of finite sets is got by multiplying the numbers of elements in each of them; it seems unreasonable to suggest that a larger product might be empty.

Despite its equivalence to the Product Theorem, the Well Ordering Principle presents a different intuition, because sets abound on which well orderings cannot be defined; there is no technique, for example, for well ordering the

power set of an arbitrary well ordered set. Such observations led to an appre-
hension that acceptance of the Axiom of Choice might introduce inconsistencies
into Set Theory. Fortunately, Gödel proved in 1938 that it does not do so. The
independence of the axiom was finally proved by Cohen in 1963, long after it
had been incorporated into modern mathematics.

We know that there are conditions on sets which do not determine sets;
and that the Subset Principle ensures that a class which is determined by a
condition and is included in a set is itself a set. We generally expect a set to
be determined by some condition. But this is not always the case; the Singleton Intersection Theorem—equivalent to the Axiom of Choice—postulates the
existence of subsets which are not determined by a condition.

Because of the discovery of the independence of the Axiom of Choice, we
are logically free to accept either it or its negation as an axiom of our system.
Within mathematical logic, acceptance of independent propositions is an important tool in testing for the independence of other propositions. But there are
many philosophical questions which can be raised concerning the acceptance
of new axioms for general use in analysis; in particular, the question of the
correspondence of analytic theory to objective reality cannot be taken lightly.
The working mathematician is most likely to be swayed by criteria deriving
from mathematics itself. If an independent proposition is shown to give sim-
pler proofs of previously known theorems; if it answers questions which appear
to be otherwise unanswerable; if it produces a wealth of new theory which has
applications to mathematics or to dependent disciplines, then the argument for
acceptance is likely to triumph. The Axiom of Choice, for all these reasons, has
taken its place as one of the fundamental assumptions of modern mathematics.

It is not unreasonable to ask why we single out the Axiom of Choice for
special treatment; why do we not discuss at length whether we are assuming
more than is justified in the Axiom of Replacement, for example, or indeed,
whether or not the Axiom of Infinity, in postulating the existence of an infinite
set, takes mathematics beyond the bounds of the real world? One reason for this
special treatment is historical, but it is also the case that there is a qualitative
difference between this axiom and the others. We should appreciate that the
Axiom of Choice is purely non-constructive. Indeed, if X is a set for which it
is unprovable in ZF that X has a choice function, then the Axiom of Choice
does not make it possible to construct such a function; it merely asserts that
one exists. As we remarked earlier, it is provable in the Predicate Calculus that
an inability to construct or define precisely any such object does not logically
prevent us from choosing one and labelling it in a proof. Having assumed the
Axiom of Choice, it is always valid within a proof to say *let f be a choice
function for X* and to make use of this f, even if it is provable that no particular
choice of f can ever be specified.

We now formally adopt the Axiom of Choice and move into the realm of ZFC. All the propositions of 1.5.2 are theorems of ZFC.

Axiom VII AXIOM OF CHOICE
Every set has a choice function.

Zorn's Lemma

Analysts regularly invoke the Axiom of Choice or an equivalent proposition without saying that they are doing so. When they do make its use explicit, it is quite often through the important lemma due to Zorn.

Definition 1.5.3
Suppose S is a set. A partial ordering on S will be called INDUCTIVE if and only if every well ordered subset of S has an upper bound in S.

Theorem 1.5.4 ZORN'S LEMMA I
Every inductively ordered set has a maximal element.

Proof
Suppose $(X, <)$ is an inductively ordered set. Then the empty set has an upper bound in X, and X is therefore not empty. Suppose X has no maximal member. By 1.4.13, let ω be an ordinal which is not equinumerous with any member of $\mathcal{P}(X)$. Let $A = \{g \in X^\alpha \mid \alpha \in \omega, g \text{ strictly increasing}\}$ and $r = \{(g, x) \in A \times X \mid \operatorname{ran}(g) \subseteq \dot{x}\}$. For each $g \in A$, $\operatorname{ran}(g)$ is well ordered because g is strictly increasing, so that the conditions on X ensure that $\operatorname{ran}(g)$ has an upper bound in $X \backslash \operatorname{ran}(g)$. Therefore $\operatorname{dom}(r) = A$. By the Recursive Choice Theorem, there exists $f : \omega \to X$ which satisfies $(f|_\alpha, f(\alpha)) \in r$ for all $\alpha \in \omega$. For each $\alpha, \beta \in \omega$ with $\beta < \alpha$, we have $f(\beta) = f|_\alpha(\beta) < f(\alpha)$, so that f is strictly increasing and $\operatorname{ran}(f)$ is similar to ω, contradicting its definition. So X has a maximal element. $\square$

Another form of Zorn's Lemma (1.5.5) is often encountered in the literature. It requires that we test all totally ordered subsets instead of just well ordered subsets and so appears to be weaker than the form given above. But each form is equivalent to the Axiom of Choice; the proof of these equivalences is left as an exercise (Q 1.5.4).

Theorem 1.5.5 ZORN'S LEMMA II

Suppose X is a partially ordered set. If every totally ordered subset of X has an upper bound in X, then X has a maximal element. If every totally ordered subset of X has a lower bound in X, then X has a minimal element.

EXERCISES

Q 1.5.1 Show that every function has an injective restriction onto its range.

Q 1.5.2 Suppose $P = \prod_{i \in I} X_i$ is a product of sets. Show that each natural projection $\pi_j \colon P \to X_j$, given by $(x_i)_{i \in I} \mapsto x_j$, is a surjective map.

Q 1.5.3 Let $(X, <)$ be a partially ordered set. Show that $<$ is inductive if and only if every totally ordered subset of X is bounded above in X.

Q 1.5.4 Show that the two forms of Zorn's Lemma are equivalent in ZF to the Axiom of Choice.

2
Counting

The ancient skill of counting is the pairing of objects in one collection with those in another in a one-to-one fashion; sheep might be paired with a particular set of pebbles, for example, to make sure that no sheep has gone astray. Whether by slow progression or in sudden illumination, our ancestors focused their minds on the central idea in this process and, abstracting it, discovered a class of *numbers*, exactly one of which can be paired with any given set. There are two important characteristics discernible in this abstraction: it exhibits *simplification* in that material objects such as pebbles are no longer needed for counting; and it exhibits *generalization* in that counting now has an infinitely wider scope of application. These two characteristics are the hallmarks of every true mathematical development.

2.1 Counting Numbers

Without the Axiom of Infinity, it cannot be proved in ZFC that there exists any LIMIT ORDINAL—an ordinal other than 0 which has no immediate predecessor amongst the ordinals. But there are certainly non-limit ordinals.

Definition 2.1.1
An ordinal α will be called a COUNTING NUMBER if and only if every member of $\alpha_+ \backslash \{0\}$ has an immediate predecessor amongst the ordinals.

41

Theorem 2.1.2

Suppose n is a counting number. Then every member of n is a counting number and n_+ is a counting number.

Proof

Suppose $m \in n_{++}$ and $p \in m_+ \backslash \{0\}$. Then either $p \in n_+ \backslash \{0\}$, in which case p has an immediate predecessor by hypothesis; or $p = m = n_+$ and n is the immediate predecessor of p. $\square$

Theorem 2.1.3

No counting number is equinumerous with any proper subset of itself. In particular, every counting number is a cardinal.

Proof

Suppose otherwise. Then, by 1.4.7, there exists a least counting number n which is in one-to-one correspondence with a proper subset S of n. Let $f \colon n \to S$ be a bijective function. Certainly, $n \neq 0$, because the empty set has no proper subset. Let $a \in n \backslash S$ and define g on n_- as follows: for $k \in n_-$, if $f(k) = n_-$ then $a \neq n_- \in S$ and we set $g(k) = a$; otherwise set $g(k) = f(k)$. Then g is injective and either $f(n_-) \in n_- \backslash \mathrm{ran}(g)$ or $a \in n_- \backslash \mathrm{ran}(g)$. So g is bijective from n_- onto a proper subset of n_-, contradicting the definition of n. $\square$

The Set of Counting Numbers

The numbers 0, 1, 2, 3, ... and subsequent successors are counting numbers. Moreover, every counting number can be realized in this way as a *remote successor* of 0; we deduce this informally as follows: if there were any counting number not so realizable, then there would be a least such number; but this number would have an immediate predecessor with the same property, giving a contradiction. The soundness of the argument depends on making the term *remote successor* precise. This can be done by saying that the remote successors of 0 are the members of the smallest set to which 0 belongs and which has the property that $x \cup \{x\}$ is a member whenever x is. The Axiom of Infinity implies that such a set exists, as we see next.

Theorem 2.1.4

There is a smallest set with respect to inclusion which satisfies the conditions of the Axiom of Infinity. Its members are precisely all the counting numbers.

Proof

The Axiom of Infinity tells us that there exists a set which has $\varnothing$ as a member and has $x \cup \{x\}$ as a member whenever x is a member. Let α be any such set. Suppose that not every counting number is a member of α. Then, by 1.4.7, there is a smallest counting number m such that $m \notin \alpha$. Certainly $m \neq 0$, so that there exists an ordinal k such that $m = k_+$. By 2.1.2, k is a counting number and, by definition of m, $k \in \alpha$. But the hypothesis on α then implies the contradiction $m = k_+ \in \alpha$. We deduce that every counting number is a member of α. Then $\{n \in \alpha \mid n$ is a counting number$\}$ is the required set. $\square$

Example 2.1.5

The set of counting numbers will be denoted by ∞. It is, in fact, the smallest ordinal which is not a counting number. The argument runs as follows: ∞ is a set of ordinals, so is well ordered by $\in$, by 1.4.8; also, since every member of a counting number is a counting number by 2.1.2, every member of ∞ is a subset of ∞. So ∞ is an ordinal. Then $\infty \notin \infty$, by 1.4.3.

Theorem 2.1.6

∞ is a cardinal.

Proof

Suppose otherwise. Then there exist $n \in \infty$ and a bijective function $f \colon \infty \to n$, whence the restriction $f|_{n_+}$ is a bijective function from n_+ onto a proper subset of n_+; since n_+ is a counting number, this contradicts 2.1.3. $\square$

Notation 2.1.7

When ∞ is regarded as a cardinal, it is frequently denoted by $\aleph_0$. This notation prepares the way for denoting by $\aleph_1$ the smallest cardinal greater than $\aleph_0$. Proceeding in this way, we get a succession of cardinals, $\aleph_0, \aleph_1, \aleph_2, \ldots$ and so on, each indexed by the appropriate ordinal. But there is no way of determining in ZFC what $\aleph_1$ actually is in terms of familiar sets; we shall return to this point at the end of the chapter.

Example 2.1.8

Not every ordinal is a cardinal: the function $\{(n, n_+) \mid n \in \infty\} \cup \{(\infty, 0)\}$ is a bijective function from ∞_+ onto ∞, demonstrating that ∞_+ is not a cardinal. Generalization of the notion of counting depends on one-to-one correspondence, and this example shows that the ordinals are not suited to such a task. The cardinals, by their definition, are tailored to it.

EXERCISES

Q 2.1.1 Why would it not have been possible to define the set ∞ (2.1.4) to be the intersection of all sets satisfying the condition of Axiom V?

Q 2.1.2 Show that ∞ is the smallest cardinal that is not a counting number.

2.2 Cardinality

The Cardinality Theorem ensures that every set S is equinumerous with exactly one cardinal. That cardinal is called the *cardinality* of S. It is clear that sets A and B have the same cardinality if and only if $A \approx B$. But we can improve considerably on this observation. In this section, we compare cardinalities by utilizing injective and surjective functions.

Definition 2.2.1

Suppose S is a set. We define the CARDINALITY of S, denoted by $|S|$, to be the unique cardinal which is equinumerous with S.

Example 2.2.2

If α is a counting number, then $|\alpha_+| = \alpha_+$ by 2.1.3. If α is an ordinal and $\infty \leq \alpha$, then the function $\{(0, \alpha)\} \cup \{(n_+, n) \mid n \in \infty\} \cup \{(\beta, \beta) \mid \infty \leq \beta < \alpha\}$ is a bijective map from α onto α_+, so that $|\alpha_+| = |\alpha|$. It follows that the only cardinals which have immediate predecessors amongst the ordinals are the counting numbers other than 0.

Theorem 2.2.3

Suppose A and B are sets. The following are equivalent:

- $|B| \leq |A|$;
- there exists an injective map from B to A;
- $B = \varnothing$ or there exists a surjective map from A onto B.

Proof

If $B = \varnothing$ then all three statements hold; so we assume $B \neq \varnothing$. Let $\phi \colon A \to |A|$ and $\psi \colon B \to |B|$ be bijective functions. If $|B| \leq |A|$ and $j \colon |B| \to |A|$ is the inclusion map, then $\phi^{-1} \circ j \circ \psi$ is an injective map from B to A. So the first assertion implies the second. If $f \colon B \to A$ is injective and $c \in B$, then the function $f^{-1} \cup \{(a, c) \mid a \in A \backslash f(B)\}$ is surjective from A onto B. So the second assertion implies the third. If $g: A \to B$ is surjective, then, by the Product Theorem, there exists $u \in \prod_{b \in B} g^{-1}\{b\}$. For all $b \in B$, we have $g(u_b) = b$, whence u is injective. Therefore $h = \phi \circ u \circ \psi^{-1}$ is an injective function from $|B|$ to $|A|$. Since $\operatorname{ran}(h)$ is a set of ordinals, there is an ordinal κ similar to $\operatorname{ran}(h)$ by 1.4.14 and, by 1.4.15, $\sup \kappa \leq \sup \operatorname{ran}(h) \leq \sup |A|$, whence $\kappa \leq |A|$. But $|B| = |\operatorname{ran}(h)| = |\kappa| \leq \kappa$, and the third assertion implies the first. $\qquad \square$

Corollary 2.2.4

Suppose A and B are sets. The following are equivalent:

- $|A| < |B|$;
- there exists no injective map from B to A;
- $B \neq \varnothing$ and there exists no surjective map from A onto B.

Corollary 2.2.5 SCHRÖDER–BERNŠTEĬN THEOREM

Suppose that A and B are sets and that each is in one-to-one correspondence with a subset of the other. Then A and B are equinumerous.

The Schröder–Bernšteĭn Theorem has been presented as a theorem of ZFC, since the Axiom of Choice was used to establish first that every set has well defined cardinality. But it is in fact a theorem of ZF and there are many proofs of it in the literature (see [12]) which do not use the Axiom of Choice.

Example 2.2.6

If A is a set and $S \subseteq A$, then $|S| \leq |A|$ because the inclusion map is injective.

Even Larger Cardinals

So far, ∞ is the only cardinal we have encountered which is not a counting number. But we do know that there are more cardinals, because, by 1.4.19, the cardinals do not form a set. The theory of cardinalities thus has wider application than the primitive theory of counting, in that it enables us to distinguish certain classes of *infinite* sets from one another.

Definition 2.2.7

Suppose A is a set. For each subset B of A, we define the CHARACTERISTIC FUNCTION of B in A to be the function $\{(x, 1) \mid x \in B\} \cup \{(x, 0) \mid x \in A \backslash B\}$ and denote it by χ_B, the superset A being understood from the context.

Theorem 2.2.8 CANTOR'S THEOREM OF POWER SETS

Suppose A is a set. Then $|A| < |\mathcal{P}(A)| = |2^A|$.

Proof

The map $S \mapsto \chi_S$ from $\mathcal{P}(A)$ to 2^A is bijective; this gives the equality. Suppose g is an arbitrary function from A to $\mathcal{P}(A)$ and consider the subset $Z = \{x \in A \mid x \notin g(x)\}$ of A. For each $z \in A$ we have $z \in g(z) \Leftrightarrow z \notin Z$, so that $Z \neq g(z)$. Therefore g is not surjective. Since $\mathcal{P}(A) \neq \varnothing$ and g is arbitrary, it follows from 2.2.4 that $|A| < |\mathcal{P}(A)|$. □

From Cantor's Theorem of Power Sets, we infer the existence of a succession of larger and larger cardinals:

$$1 < 2 < 3 < \ldots < \infty < |2^\infty| < \left|2^{2^\infty}\right| < \ldots.$$

Indeed, for each cardinal $\aleph$, there is a strictly larger cardinal $|2^\aleph|$. These larger cardinals behave well, in the sense that sets a and b have the same cardinality if and only if the corresponding sets 2^a and 2^b have the same cardinality, as 2.2.9 below shows.

Theorem 2.2.9

Suppose A and B are sets. Then $|A| < |B| \Leftrightarrow |\mathcal{P}(A)| < |\mathcal{P}(B)|$. Consequently also $|A| = |B| \Leftrightarrow |\mathcal{P}(A)| = |\mathcal{P}(B)|$.

Proof

Suppose $f: A \to B$. Let $\tilde{f}: \mathcal{P}(A) \to \mathcal{P}(B)$ be $S \mapsto \{f(x) \mid x \in S\}$. If f is surjective, then, for each $C \in \mathcal{P}(B)$, $\tilde{f}(f^{-1}(C)) = C$, whence $\tilde{f}$ is also surjective. If f is injective and $U, V \in \mathcal{P}(A)$ and $\tilde{f}(U) = \tilde{f}(V)$, then we have $U = f^{-1}(\tilde{f}(U)) = f^{-1}(\tilde{f}(V)) = V$, so that $\tilde{f}$ is also injective. □

EXERCISES

Q 2.2.1 Suppose X is a set and α is a cardinal with $\alpha \leq |X|$. Prove that X has a subset whose cardinality is α.

Q 2.2.2 Suppose S is a set. Show that there exists a well ordering $<$ on S such that $(S, <)$ is similar to $(|S|, \in)$.

2.3 Enumeration

Finite Sets

Definition 2.3.1

Suppose S is a set. S is said to be FINITE if and only if $|S| < \infty$; S is said to be INFINITE otherwise. These terms are used also to describe families, products, unions and so on, where the meaning should be apparent from the context; so we might use the term *finite product* for a product in which the indexing set is finite, or *finite union* for the union of the members of a finite set.

Example 2.3.2

It is a consequence of the Axiom of Choice that every set admits a well ordering. But a set S with $|S| \leq \infty$ can be characterized by being equinumerous with a counting number or with ∞. In either case, any bijective function from such a number onto S determines a well ordering on S without recourse to the Axiom of Choice.

Theorem 2.3.3

Suppose S is a set and $a \notin S$. If S is infinite, then $|S \cup \{a\}| = |S|$. If S is finite, then $|S \cup \{a\}| = |S|_+$, so that, in particular, $S \cup \{a\}$ is also finite.

Proof

Suppose $\psi \colon S \to |S|$ is bijective; then $\psi \cup \{(a, |S|)\}$ is a bijective function from $S \cup \{a\}$ onto $|S|_+$. So $|S \cup \{a\}| = \big||S|_+\big|$. The result follows from 2.2.2. $\quad\square$

Corollary 2.3.4

Suppose X is a set and $S \subset X$. If S is finite, then $|S| < |X|$.

Proof

Let $a \in X \backslash S$. Then $S \cup \{a\} \subseteq X$ and $\mathrm{inc} \colon S \cup \{a\} \to X$ is injective. Therefore $|S| < |S \cup \{a\}| \leq |X|$ by 2.3.3 and 2.2.3. $\quad\square$

Theorem 2.3.5

Suppose S is a non-empty set. The following are equivalent:

- S is finite;
- every total ordering on S determines a maximum element;
- every total ordering on S determines a minimum element.

Proof

Suppose firstly that S is infinite and let $\phi: S \to |S|$ be a bijective function. By 2.2.2, $|S|$ has no immediate predecessor amongst the ordinals; so there is no maximum member of S when it is ordered by $\{(a,b) \in S \times S \mid \phi(a) < \phi(b)\}$ and no minimum element with the ordering $\{(b,a) \in S \times S \mid \phi(a) < \phi(b)\}$. So the first assertion is implied by either of the other two. Towards the converse, suppose there exists a non-empty finite set which can be totally ordered such that it has either no maximum member or no minimum member; invoking 1.4.7, let S be such a set with the least possible cardinality; let $<$ be a total ordering on S which manifests the stated deficiency. Let $a \in S$. Certainly S is not a singleton set, so that $S \backslash \{a\}$ is a non-empty set of strictly smaller cardinality than S, by 2.3.4; therefore $S \backslash \{a\}$ has a maximum element s and a minimum element l with respect to $<$. But then either s or a is maximum in S and either l or a is minimum in S, contradicting the hypothesis. $\square$

Enumerative Orderings

Countable sets admit some total orderings which are almost, but not quite, well orderings.

Definition 2.3.6

Suppose S is a set. A total ordering $<$ on S will be called an ENUMERATIVE ordering if and only if $\acute{x} \cap \grave{y}$ is finite for all $x, y \in S$.

Theorem 2.3.7

Suppose $(S, <)$ is an enumeratively ordered set. Then every $a \in S$ which is not maximum in S has an immediate successor in S; and every $a \in S$ which is not minimum in S has an immediate predecessor in S.

Proof

Suppose $a \in S$. If $\grave{a} \neq \varnothing$ then there exists $c \in \grave{a}$ and, by hypothesis, $\acute{c} \cap \grave{a}$ is finite. If $\acute{c} \cap \grave{a}$ is empty, then $c = a_-$; if not, $\acute{c} \cap \grave{a}$ has a maximum member, by 2.3.5, which is the required predecessor of a in S. In either case, a has an immediate predecessor in S. A similar argument shows that if $\acute{a} \neq \varnothing$ then a has an immediate successor in S. $\square$

Corollary 2.3.8

Suppose $(S, <)$ is an enumeratively ordered set with a minimum element. Then $<$ is a well ordering on S.

Proof

Let $m = \min S$. Suppose A is a non-empty subset of S and let L denote the set of lower bounds for A in S. If $A \subseteq L$, then A is a singleton set, so has a minimum element. Suppose there exists $z \in A \backslash L$. Then $L \subseteq \check{z} = \{m\} \cup (\acute{m} \cap \check{z})$, which is finite by 2.3.3. So L is finite by 2.2.6. But $L \neq \varnothing$ because $m \in L$; so L has a maximum element s by 2.3.5. Then $s \neq \max S$ because $s < z$. So s has immediate successor s_+ in S by 2.3.7, whence $s_+ \notin L$ and $s = \min A$. $\square$

Corollary 2.3.9

Suppose S is a non-empty set and $<$ is a well ordering on S. Then $<$ is enumerative if and only if every member of S other than $\min S$ has an immediate predecessor with respect to $<$.

Proof

If $<$ is enumerative, then the condition is satisfied by 2.3.7. Towards the converse, suppose that the condition is satisfied and let $p, q \in S$. Note that $q \in \{x \in S \mid |\acute{x} \cap \acute{q}| < \infty\}$ and let z be the minimum member of this set; if z had an immediate predecessor y in S, then, since $\acute{y} \cap \acute{q} \subseteq \{z\} \cup (\acute{z} \cap \acute{q})$, we should have $\acute{y} \cap \acute{q}$ finite by 2.3.3 and 2.2.6, contradicting the definition of z. So $z = \min S$. Then $\acute{p} \cap \acute{q}$, being a subset of $\acute{z} \cap \acute{q}$, is finite by 2.2.6. $\square$

Example 2.3.10

The converse to 2.3.7 is not true. For each non-zero counting number n, let $-n$ denote the set $\infty \backslash n$. Let $-\infty$ denote the set $\{-n \mid n \in \infty\}$. The standard ordering on $-\infty$ is $\{(-n, -m) \mid m < n\}$. Now consider the non-standard ordering on $-\infty \cup \infty$ which includes the standard orderings on ∞ and on $-\infty$ with the set $\infty \times (-\infty \backslash \{0\})$. This is not enumerative but has the properties of 2.3.7.

Countable Sets

Definition 2.3.11

Suppose S is a set. S is said to be

- COUNTABLE if and only if $|S| \leq \infty$;
- DENUMERABLE or COUNTABLY INFINITE if and only if $|S| = \infty$;
- UNCOUNTABLE if and only if $|S| > \infty$.

Theorem 2.3.12

Suppose S is a set. Then the following are equivalent:

- S is countable;
- there exists an enumerative well ordering on S;
- there exists an enumerative ordering on S.

Proof

If $h: S \to |S|$ is bijective, then $\{(x, y) \in S \times S \mid h(x) < h(y)\}$ is a well ordering on S. And, if S is countable and $a, b \in S$, then $|\acute{a} \cap \grave{b}| \leq |\grave{b}| \leq h(b) < \infty$; so the ordering is enumerative. To show that the third assertion implies the first, we suppose that S is enumeratively ordered by $<$. Let $m \in S$. Then $<$ is a well ordering on $\{m\} \cup \acute{m}$ by 2.3.8; moreover $>$ is a well ordering on $\{m\} \cup \acute{m}$, also by 2.3.8. Let s_1 and s_2 be similarity maps from $\{m\} \cup \acute{m}$ and from $\{m\} \cup \grave{m}$ to the appropriate ordinals (1.4.14). Note that $s_1(m) = 0 = s_2(m)$, so that $f = s_1 \cup s_2$ is a function from S onto an ordinal. It is easy to check that $\{(a, b) \in S \times S \mid f(a) < f(b) \text{ or } (f(a) = f(b) \text{ and } a < b)\}$ is an enumerative ordering on S; it has minimum element m, so is a well ordering by 2.3.8; then, by 2.3.9, every member of S other than m has an immediate predecessor, and the same is true of the ordinal which is similar to S with this ordering; so ∞ is not a member of that ordinal. Therefore S is countable. $\square$

Example 2.3.13

Every total ordering on a finite set is enumerative (Q 2.3.1); but this is not true for denumerable sets. Indeed, let S be a denumerable set and $s \in S$; then $S \backslash \{s\}$ is also denumerable; so there exists a bijective map $\phi: \infty \to S \backslash \{s\}$. Then $\{(\phi(m), \phi(n)) \mid m, n \in \infty; m < n\} \cup ((S \backslash \{s\}) \times \{s\})$ is a well ordering of S in which s has no immediate predecessor.

Sequences

Definition 2.3.14

Suppose X is a set. A family $(x_i)_{i \in I}$ in X is called a SEQUENCE in X if and only if I is a non-empty countable set with an enumerative ordering; if I has a minimum element, then this is a well ordering by 2.3.8. We shall assume henceforth that I is infinite and well ordered unless it is clear from the context that it may not be; if I is not well ordered, then the sequence is called a DOUBLE SEQUENCE; if I is finite then the sequence will be called a FINITE SEQUENCE and $|I|$ will be called its LENGTH. For each $i \in I$, the set $t_i = \{x_j \mid i \leq j\}$ is called the ith TAIL of the sequence; a sequence is said to be EVENTUALLY CONSTANT if and only if it has a tail which is a singleton set. A bijective well ordered sequence in X is called an ENUMERATION of X.

Definition 2.3.15

Suppose X is a non-empty set, $(x_i)_{i \in I}$ is a sequence in X and $(t_j)_{j \in J}$ is a strictly increasing sequence in I. Then $x \circ t$ is itself a sequence in X; it is called a SUBSEQUENCE of x and will generally be denoted by $(x_{t_j})_{j \in J}$ or more simply by (x_{t_j}).

Example 2.3.16

Suppose X is a set. A family $(a_{i,j})_{(i,j) \in I \times J}$ in X is called a MATRIX with ENTRIES in X if and only if both I and J are non-empty countable sets with enumerative orderings—the word *entry* is preferred to the word *term* to describe a value of such a family. For each $r \in I$, the matrix $(a_{r,j})_{(r,j) \in \{r\} \times J}$ is called the ROW of the matrix $(a_{i,j})$ corresponding to the index r; for each $s \in J$, $(a_{i,s})_{(i,s) \in I \times \{s\}}$ is the COLUMN corresponding to the index s; then $a_{r,s}$ is the unique entry in row r and column s. If I and J are finite, the SIZE of such a FINITE MATRIX is $|I| \times |J|$; if $|I| = |J|$, then the matrix is called a SQUARE MATRIX. If $m, n \in \infty \backslash \{0\}$, then the set of $m \times n$ matrices with entries in X (and indexed by an appropriate set $I \times J$) is denoted by $\mathcal{M}_{m \times n}(X)$; it is, in fact, $X^{I \times J}$.

Finite Induction and Recursion

The Principle of Induction applies to any well ordered set. But countable sets have the special property that they admit enumerative well orderings; this fact gives us, for such sets, special forms of the Principle of Induction and of the Recursion Theorems. We state them below for a denumerable set W; obvious modifications may be made for finite W. These theorems can be proved as corollaries of their sister theorems for arbitrary well ordered sets.

Theorem 2.3.17 PRINCIPLE OF FINITE INDUCTION

Suppose W is a denumerable set with an enumerative well ordering. Suppose that S is a subset of W, that $\min W \in S$ and that for each $x \in S$, we have also $x_+ \in S$. Then $S = W$.

Theorem 2.3.18 FINITE RECURSION THEOREM

Suppose that W is a denumerable set with an enumerative well ordering. Suppose that X is a non-empty set and that $r \colon X \rightarrowtail X$ is a function which satisfies $\operatorname{ran}(r) \subseteq \operatorname{dom}(r)$. Then, for each $a \in \operatorname{dom}(r)$, there exists a unique sequence $(x_n)_{n \in W}$ in X such that $x_{\min W} = a$ and $x_{n_+} = r(x_n)$ for all $n \in W$.

Theorem 2.3.19 FINITE RECURSIVE CHOICE THEOREM

Suppose that W is a denumerable set with an enumerative well ordering. Suppose that X is a non-empty set and that r is a relation on X which satisfies $\operatorname{ran}(r) \subseteq \operatorname{dom}(r)$. Then, for each $a \in \operatorname{dom}(r)$, there exists a sequence $(x_n)_{n \in W}$ in X such that $x_{\min W} = a$ and $(x_n, x_{n_+}) \in r$ for all $n \in W$.

Example 2.3.20

It cannot be proved in ZF that every infinite set has a denumerable subset. We can, however, by invoking the Well Ordering Principle, provide a proof which involves recursive definition. In mathematical texts, statements of recursive definition are usually quite terse, their justification by one of the recursion theorems being generally omitted; in this case, we might write the following: *suppose S is an infinite set; impose a well ordering on S; define recursively an injective sequence (a_n) in S by setting $a_0 = \min S$ and, for each $n \in \infty$, letting a_{n_+} be the immediate successor of a_n in S; then $\{a_n \mid n \in \infty\}$ is a denumerable subset of S.* A justification of the inductive definition involved here might run as follows: let $r \colon S \rightarrowtail S$ be defined by $r(x) = x_+$ on the subset $\{x \in S \mid |\acute{x}| \geq \infty\}$ of S. Then $\operatorname{ran}(r) \subseteq \operatorname{dom}(r)$ and, by the Finite Recursion Theorem, there exists a sequence $(a_n)_{n \in \infty}$ in S for which $a_0 = \min S$ and $a_{n_+} = (a_n)_+$ for each $n \in \infty$.

Example 2.3.21

It is often easier to use the general Recursion Theorem rather than its finite counterpart even when dealing with sequences. For example, we define FACTORIALS as follows: set $0! = 1$ and, for each $n \in \infty \backslash \{0\}$, recursively define $n! = |n \times (n_-)!|$. To show that each $n!$ is a counting number requires 2.4.3 which we shall prove shortly; but, assuming 2.4.3, the assertion that this process defines $n!$ for all $n \in \infty$ is validated by an application of the Recursion Theorem (1.3.24) to the function $r \colon \bigcup \{\infty^n \mid n \in \infty\} \to \infty$ defined by setting $r(\varnothing) = 1$ and $r(f) = |n \times f(n_-)|$ for each $f \in \infty^n$ and each $n \in \infty \backslash \{0\}$. We usually omit such arguments; but it is important to know that there is a justification for the bald statements of recursive definition which we make from time to time.

EXERCISES

Q 2.3.1 Suppose $(S, <)$ is a finite totally ordered set. Show that $<$ is an enumerative well ordering.

Q 2.3.2 Suppose that S is a totally ordered set and suppose that every countable subset of S is well ordered. Show that S itself is well ordered.

Q 2.3.3 Show that a set is infinite if and only if it is in one-to-one correspondence with a proper subset of itself.

Q 2.3.4 (FINITE RECURSION THEOREM II) Suppose X is a non-empty set,
$r: X \rightarrowtail X$ is an injective function and $\operatorname{ran}(r) = \operatorname{dom}(r)$. Suppose W is
a denumerable set with an enumerative ordering. Show that, for each
$a \in \operatorname{dom}(r)$ and $w \in W$, there exists a unique sequence $(x_n)_{n \in W}$ in X
such that $x_w = a$ and $x_{n_+} = r(x_n)$ for all $n \in W$.

2.4 Cardinality of Unions and Products

Finite Sets

Theorem 2.4.1
Suppose $\mathcal{C}$ is a finite collection of finite sets. Then $\bigcup \mathcal{C}$ is finite.

Proof
If $\mathcal{C} \backslash \{\varnothing\} = \varnothing$, then $\bigcup \mathcal{C} = \varnothing$ which is finite; so suppose $\mathcal{C} \backslash \{\varnothing\} \neq \varnothing$. Suppose
$<$ is a total ordering on $\bigcup \mathcal{C}$. Since each $A \in \mathcal{C} \backslash \{\varnothing\}$ is non-empty and finite,
it has a maximum member with respect to the restriction of $<$, by 2.3.5. Let
$S = \{\max A \mid A \in \mathcal{C} \backslash \{\varnothing\}\}$; the map $A \mapsto \max A$ from $\mathcal{C} \backslash \{\varnothing\}$ to S is surjective; so S is finite, by 2.2.3, and has a maximum member with respect to the
restriction of $<$; this is a maximum for $\bigcup \mathcal{C}$, which is then finite by 2.3.5.   □

Corollary 2.4.2
Suppose X is a finite set. Then $\mathcal{P}(X)$ is also finite.

Proof
Let $N = \{n \in \infty \mid \text{for every set } A, (|A| = n \Rightarrow |\mathcal{P}(A)| < \infty)\}$. Then, because
$\mathcal{P}(\varnothing) = \{\varnothing\}$, we certainly have $0 \in N$. Suppose $m \in \infty \backslash \{0\}$ and $\dot{m} \subseteq N$; let S
be a set with $|S| = m$. Let $a \in S$. Then $|S \backslash \{a\}| < m$ by 2.3.4, so that $\mathcal{P}(S \backslash \{a\})$
is finite by hypothesis. Now $\{C \cup \{a\} \mid C \in \mathcal{P}(S \backslash \{a\})\}$ is clearly equinumerous
with $\mathcal{P}(S \backslash \{a\})$; so it too is a finite set. Then $\mathcal{P}(S)$, being the union of these two
sets, is finite by 2.4.1. Therefore $N = \infty$ by the Principle of Finite Induction
(2.3.17).   □

Theorem 2.4.3

Suppose $(A_i)_{i \in I}$ is a finite family of finite sets. Then $|\prod_i A_i| = |\prod_i |A_i|| < \infty$.

Proof

$\bigcup(A_i)$ is finite by 2.4.1. For each $j \in I$, $\bigcup(A_i)$ is equinumerous with $\{j\} \times \bigcup(A_i)$, which is therefore also finite. Then $I \times \bigcup(A_i) = \bigcup\{\{j\} \times \bigcup(A_i) \mid j \in I\}$ is a finite union of finite sets, and is finite by 2.4.1. Since $\prod_i A_i \subseteq \mathcal{P}(I \times \bigcup(A_i))$, it too is finite by 2.4.2 and 2.2.6. Lastly, for each $j \in I$, let $f_j : A_j \to |A_j|$ be a bijective map; then the map $(a_i) \mapsto (f_i(a_i))$ from $\prod_i A_i$ to $\prod_i |A_i|$ is clearly also bijective, so that $|\prod_i A_i| = |\prod_i |A_i||$.   □

Infinite Sets

Lemma 2.4.4

Suppose α is an infinite cardinal. Then $|\alpha \times \alpha| = \alpha$.

Proof

Let $\mathcal{S} = \{f : \gamma \times \gamma \to \gamma \mid \gamma \text{ is an ordinal}, \infty \leq \gamma \leq \alpha, f \text{ injective}\}$. $\infty \times \infty$ is well ordered by the ordering of 1.3.14 induced from that of ∞; it is easily checked that every member except $(0, 0)$ has an immediate predecessor in that ordering, so that it is enumerative by 2.3.9; then $\infty \times \infty$ is countable by 2.3.12. Therefore $\mathcal{S} \neq \varnothing$. We order $\mathcal{S}$ by inclusion and observe that, for any nest $\mathcal{U} \subseteq \mathcal{S}$, the function $\bigcup \mathcal{U}$ is an injective function from $\delta \times \delta$ to δ, where δ is a union of ordinals and therefore an ordinal by 1.4.8. So either $\mathcal{U} = \varnothing$ or $\bigcup \mathcal{U} \in \mathcal{S}$. In either case, $\mathcal{U}$ has an upper bound in $\mathcal{S}$; so Zorn's Lemma ensures that there exists a maximal member h of $\mathcal{S}$. Let β be the ordinal for which $\mathrm{dom}(h) = \beta \times \beta$. Then $\beta \leq \alpha$. We claim that $\beta = \alpha$. In order to show this, we let ρ be the ordinal similar to $\alpha \backslash \beta$ and $s : \alpha \backslash \beta \to \rho$ be the similarity map. It is not possible that $\beta \leq \rho$, for in that case, we set $\beta' = \{s^{-1}(\gamma) \mid \gamma \in \beta\}$, note that $\beta \cup \beta'$ is an ordinal, and, letting f denote the function

$$(\eta, \xi) \mapsto \begin{cases} (h(\eta, s(\xi)), 0) & \text{if } \eta \in \beta \text{ and } \xi \in \beta' \\ (h(s(\eta), \xi), 1) & \text{if } \eta \in \beta' \text{ and } \xi \in \beta \\ (h(s(\eta), s(\xi)), 2) & \text{if } \eta, \xi \in \beta', \end{cases}$$

observe that $h \cup (s^{-1} \circ h \circ f)$ is an injective function from $(\beta \cup \beta') \times (\beta \cup \beta')$ to $\beta \cup \beta'$, contradicting the maximality of h. So $\rho < \beta$. Then the function

$$\gamma \mapsto \begin{cases} h(\gamma, 0) & \text{if } \gamma < \beta \\ h(s(\gamma), 1) & \text{if } \beta \leq \gamma < \alpha \end{cases}$$

is injective from α to β. So $\alpha = |\alpha| \leq |\beta|$. But $|\beta| \leq \beta \leq \alpha$, so that $\alpha = \beta$ as claimed. Therefore h is injective from $\alpha \times \alpha$ to α and, since $\xi \mapsto (\xi, 0)$ is injective from α to $\alpha \times \alpha$, the Schröder–Bernšteĭn Theorem gives the result. $\qquad \square$

Theorem 2.4.5

Suppose $\mathcal{C}$ is a non-empty set of sets. Let $\gamma = \sup\{|A| \mid A \in \mathcal{C}\}$ and suppose that either $\mathcal{C}$ or γ is infinite. Then $|\bigcup \mathcal{C}| = \gamma$ if $|\mathcal{C}| \leq \gamma$; and $|\bigcup \mathcal{C}| \leq |\mathcal{C}|$ otherwise, with equality if the members of $\mathcal{C}$ are non-empty and mutually disjoint.

Proof

For each $A \in \mathcal{C}$, we have $A \subseteq \bigcup \mathcal{C}$, so that $|A| \leq |\bigcup \mathcal{C}|$; therefore $\gamma \leq |\bigcup \mathcal{C}|$. Since $|A| \leq |\gamma| = |\{A\} \times \gamma|$ for each $A \in \mathcal{C}$, it follows from the Product Theorem that there exists a family $(g_A)_{A \in \mathcal{C}}$ of injective functions $g_A \colon A \to \{A\} \times \gamma$. The union of these functions is a relation with domain $\bigcup \mathcal{C}$ and range included in $\mathcal{C} \times \gamma$; and it is clearly injective. The Included Function Theorem now ensures the existence of an injective function from $\bigcup \mathcal{C}$ into $\mathcal{C} \times \gamma$ and hence one from $\bigcup \mathcal{C}$ into $|\mathcal{C}| \times |\gamma|$. Since $|\mathcal{C}| \times |\gamma|$ is a subset either of $|\mathcal{C}| \times |\mathcal{C}|$ or of $|\gamma| \times |\gamma|$, it follows from 2.4.4 that $|\bigcup \mathcal{C}| \leq \max\{|\mathcal{C}|, |\gamma|\}$. If $|\mathcal{C}| \leq \gamma$ then $|\mathcal{C}| \leq |\gamma|$, so that $\gamma \leq |\bigcup \mathcal{C}| \leq |\gamma| \leq \gamma$, whence $|\bigcup \mathcal{C}| = \gamma$. Otherwise, $|\bigcup \mathcal{C}| \leq |\mathcal{C}|$. Finally, suppose that the members of $\mathcal{C}$ are non-empty and mutually disjoint; then, by the Axiom of Choice, there exists a choice function for $\mathcal{C}$. Any such function is injective from $\mathcal{C}$ to $\bigcup \mathcal{C}$, so that $|\mathcal{C}| \leq |\bigcup \mathcal{C}|$. This, combined with the inequality $|\bigcup \mathcal{C}| \leq |\mathcal{C}|$, gives us the required result. $\qquad \square$

Corollary 2.4.6

Suppose I is a non-empty finite set and $(A_i)_{i \in I}$ is a family of non-empty sets at least one of which is infinite. Then $|\prod_i A_i| = \max\{|A_i| \mid i \in I\}$.

Proof

I is finite; so $\{|A_i| \mid i \in I\}$ is finite and has a maximum member γ by 2.3.5; let $k \in I$ be such that $|A_k| = \gamma$. The natural projection $\pi_k \colon \prod_i A_i \to A_k$ is surjective by Q 1.5.2; so $\gamma \leq |\prod_i A_i|$ by 2.2.3. Towards the reverse inequality: for each $j \in I$, we have $|A_j| \leq \gamma$; so there exists an injective function $g_j \colon A_j \to \gamma$; then $(a_i)_{i \in I} \mapsto (g_i(a_i))_{i \in I}$ is an injective map from $\prod_i A_i$ to γ^I, and therefore $|\prod_i A_i| \leq |\gamma^I|$. That $|\gamma^I| = \gamma$ is proved using finite induction and 2.4.4. $\qquad \square$

Corollary 2.4.7

Suppose X is a set. Denote by $\mathcal{F}$ the collection of all finite subsets of X. Then $|\mathcal{F}| = |2^X|$ if X is finite and $|\mathcal{F}| = |X|$ otherwise.

Proof

If X is finite, then $\mathcal{F} = \mathcal{P}(X)$, and the result follows from 2.2.8. Suppose therefore that X is infinite. Each $S \in \mathcal{F}$ is in one-to-one correspondence with a counting number; it follows that the map $f \mapsto \operatorname{ran}(f)$ from $\bigcup\{X^n \mid n \in \infty\}$ to $\mathcal{F}$ is surjective. So $|\mathcal{F}| \leq |\bigcup(X^n)|$ by 2.2.3. But $|X^n| = |X|$ by 2.4.6; so $|\bigcup(X^n)| = |X|$ by 2.4.5. Therefore $|\mathcal{F}| \leq |X|$; and the reverse inequality holds because $\{\{x\} \mid x \in X\} \subseteq \mathcal{F}$.   □

In 2.4.6 we calculated the cardinality of finite products. It is natural to ask whether or not the finiteness condition can be relaxed. More particularly, if α and β are cardinals with $\infty \leq \beta < \alpha$, is it necessarily the case that $|\alpha^\beta| = \alpha$? In fact, it is not difficult to show, using Q 2.4.4, that $\alpha \leq |\alpha^\beta| \leq |2^\alpha|$. This observation leads naturally to a most interesting question: is there any cardinal γ such that $\alpha < \gamma < |2^\alpha|$? Cohen's work has shown that this question cannot be resolved in ZFC. Cantor had much earlier put forward the weaker CONTINUUM HYPOTHESIS that $|2^\infty|$ is $\aleph_1$, the smallest cardinal strictly greater than $\aleph_0$, which is, of course, ∞ (2.1.7). This too Cohen has shown to be independent of ZFC. There is no particular reason for us to accept either of these hypotheses (or, indeed, their negations) as axioms and we merely state them here for the interested reader.

EXERCISES

Q 2.4.1 Suppose S is a non-empty set. Show that S is finite if and only if there exists an enumerative well ordering on S in respect of which there exists a maximum member of S.

Q 2.4.2 Suppose $(X_i)_{i \in I}$ is a family and $2 \leq |X_j| \leq |I|$ for all $j \in I$. Show that $|\prod_i X_i| = |2^{|I|}|$.

Q 2.4.3 Suppose $(X_i)_{i \in I}$ is a countable family of countable non-empty sets. Show that $\prod_i X_i$ has cardinality $|2^\infty|$ unless all except a finite number of the co-ordinate sets are singleton sets, in which case it is countable.

Q 2.4.4 Suppose X is an infinite set. Show that $|X^X| = |2^X|$.

3

Algebraic Structure

What sport shall we devise here in this garden
To drive away the heavy thought of care?   *Richard II, III,iv.*

Modern analysis makes use of sets which exhibit various degrees of algebraic structure. In this chapter we shall describe the most elementary forms of that structure. Our chief concern is to articulate and formalize the algebraic properties we naturally associate with the *real* and *complex* number systems.

3.1 Elementary Algebraic Structures

In this section, we define *semigroups, groups, rings, integral domains* and *fields* and examine some of their most elementary structure. In particular, we introduce *finite sums* and *series*, and *finite products* and *powers*.

Binary Operations

Definition 3.1.1
A function is called a BINARY OPERATION if and only if its domain is a set of ordered pairs. For any set S, a function $\star$ whose domain is $S \times S$ and whose range is included in S will be called a binary operation on S.

It is usual to employ suggestive notation for many of the binary operations we use. Accordingly, the image of an ordered pair (a, b) under a binary operation will be denoted by $a + b$, $a \cdot b$, ab, $a \circ b$, $a \times b$, $a \star b$ or otherwise as may be appropriate for the intended interpretation. If $\star$ is a binary operation on a set

S, then a subset A of S is said to be CLOSED UNDER $\star$ if and only if $a \star b \in A$ for all $a, b \in A$; in this case, the restriction of $\star$ to $A \times A$ is a binary operation on A, which will habitually be denoted by the same symbol $\star$.

Associativity, Commutativity and Distribution

Definition 3.1.2
Suppose $\star$ is a binary operation. Then $\star$ is said to

- be COMMUTATIVE if and only if $b \star a = a \star b$ whenever $(a, b) \in \mathrm{dom}(\star)$ or $(b, a) \in \mathrm{dom}(\star)$;
- be ASSOCIATIVE if and only if $(a \star b) \star c = a \star (b \star c)$ whenever either of these expressions is defined;
- DISTRIBUTE over a binary operation $\circ$ if and only if $(a \star b) \circ (a \star c) = a \star (b \circ c)$ and $(b \star a) \circ (c \star a) = (b \circ c) \star a$ whenever any of these expressions is defined.

Example 3.1.3
Suppose α and β are cardinals; we define the PRODUCT $\alpha\beta$ to be $|\alpha \times \beta|$ and the SUM $\alpha + \beta$ to be $|(\alpha \times \{0\}) \cup (\beta \times \{1\})|$. If α and β are finite then so are both $\alpha + \beta$ and $\alpha\beta$ by 2.4.3 and 2.4.1. If not both α and β are finite, it follows from 2.4.5 that the sum is $\max\{\alpha, \beta\}$ and from 2.4.6 that the product is also $\max\{\alpha, \beta\}$. So, given any infinite cardinal γ, these operations, called ADDITION and MULTIPLICATION respectively, are binary operations on the set of cardinals in γ. A little calculation shows that they are both commutative and associative and that multiplication distributes over addition. This is all true, in particular, for $\gamma = \infty$; and, for each $n \in \infty$, we have $n + 1 = n_+$, $n + 2 = n_{++}$ and so on.

Example 3.1.4
Suppose S is a set. The maps $(A, B) \mapsto A \cup B$ and $(A, B) \mapsto A \cap B$ and $(A, B) \mapsto A \backslash B$ are binary operations on $\mathcal{P}(S)$, which we might denote by $\cup$ and $\cap$ and $\backslash$ respectively. $\cup$ and $\cap$ are commutative and associative and distribute over each other. If $S \neq \varnothing$, then $\backslash$ is neither commutative nor associative.

Generalized Associativity, Commutativity and Distribution

Suppose S is a set and $\star$ is a binary operation on S. If $\star$ is associative, then it is not necessary to use parentheses to indicate the order of application of the operation $\star$ in expressions such as $a \star b \star c \star d \star e$, because the result is independent of the particular order. This *Finite Associative Law* is more or less obvious and is easy to prove by induction once it has been stated; but it is a tricky enough matter to state it precisely (3.1.7). The *Finite Commutative Law* and the *Finite Distributive Law* can be proved similarly by induction. Most of the work lies in the rather cumbersome Definition 3.1.5, which is followed by examples (3.1.6) to help to clarify its purpose. Also for clarity, when A is totally ordered with a maximum element, we shall, in this section, denote the set $A \backslash \{\max A\}$ by $\underline{A}$.

Definition 3.1.5

Suppose S is a set and $\star$ is a binary operation on S. For every finite sequence $(a_i)_{i \in I}$ in S and every permutation v of $\underline{I}$, we define $\star_{i \in I}^v [a_i]$ recursively as follows: if $|I| = 1$, then $\star_{i \in I}^v [a_i]$ is the unique term of the sequence $(a_i)_{i \in I}$; if $|I| = n \in \infty \backslash \{0, 1\}$ and $\star_{i \in J}^u [b_i]$ has been defined for every sequence $(b_i)_{i \in J}$ in S of length strictly less than n and every permutation u of $\underline{J}$, then we define $\star_{i \in I}^v [a_i]$ to be $(\star_{i \in I \backslash \acute{m}}^{w_1} [a_i]) \star (\star_{i \in \acute{m}}^{w_2} [a_i])$, where $m = v(\max \underline{I})$, $w_1 = s \circ v|_{\acute{m}}$ and $w_2 = t \circ v|_{\underline{\acute{m}}}$ where s is the similarity map from $v(\acute{m})$ onto $\acute{m}$ and t is the similarity map from $v(\underline{\acute{m}})$ onto $\underline{\acute{m}}$. In the case where v is the identity map on $\underline{I}$, we shall write $\star_{i \in I} [a_i]$ rather than $\star_{i \in I}^v [a_i]$. If $\star$ is both associative and commutative, then 3.1.8 below allows us to extend unambiguously the definition of $\star_{i \in I} [a_i]$ to finite sets I with no specified order.

Example 3.1.6

Suppose $I = \{1, 2, 3, 4\}$, ordered as usual, and $v = \{(1, 3), (2, 1), (3, 2)\}$ in 3.1.5. Then $m = 2$, $s = \{(3, 1)\}$ and $t = \{(2, 3)\}$; so $w_1 = \{(1, 1)\}$ and $w_2 = \{(3, 3)\}$. Then $\star_{i \in I}^v [a_i] = (\star_{i \in \{1, 2\}} [a_i]) \star (\star_{i \in \{3, 4\}} [a_i]) = (a_1 \star a_2) \star (a_3 \star a_4)$. If, on the other hand, $v = \{(1, 3), (2, 2), (3, 1)\}$, then $m = 1$, $s = \varnothing$ and $t = \{(1, 2), (2, 3)\}$, yielding $w_1 = \varnothing$ and $w_2 = \{(2, 3), (3, 2)\}$. Then $\star_{i \in I}^v [a_i] = a_1 \star (\star_{i \in \{2, 3, 4\}}^{w_2} [a_i])$; the latter expression is similarly calculated to be $a_2 \star (a_3 \star a_4)$, and it follows that $\star_{i \in I}^v [a_i] = a_1 \star (a_2 \star (a_3 \star a_4))$.

Theorem 3.1.7 FINITE ASSOCIATIVE LAW

Suppose S is a set and $\star$ is an associative binary operation on S. For every finite sequence $(a_i)_{i \in I}$ in S and every permutation v of $\underline{I}$, we have $\star_{i \in I}^v [a_i] = \star_{i \in I} [a_i]$.

Proof

This is true by definition for sequences of length 1 in S; let $n \in \infty \backslash \{0, 1\}$ and suppose it is true for all sequences of length strictly less than n in S. Suppose $(a_i)_{i \in I}$ is a sequence in S of length n and v is a permutation of $\underline{I}$. Let $m = v(\max \underline{I})$. Then, applying the inductive hypothesis to Definition 3.1.5, we have $\star^{v}_{i \in I}[a_i] = (\star_{i \in I \setminus \dot{m}}[a_i]) \star (\star_{i \in \dot{m}}[a_i])$. If $m = \max \underline{I}$, then this expression is $(\star_{i \in \underline{I}}[a_i]) \star a_{\max I}$ which is precisely $\star_{i \in I}[a_i]$; otherwise, by associativity of $\star$, the expression equals $((\star_{i \in I \setminus \dot{m}}[a_i]) \star (\star_{i \in \dot{m}}[a_i])) \star a_{\max I}$ which, by the inductive hypothesis, equals $(\star_{i \in \underline{I}}[a_i]) \star a_{\max I}$ which again is $\star_{i \in I}[a_i]$.   □

Theorem 3.1.8 FINITE COMMUTATIVE LAW

Suppose S is a set and $\star$ is a commutative associative binary operation on S. Then $\star^{v}_{i \in I}[a_{u(i)}] = \star_{i \in I}[a_i]$ for every finite sequence $(a_i)_{i \in I}$ in S and every pair of permutations u of I and v of $\underline{I}$.

Theorem 3.1.9 FINITE DISTRIBUTIVE LAW

Suppose S is a set, $\circ$ and $\star$ are binary operations on S and $\star$ distributes over $\circ$. Suppose that $((a_{i,j})_{j \in J_i})_{i \in I}$ is a finite sequence of finite sequences in S, that v is a permutation of $\underline{I}$ and that, for each $i \in I$, u_i is a permutation of J_i. Then $\star^{v}_{i \in I}[\circ^{u_i}_{j \in J_i}[a_{i,j}]] = \circ^{w}_{p \in \prod_i J_i}[\star^{v}_{i \in I}[a_{i,p_i}]]$ for some total ordering on $\prod_i J_i$ and some permutation w of $\underline{\prod_i J_i}$.

Distributive Laws for Union and Intersection

The commutativity and associativity of union and intersection at the most general level are ensured by their definitions. The case for distribution is altogether different. The generalized distributive laws given below (3.1.10) make reference to arbitrary products of sets, and it is clear that an empty product would render them generally false. Since these laws do hold in ZFC, they are, in fact, equivalent in ZF to the Axiom of Choice.

Theorem 3.1.10 DISTRIBUTIVE LAWS FOR UNION AND INTERSECTION

Suppose $((A_{i,j})_{j \in J_i})_{i \in I}$ is a family of families of sets. Then the following equations both hold:

$$\bigcup \left\{ \bigcap \{A_{i,j} \mid j \in J_i\} \,\middle|\, i \in I \right\} = \bigcap \left\{ \bigcup \{A_{i,p_i} \mid i \in I\} \,\middle|\, p \in \prod_i J_i \right\}$$

$$\bigcap \left\{ \bigcup \{A_{i,j} \mid j \in J_i\} \,\middle|\, i \in I \right\} = \bigcup \left\{ \bigcap \{A_{i,p_i} \mid i \in I\} \,\middle|\, p \in \prod_i J_i \right\}.$$

Proof

Suppose firstly that $x \in \bigcup \{\bigcap \{A_{i,j} \mid j \in J_i\} \mid i \in I\}$. Then there exists $k \in I$ such that, for all $j \in J_k$, we have $x \in A_{k,j}$; so, for each $p \in \prod_i J_i$, we have

$x \in A_{k,p_k}$, whence $x \in \bigcap\{\bigcup\{A_{i,p_i} \mid i \in I\} \mid p \in \prod_i J_i\}$. Conversely, suppose $x \notin \bigcup\{\bigcap\{A_{i,j} \mid j \in J_i\} \mid i \in I\}$. Then, for every $i \in I$, there exists $j \in J_i$ such that $x \notin A_{i,j}$; so the relation $\{(i,j) \mid i \in I;\ j \in J_i;\ x \notin A_{i,j}\}$ has domain I. By the Included Function Theorem, there exists $q \in \prod_i J_i$ such that, for every $i \in I$, $x \notin A_{i,q_i}$. Then $x \notin \bigcap\{\bigcup\{A_{i,p_i} \mid i \in I\} \mid p \in \prod_i J_i\}$. These calculations establish the first equation; the second can be proved just as easily. □

Idempotents and Identities

Definition 3.1.11
Suppose S is a non-empty set, $\star$ is a binary operation on S and $e \in S$. Then e is called an IDEMPOTENT of S with respect to $\star$ if and only if $e \star e = e$, and an IDENTITY of S with respect to $\star$ if and only if $s \star e = s = e \star s$ for all $s \in S$.

Example 3.1.12
Suppose S is a set. There is at most one identity element with respect to a given binary operation $\star$ on S, for if e and f are identities, then $e = e \star f = f$. With reference to 3.1.4, $\varnothing$ is the identity with respect to the binary operation $\cup$ on $\mathcal{P}(S)$, and S is the identity with respect to $\cap$ on $\mathcal{P}(S)$.

Semigroups

Definition 3.1.13
An ordered pair $(S, \star)$, where S is a non-empty set and $\star$ is a binary operation on S, is called a SEMIGROUP if and only if $\star$ is associative. Usually we shall say simply that S is a semigroup (with operation $\star$). S will be called a COMMUTATIVE SEMIGROUP if and only if $\star$ is commutative.

Example 3.1.14
∞ is a commutative semigroup both under addition and under multiplication of cardinals, and multiplication distributes over addition (3.1.3). 0 is the identity with respect to addition, 1 with respect to multiplication.

Example 3.1.15
There is an important partition $\{\{0,1\}, P, C\}$ of ∞, where C is the set $\{ab \in \infty \mid a, b \in \infty\backslash\{0,1\}\}$ of COMPOSITE counting numbers and P, its complement in $\infty\backslash\{0,1\}$, is the set of PRIME counting numbers. C is a semigroup without an identity under cardinal multiplication (3.1.3).

Definition 3.1.16

Suppose that $(S, \star)$ is a semigroup with identity e and that $x \in S$. An element y of S is called a LEFT INVERSE for x in S if and only if $y \star x = e$; an element z of S is called a RIGHT INVERSE for x in S if and only if $x \star z = e$. An element of S which is both a left and a right inverse for x in S is called an INVERSE of x in S; if such an inverse exists, then x is said to be INVERTIBLE in S.

Theorem 3.1.17

Suppose that $(S, \star)$ is a semigroup with identity e and that $x \in S$. If x has a left inverse and a right inverse in S, then these are equal. In particular, x has at most one inverse in S.

Proof

If $y \star x = e = x \star z$, then $z = e \star z = (y \star x) \star z = y \star (x \star z) = y \star e = y$. $\square$

Additive and Multiplicative Semigroups

When the binary operation of a semigroup S is denoted by $+$, we generally interpret it additively and call it ADDITION; and an operation which is interpreted additively is usually denoted by $+$. We call its values SUMS and, if S has an identity, call the identity ZERO and denote it by 0, any ambiguity being dispelled by the context. Then, if an element x of S has an inverse, it is denoted by $-x$ and, for each $y \in S$, the member $y + (-x)$ of S is written $y - x$; the binary operation $(a, b) \mapsto a - b$ thus defined on the appropriate subset of $S \times S$ is called SUBTRACTION. If A and B are subsets of S, then the set $\{a + b \mid a \in A,\ b \in B\}$ is denoted by $A + B$; if the members of B all have inverses, then $-B$ will denote the set $\{-b \mid b \in B\}$ and $A - B$ the set $\{a - b \mid a \in A,\ b \in B\}$. For each $z \in S$, the maps $s \mapsto s + z$ and $s \mapsto z + s$ defined on S are called TRANSLATIONS and the images of A under these maps, namely $A + \{z\}$ and $\{z\} + A$ are called the TRANSLATES of A by z. A commutative additive semigroup is usually called an ABELIAN semigroup rather than a commutative one.

On the other hand, when the binary operation of a semigroup S is interpreted multiplicatively its values are called PRODUCTS and it is usually denoted by $\cdot$, by $\times$, by $\circ$ or by juxtaposition; in practice, we shall often drop the symbol in favour of juxtaposition if there can be no doubt about the operation in question; we write, for example, xy rather than $x \cdot y$ or $x \times y$. If S has an identity, we shall name the identity UNITY and in many circumstances denote it by 1, any ambiguity being dispelled by the context. Suppose $x \in S$ and $A \subseteq S$; then we shall write xA for the set $\{xa \mid a \in A\}$ and Ax for the set $\{ax \mid a \in A\}$. If x has an inverse, it is denoted by x^{-1}; then, for each $y \in S$, the member yx^{-1} of S, called the QUOTIENT of y by x, is often written y/x and the binary operation $(a, b) \mapsto a/b$ thus defined on the appropriate subset of $S \times S$ is called DIVISION. The set of members of S which are invertible in S will be denoted by $\mathrm{inv}(S)$.

Finite Sums and Products. Series

Definition 3.1.18

Suppose $(S, +)$ is an additive semigroup, and $(x_i)_{i \in I}$ is a sequence in S. We define the FINITE SUM $\sum_{i=m}^{n} x_i$ to be $+_{i \in \{k \in I \mid m \leq k \leq n\}} [x_i]$ if $m, n \in I$ and $m \leq n$; otherwise, provided S has a zero, we define it to be 0 (this covers the cases where, for one reason or another, the notation hides the fact that we are trying to sum over an empty set). The Finite Associative Law ensures that the order in which the operations are performed is irrelevant. If I is well ordered, the sequence $(x_i)_{i \in I}$ spawns another sequence $(\sum_{j=\min I}^{i} x_j)_{i \in I}$ which we call a SERIES and denote by $\sum_{i \in I} x_i$, or by $\sum_i x_i$ when the indexing set is clear from the context; we may use some other suitable suffix for clarification. Each term of this sequence is called a PARTIAL SUM of the series $\sum_i x_i$ rather than a term, and it is the individual x_j which are called the TERMS of the series. If I is finite, or indeed if the set N of non-zero terms of (x_i) is finite, we might use the notation $\sum_{i \in I} x_i$ or simply $\sum_i x_i$ also for the finite sum $\sum_{i=\min N}^{\max N} x_i$; whether such expressions stand for series or for finite sums should be apparent from the context. Moreover, if also $+$ is commutative, we shall use this notation for finite sums even when no ordering has been specified for the finite set I—this is justified by the Finite Commutative Law. We shall also employ notation of the type $\sum_{i \in I} f(i)$ for a sum of members of the range of a function f whose domain includes the finite set I and whose range is an Abelian semigroup; thus—assuming I and f to have the necessary properties—when f is the identity function, $\sum_{i \in I} i$ is the sum of the members of I and, when f is a constant function with value v, $\sum_{i \in I} v$ is the sum of $|I|$ copies of v.

Suppose $(S, \times)$ is a multiplicative semigroup and $(x_i)_{i \in I}$ is a sequence in S. We define the FINITE PRODUCT $\prod_{i=m}^{n} x_i$ to be $\times_{i \in \{k \in I \mid m \leq k \leq n\}} [x_i]$ if $m, n \in I$ and $m \leq n$; otherwise, provided S has a unity, we define it to be 1. The Finite Associative Law ensures that the order in which the operations are performed is irrelevant. If I is finite, or if the set $N = \{i \in I \mid a_i \neq 1\}$ is finite, we might use the notation $\prod_{i \in I} a_i$ or simply $\prod_i a_i$ also for the finite product $\prod_{i=\min N}^{\max N} a_i$; it should be clear from the context whether such notation refers to a Cartesian product or to a finite product. Moreover, if also $\times$ is commutative, we shall use this notation for finite products even when no ordering has been specified for the finite set I; as in the case of sums, this is justified by the Finite Commutative Law. We shall also employ notation of the type $\prod_{i \in I} f(i)$ in a similar way to that specified above for summation. Suppose $z \in S$; then, for each $n \in \infty \backslash \{0\}$, earlier notation notwithstanding, we define the nth POWER z^n to be the finite product $\prod_{i \in n} z$; if the semigroup has a unity 1, then z^0 is defined to be 1; if also z has an inverse in S, then it follows by finite induction that $(z^{-1})^n$, written as z^{-n}, is inverse to z^n for each $n \in \infty$.

Example 3.1.19

Suppose S is a set. Composition of maps is clearly an associative binary operation. So $(S^S, \circ)$ is a semigroup with identity ι_S; it is not commutative unless $|S| \leq 1$. It is treated as being multiplicative, so that for $f \in S^S$, f^0 is ι_S and, for $n \in \infty \backslash \{0\}$, f^n denotes the iterated composition of n copies of f. If f is bijective, then f has inverse $f^{-1} \in S^S$. The inverse was earlier defined (1.2.11) for every injective function; but if $f \in S^S$ is not also surjective, then f^{-1} is not a member of S^S. Iterated compositions of the identity map, namely functions of the type $z \mapsto z^n$ defined on S for $n \in \infty$, are called POWER FUNCTIONS; it is usual to denote the function $z \mapsto z^n$ simply by z^n (z may be replaced by some other symbol), rather than by ι_S^n, any confusion being dispelled by the context.

Example 3.1.20

The finite laws of associativity and of commutativity allow us to change the order of summation, or indeed of products. Specifically, suppose $(S, +)$ is a semigroup and I and J are non-empty finite well ordered sets (the same can be done with any finite number of indexing sets, but two will serve to illustrate the point); suppose $(a_{i,j})$ is a family indexed by $I \times J$. Let $\triangleleft$ denote the well ordering on $I \times J$ given by specifying that $(i, j) \triangleleft (k, l)$ if and only if $i < k$ or ($i = k$ and $j < l$); let $\vartriangleleft$ denote the well ordering on $I \times J$ given by setting $(i, j) \vartriangleleft (k, l)$ if and only if $j < l$ or ($j = l$ and $i < k$). Then the Finite Associative Law tells us that $\sum_{i \in I} \sum_{j \in J} a_{i,j}$ is identical to $\sum_{(i,j) \in I \times J} a_{i,j}$ where $I \times J$ is ordered by $\triangleleft$. Similarly $\sum_{j \in J} \sum_{i \in I} a_{i,j}$ is identical to $\sum_{(i,j) \in I \times J} a_{i,j}$ where $I \times J$ is ordered by $\vartriangleleft$. If $+$ is commutative, the Finite Commutative Law ensures that these two sums are equal.

Groups

Definition 3.1.21

Suppose $(G, \star)$ is a semigroup. Then $(G, \star)$ is called a GROUP if and only if G has an identity and every member of G has an inverse with respect to $\star$. Usually we shall say simply that G is a group (with operation $\star$). Suppose $(G, \star)$ is a group and $S \subseteq G$. Then $(S, \star)$, or simply S, will be called a SUBGROUP of $(G, \star)$ if and only if $(S, \star)$ is itself a group. A group $(G, \star)$ is called a TRIVIAL GROUP if and only if G is a singleton.

Example 3.1.22

Suppose S is a multiplicative semigroup with unity. Then $\text{inv}(S)$ is a group.

Example 3.1.23

Suppose that $(G, +)$ is a group and that $m, n \in \infty\backslash\{0\}$. For each $A = (a_{i,j})$ and $B = (b_{i,j})$ in $\mathcal{M}_{m \times n}(G)$, we define $A + B$ to be $(a_{i,j} + b_{i,j}) \in \mathcal{M}_{m \times n}(G)$. Endowed with this addition, $\mathcal{M}_{m \times n}(G)$ is clearly a group which is Abelian if G is Abelian; its zero is the matrix whose entries are all zero.

Example 3.1.24

Recalling 2.3.10, we extend cardinal addition (3.1.3) from ∞ to $-\infty \cup \infty$ as follows: for each $m, n \in \infty\backslash\{0\}$, set both $(-m) + 0$ and $0 + (-m)$ to be $-m$; set $(-m) + (-n)$ to be $-(m + n)$; and set both $m + (-n)$ and $(-n) + m$ to be $|m\backslash n|$ if $n \leq m$ and to be $-|n\backslash m|$ otherwise; these are in $-\infty \cup \infty$ by 2.2.6. It is easily verified that $-\infty \cup \infty$ is an Abelian group under addition with 0 as its zero and, for each $n \in \infty\backslash\{0\}$, $-n$ fulfilling the rôle notationally ascribed to it as the additive inverse for n.

Rings and Fields

Definition 3.1.25

An ordered triple $(R, +, \times)$ is called a RING if and only if $(R, +)$ is an Abelian group, $(R, \times)$ is a semigroup, and multiplication, $\times$, distributes over addition, $+$. We say simply that R is a ring (with operations $+$ and $\times$). It is called

- a COMMUTATIVE RING if and only if multiplication is commutative;
- a UNITAL RING if and only if it is non-trivial and unital;
- a DIVISION RING if and only if it is unital and every non-zero element has a multiplicative inverse;
- a FIELD if and only if it is a commutative division ring.

Suppose $(R, +, \times)$ is a ring and $S \subseteq R$. Then $(S, +, \times)$, or simply S, is called a SUBRING or a SUBFIELD of $(R, +, \times)$ if and only if $(S, +, \times)$ is itself a ring or a field respectively. If R is unital, a subring S of R is described as a UNITAL SUBRING of R if and only if the unity of R is a member of S.

Example 3.1.26

Suppose that R is a ring and that $x \in R$. Then the simple calculation $0 + 0x = (-x^2 + x^2) + 0x = -x^2 + (x^2 + 0x) = -x^2 + (x + 0)x = -x^2 + x^2$ yields

$0x = 0$; similarly $x0 = 0$. It follows incidentally that $(-a)b = -(ab) = a(-b)$ for all $a, b \in R$ and that $1 \neq 0$ in a unital ring. But the property $0x = 0 = x0$ does not necessarily determine 0; indeed, given any Abelian group $(G, +)$, if we define multiplication on G by setting $ab = 0$ for all $a, b \in G$, then G is a ring in which every product is zero.

Example 3.1.27

A non-unital ring may have unital subrings, and a unital ring may have subrings with different unities. Such phenomena occur quite naturally, but they always depend on one contingency—the existence of non-trivial multiplicative idempotents. The zero of a ring R is a multiplicative idempotent by 3.1.26, and, if it exists, so is the unity; if there is any other $e \in R$ such that $e^2 = e$, then the subring $eRe = \{ere \mid r \in R\}$ has unity e. No such idempotents exist if R is a field; so a subfield necessarily has the same identity as the original field.

Example 3.1.28

Suppose R is a ring. Suppose $n, p, q \in \infty \backslash \{0\}$ and $C = (c_{i,j}) \in \mathcal{M}_{q \times p}(R)$ and $D = (d_{i,j}) \in \mathcal{M}_{p \times n}(R)$. CD is defined to be the $q \times n$ matrix $(\sum_{k=1}^{p} c_{i,k} d_{k,j})_{i,j}$. This multiplication is clearly not commutative even when R is commutative, except in trivial cases. That it is associative is shown as follows. If $m \in \infty \backslash \{0\}$ and $G = (g_{i,j}) \in \mathcal{M}_{m \times q}(R)$, then $G(CD) = (e_{i,j})$ and $(GC)D = (f_{i,j})$ are $m \times n$ matrices. For $1 \leq i \leq m$ and $1 \leq j \leq n$, we have $e_{i,j} = \sum_{l=1}^{q} g_{i,l}(\sum_{k=1}^{p} c_{l,k} d_{k,j})$. By the Finite Distributive Law, this equals $\sum_{l=1}^{q} \sum_{k=1}^{p} g_{i,l} c_{l,k} d_{k,j}$. A similar calculation shows $f_{i,j} = \sum_{k=1}^{p} \sum_{l=1}^{q} g_{i,l} c_{l,k} d_{k,j}$. But, since $+$ is commutative, the order of summation can be reversed as in 3.1.20 and we have $e_{i,j} = f_{i,j}$. It is easy to check that, for appropriately sized matrices, multiplication distributes over addition. So, for $n \in \infty \backslash \{0\}$, $\mathcal{M}_{n \times n}(R)$ is a ring; it is unital if R is unital, with unity $(\delta_{i,j})$, where the DIAGONAL ENTRIES $\delta_{i,i}$ are 1 and the rest are 0.

Definition 3.1.29

Suppose R is a ring and $x \in R \backslash \{0\}$. Then x is called a LEFT DIVISOR OF ZERO in R if and only if there exists $z \in R \backslash \{0\}$ such that $xz = 0$; x is called a RIGHT DIVISOR OF ZERO in R if and only if there exists $z \in R \backslash \{0\}$ such that $zx = 0$. A unital ring is called an INTEGRAL DOMAIN if and only if it is commutative and has no zero divisors of either type.

Example 3.1.30

Suppose R is a unital ring. Clearly a left divisor of zero cannot be left invertible in R and a right divisor of zero cannot be right invertible in R. But, if e is a multiplicative idempotent in R and $0 \neq e \neq 1$, then $e(1 - e) = 0 = (1 - e)e$; and (3.1.27) e is the unity of a subring of R.

Example 3.1.31

$-\infty \cup \infty$ is made into a ring by extending the multiplication of 3.1.3 from ∞ as follows: for each $m, n \in \infty \backslash \{0\}$, set both $(-m)0$ and $0(-m)$ to be 0; set $(-m)(-n)$ to be mn; and set both $(-m)n$ and $m(-n)$ to be $-(mn)$. That this ring is an integral domain is evident from the definition.

Example 3.1.32

A member of a unital ring cannot be both invertible and a divisor of zero, but it can be neither; indeed the ring $-\infty \cup \infty$ (3.1.31) has no zero divisors and only two invertible elements. And there are conditions under which a non-zero member of a unital ring must be either invertible or a zero divisor; this is so, in particular, for every member of a finite unital ring. Slightly more generally, if R is a unital ring and $a \in R \backslash \{0\}$ and if the set $\{a^n \mid n \in \infty\}$ of powers of a is finite, then there exist $m, k \in \infty$ with $m < k$ such that $a^m = a^k$; so $a^m(a^{k-m} - 1) = 0$, and, by induction, either $a^{k-m} = 1$ or a is a divisor of zero.

Example 3.1.33

The calculation in 3.1.32 ensures that every finite integral domain is a field. Let $n \in \infty \backslash \{0, 1\}$; for each $k \in \infty$, let $\text{rem}_n(k) = k - \max\{an \mid a \in \infty, \, an \leq k\}$ be the REMAINDER when k is divided by n. Define $+_n$ and $\times_n$ on n by the equations $a +_n b = \text{rem}_n(a + b)$ and $a \times_n b = \text{rem}_n(ab)$. Then $(n, +_n, \times_n)$ is a unital commutative ring; moreover, it is easily shown to be an integral domain if and only if n is prime and, in that case, since it is finite, it is in fact a field.

Ideals

Definition 3.1.34

Suppose R is a ring. A subring S of R is called a LEFT IDEAL of R if and only if $\{rs \mid r \in R, \, s \in S\} \subseteq S$; it is called a RIGHT IDEAL of R if and only if $\{sr \mid r \in R, \, s \in S\} \subseteq S$ and an IDEAL or a TWO-SIDED IDEAL of R if and only if it is both a left and a right ideal of R. An ideal I of R (of any type) is deemed to be PROPER if and only if $I \neq R$. The term MAXIMAL applied to each type of ideal refers to maximality with respect to inclusion amongst the proper ideals of that type. The ideal $\{0\}$ is called the TRIVIAL IDEAL. R is said to be SIMPLE if and only if it has no proper non-trivial two-sided ideal.

Theorem 3.1.35

Suppose R is a unital ring and J is a left ideal of R. Then $J = R$ if and only if some member of J has a left inverse in R. A similar result holds for right ideals and right inverses.

Proof

Suppose $z \in J$, $y \in R$ and $yz = 1$. Then we have $a = ayz \in ayJ \subseteq J$ for each $a \in R$, whence $J = R$. Conversely, if $J = R$, then $1 \in J$.   □

Theorem 3.1.36

Suppose R is a ring. The collection of left ideals of R is closed under non-trivial intersections and under non-trivial nested unions. Similar statements apply to right ideals and to two-sided ideals.

Proof

Suppose $\mathcal{N}$ is a non-trivial nest of left ideals of R. Suppose $a, b \in \bigcup \mathcal{N}$ and $x \in R$. Then there exist $A, B \in \mathcal{N}$ such that $a \in A$ and $b \in B$; either $A \subseteq B$ or $B \subseteq A$, so that $a + b$ is in the larger and therefore is in $\mathcal{N}$. Also $xa \in A \subseteq \bigcup \mathcal{N}$. So $\bigcup \mathcal{N}$ is a left ideal of R. The other proofs are just as easy.   □

Theorem 3.1.37

Suppose R is a unital ring. Then every proper left ideal of R is included in a maximal left ideal of R; every proper right ideal of R is included in a maximal right ideal of R; and every proper ideal of R is included in a maximal ideal of R. In particular, R has at least one maximal ideal of each type.

Proof

We prove the result for left ideals; the other two proofs are similar. Suppose J is a proper left ideal of R. Let $\mathcal{I}$ denote the collection of all proper left ideals of R which include J. Order $\mathcal{I}$ by inclusion and suppose $\mathcal{N} \subseteq \mathcal{I}$ is a nest. $\bigcup \mathcal{N}$ is a left ideal of R by 3.1.36; it includes J and includes every member of $\mathcal{N}$; moreover, $1 \notin \bigcup \mathcal{N}$ by 3.1.35. So, by Zorn's Lemma, $\mathcal{I}$ has a maximal member, which is then a maximal left ideal of R. Since $\{0\}$ is a proper ideal of R, maximal ideals of each type exist.   □

Theorem 3.1.38

Suppose R is a unital ring. Let $\mathcal{L}$ denote the set of maximal left ideals of R and $\mathcal{R}$ the set of maximal right ideals of R. Then $\bigcap \mathcal{L} = \bigcap \mathcal{R}$; and this is a two-sided ideal.

Proof

Suppose $a \in \bigcap \mathcal{L}$ and $x \in R$. Then $xa \in \bigcap \mathcal{L}$ and, using 3.1.35, $xa - 1 \notin \bigcup \mathcal{L}$, so that, by 3.1.37, $R(xa - 1) = R$ and $xa - 1$ has a left inverse in R. Let $r \in R$ be such that $r(xa - 1) = 1$. Now $rxa \in \bigcap \mathcal{L}$, and a similar argument shows that $rxa - 1$ has a left inverse in R. But $rxa - 1 = r$; and, since r has a right inverse, namely $xa - 1$, in R, r is invertible in R with inverse $xa - 1$, by 3.1.17. So $xar = 1 + r$, whence $(1 - ax)(1 - arx) = 1$, so that $1 - ax$ is right invertible. Since x is arbitrary in R, this implies that, for every $M \in \mathcal{R}$, $1 \notin M + aR$, and therefore that the right ideal $M + aR$ is proper, whence $a \in M$. Since M is arbitrary, $a \in \bigcap \mathcal{R}$. So $\bigcap \mathcal{L} \subseteq \bigcap \mathcal{R}$. The reverse inclusion is got similarly. This is both a left and a right ideal of R by 3.1.36, and is therefore two-sided.   □

Definition 3.1.39

The intersection of the maximal left (or right) ideals of a unital ring is called the JACOBSON RADICAL of the ring. A unital ring is said to be SEMI-SIMPLE if and only if its Jacobson radical is trivial.

EXERCISES

Q 3.1.1 Suppose $\mathcal{C}$ is a finite disjointed set. Show that $|\bigcup \mathcal{C}| = \sum_{A \in \mathcal{C}} |A|$.

Q 3.1.2 Suppose R is a ring, $x \in R \backslash \{0\}$ and $|xR| < |R|$. Show that x is a zero divisor.

Q 3.1.3 Suppose R is a unital ring, $a, b, z \in R$, $az = za$ and $z \in \text{inv}(R)$. Show that, if $z - ab$ is left invertible in R, then so also is $z - ba$; and that, if $z - ab$ is right invertible in R, then so also is $z - ba$.

Q 3.1.4 Suppose that R is a commutative unital ring. Show that R is a field if and only if R is simple.

3.2 Vector Spaces

In this section, we shall discuss *vector spaces* over arbitrary fields. Our chief interest for analysis is, of course, in vector spaces over the fields of *real* and *complex* numbers which we have yet to define. We shall give some interesting examples of them in Chapter 5.

Definition 3.2.1

Suppose that $(V, +)$ is an Abelian group, that F is a field with operations denoted by $+$ and by juxtaposition and that $\cdot$ is a binary operation from $F \times V$ to V, denoted here also by juxtaposition. Then the ordered triple $(V, +, \cdot)$ is called a VECTOR SPACE over F if and only if, for each $x, y \in V$ and $\alpha, \beta \in F$

- $\alpha(x + y) = \alpha x + \alpha y$;
- $(\alpha + \beta)x = \alpha x + \beta x$;
- $(\alpha\beta)x = \alpha(\beta x)$;
- $1x = x$, where 1 denotes the unity of the field F.

We usually say simply that V is a vector space; the binary operation from $F \times V$ to V is called SCALAR MULTIPLICATION, the elements of V are called VECTORS and the elements of F are called SCALARS. The symbol 0 will habitually be used to denote both the zero of the field F and the zero vector of V, the context indicating which is intended. The zero vector will also be called the ORIGIN of V; if it is the only vector in V, then V is said to be TRIVIAL. For $\alpha \in F$, $v \in V$, $S \subseteq V$ and $G \subseteq F$, we denote the set $\{\alpha s \mid s \in S\}$ by αS, the set $\{\alpha g \mid g \in G\}$ by αG, the set $\{\beta v \mid \beta \in G\}$ by Gv and the set $\{\beta s \mid \beta \in G, s \in S\}$ by GS. The members of the set Fv are called the SCALAR MULTIPLES of v. S is said to be CLOSED UNDER SCALAR MULTIPLICATION if and only if $FS = S$.

Vector Subspaces

Definition 3.2.2

Suppose $(V, +, \cdot)$ is a vector space over a field F and $S \subseteq V$. Then $(S, +, \cdot)$, or simply S, is called a VECTOR SUBSPACE of $(V, +, \cdot)$ if and only if $(S, +, \cdot)$ is itself a vector space over F. A subspace S of V is said to be PROPER if and only if $S \neq V$. A subspace of V which is not properly included in any proper subspace is called a MAXIMAL SUBSPACE of V.

Example 3.2.3

Suppose that V is a vector space over a field F and that A and B are subspaces of V. Then $A + B$ is also a subspace of V. If $A \cap B = \{0\}$, then the space $A + B$ is called the DIRECT SUM of A and B and is usually written $A \oplus B$ or $B \oplus A$. If $V = A \oplus B$, then this is called a DECOMPOSITION of V, and A and B are called ALGEBRAIC COMPLEMENTS of each other in V. Suppose now that W is also a vector space over F. Then $V \times W$ is a vector space over F when the operations are defined POINTWISE: $((a, c), (b, d)) \mapsto (a + b, c + d)$ and $(\lambda, (a, c)) \mapsto (\lambda a, \lambda c)$, for $(a, c), (b, d) \in V \times W$ and $\lambda \in F$. It too is called a DIRECT SUM and is denoted by $V \oplus W$. In practice, this duplication of notation is unlikely to cause problems because the latter space is the direct sum, in the earlier sense, of the subspaces $V \times \{0\}$ and $\{0\} \times W$ of $V \times W$—and the spaces V and $V \times \{0\}$ are *identical in form* as also are $\{0\} \times W$ and W, the maps $v \mapsto (v, 0)$ on V and $w \mapsto (0, w)$ on W performing the noted identification. Spaces which are identical in form are said to be *isomorphic* and for many practical purposes we do not distinguish between them; this phenomenon will be discussed in Section 3.4.

Example 3.2.4

Let V be a vector space. It is readily shown that a non-empty subset S of V is a subspace of V if and only if S is closed under addition and under scalar multiplication. If $\mathcal{C}$ is a non-empty collection of subspaces of V, then $0 \in \bigcap \mathcal{C}$ and $\bigcap \mathcal{C}$ is closed under the operations; so $\bigcap \mathcal{C}$ is a subspace of V, clearly the largest subspace of V which is included in each member of $\mathcal{C}$. Note that *largest* in this context means *the maximum member of the set, partially ordered by inclusion, of all subspaces of V which are included in every member of $\mathcal{C}$*. In contrast, $\bigcup \mathcal{C}$ need not be closed under addition. Consider, for example, the vector space $F \oplus F$ over a field F. $A = \{(a, 0) \mid a \in F\}$ and $B = \{(0, b) \mid b \in F\}$ are subspaces of $F \oplus F$, but $A \cup B$ is not because $(1, 0) + (0, 1) = (1, 1) \notin A \cup B$. If, however, $\mathcal{C}$ is a non-trivial nest of subspaces of V, then $\bigcup \mathcal{C}$ is closed under the operations and is therefore a subspace of V.

Linear Independence and Spanning

Definition 3.2.5

Suppose V is a vector space over a field F, $v \in V$ and $S \subseteq V$. The intersection of all subspaces of V which include S is called the LINEAR SPAN of S and will be denoted by $\langle S \rangle$; it is evidently the smallest subspace of V which includes S. We say that S SPANS V or that S is a SPANNING SET for V if and only if $\langle S \rangle = V$. The vector v is called a LINEAR COMBINATION of members of S if and only if there exists a finite subset A of S and a family of COEFFICIENTS $(\alpha_a)_{a \in A}$ in F such that $v = \sum_{a \in A} \alpha_a a$. We say that S is LINEARLY DEPENDENT, or that the members of S are linearly dependent, if and only if there exists a finite non-empty subset A of S and a family $(\alpha_a)_{a \in A}$ in $F \backslash \{0\}$ such that $\sum_{a \in A} \alpha_a a = 0$; otherwise we say that S is LINEARLY INDEPENDENT. A family $(v_i)_{i \in I}$ in V is said to be LINEARLY DEPENDENT if and only if there exists a finite non-empty subset J of I and a family $(\xi_i)_{i \in J}$ in $F \backslash \{0\}$ such that $\sum_{i \in J} \xi_i v_i = 0$; otherwise it is LINEARLY INDEPENDENT.

The empty set is linearly independent and, clearly, every subset of a linearly independent subset of V is linearly independent; and a subset S of V is linearly independent if and only if every finite subset of it is linearly independent. On the other hand, every superset of a spanning set of V is also a spanning set of V. The vector 0 belongs to the linear span of every subset of V, but does not belong to any linearly independent subset of V.

Theorem 3.2.6

Suppose that V is a vector space over a field F and that S is a subset of V. Then $\langle S \rangle$ is the set of all linear combinations of members of S.

Proof

If $S = \varnothing$, then $\langle S \rangle = \{0\}$, and 0 is, by definition, the trivial sum. Suppose $S \neq \varnothing$; denote by L the set of all linear combinations of members of S. Clearly $S \subseteq L$, and, by induction, $L \subseteq \langle S \rangle$. We claim that L is closed under the two operations. Suppose $u, v \in L$; then there are finite subsets A, B of S and families $(\alpha_a)_{a \in A}$ and $(\beta_b)_{b \in B}$ in F with $u = \sum_a \alpha_a a$ and $v = \sum_b \beta_b b$. Set $\gamma_c = \alpha_c$ if $c \in A \backslash B$, $\gamma_c = \beta_c$ if $c \in B \backslash A$ and $\gamma_c = \alpha_c + \beta_c$ if $c \in A \cap B$; then $u + v = \sum_{c \in A \cup B} \gamma_c c$. So $u + v \in L$; moreover, for each $\lambda \in F$, we have $\lambda u = \sum_a (\lambda \alpha_a) a$. So L is closed under the two operations and is therefore a subspace of V, whence $\langle S \rangle = L$.   □

Bases

A linearly independent subset which spans a vector space is called a *basis*. It is a consequence of the Axiom of Choice that every vector space has a basis.

Definition 3.2.7

Suppose V is a vector space over a field F. A subset S of $V \backslash \{0\}$ is called a HAMEL BASIS for V if and only if, for each vector $v \in V$, there exist a unique finite subset A of S and a unique family $(\alpha_a)_{a \in A}$ in $F \backslash \{0\}$ with $v = \sum_{a \in A} \alpha_a a$.

Note that Definition 3.2.7 covers trivial spaces, whose only basis is the empty set. A spanning set for V which does not properly include any other spanning set for V is called a MINIMAL SPANNING SET for V; a linearly independent subset of V which is not properly included in any other linearly independent subset of V is called a MAXIMAL LINEARLY INDEPENDENT set. With these concepts in mind, the following proposition gives equivalent formulations for the definition of a basis; its proof is straightforward and is left for the reader.

Theorem 3.2.8

Suppose V is a vector space and $B \subseteq V$. Then the following are equivalent:

- B is a Hamel basis for V;
- B is linearly independent and spans V;
- B is a maximal linearly independent subset of V;
- B is a minimal spanning set for V.

Theorem 3.2.9

Suppose V is a vector space and A is a linearly independent subset of V. Then V has a Hamel basis which includes A. In particular, every vector space has a Hamel basis.

Proof

Consider the set S of all linearly independent subsets of V which include A. This is non-empty because $A \in S$. Partially order S by inclusion. Suppose $\mathcal{N} \subseteq S$ is a non-trivial nest. Every member of $\mathcal{N}$ is a subset of $\bigcup \mathcal{N}$ and $A \subseteq \bigcup \mathcal{N}$. Moreover, if B is any finite subset of $\bigcup \mathcal{N}$, then the total ordering of $\mathcal{N}$ ensures that there is a member of $\mathcal{N}$ which includes B; so B is linearly independent. Therefore $\bigcup \mathcal{N}$ is linearly independent. It follows, using Zorn's Lemma, that S has a maximal element, which, by 3.2.8, is a basis for V which includes A. For the particular case, note that $\varnothing$ is linearly independent. □

Example 3.2.10

Every subspace S of a vector space V has an algebraic complement in V: let P be a basis for S and let B be a basis for V which includes P; then $C = \langle B \backslash P \rangle$ is such a complement.

Example 3.2.11

Every proper subspace S of a non-trivial vector space V is included in a maximal subspace of V. Let P be a basis of S; let B be a basis for V with $P \subseteq B$. Then $P \subset B$ because S is a proper subspace of V. Let $a \in B \backslash P$. Then $\langle B \backslash \{a\} \rangle$ is a subspace of V which includes S; it is easy to check that it is maximal.

Dimension

There may be many different bases for a given vector space, but all have the same cardinality, enabling us to define its *dimension* to be that cardinality. The proof is easier for infinite dimensional spaces than for finite dimensional ones.

Lemma 3.2.12

Suppose V is a vector space over a field F. If B and U are finite linearly independent subsets of V with $|U| = |B|$ and $U \subseteq \langle B \rangle$, then $\langle U \rangle = \langle B \rangle$.

Proof

We use induction. If $B = \varnothing = U$, the assertion certainly holds. Suppose that $k \in \infty$ and that the result holds for every pair of linearly independent subsets of V of cardinality k. Suppose B and U are linearly independent subsets of V and $|B| = k + 1 = |U|$. Suppose $U \subseteq \langle B \rangle$ and $c \in B$; we want to show that $c \in \langle U \rangle$. For each $u \in U$, there exists a family $(\alpha_{u,b})_{b \in B}$ in F such that $u = \sum_{b \in B} \alpha_{u,b} b$. For each $u \in U$, let $u' = u - \alpha_{u,c} c$ and set $U' = \{u' \mid u \in U\}$. We consider two cases. Firstly, if $|U'| \neq k + 1$, then there exist distinct $v, w \in U$ such that $v' = w'$, so that $\alpha_{v,c} \neq \alpha_{w,c}$ and therefore $c = (\alpha_{v,c} - \alpha_{w,c})^{-1}(v - w) \in \langle U \rangle$, as required. Secondly, if $|U'| = k + 1$, then, given $v \in U$, $\langle B \backslash \{c\} \rangle = \langle U' \backslash \{v'\} \rangle$ by the inductive hypothesis because $|U' \backslash \{v'\}| = k$; and, since $v' \in \langle B \backslash \{c\} \rangle$, we have $v' \in \langle U' \backslash \{v'\} \rangle$, so that U' is linearly dependent. So there exists a family $(\gamma_u)_{u \in U}$ in F, not all of whose terms are zero, such that $\sum_{u \in U} \gamma_u u' = 0$; it follows that $\sum_{u \in U} \gamma_u u = \lambda c$ for some $\lambda \in F$, and that $\lambda \neq 0$ because U is linearly independent. Then $c = \sum_{u \in U} \lambda^{-1} \gamma_u u \in \langle U \rangle$, as required. In either case, $c \in \langle U \rangle$, and, since c is arbitrary in B, it follows that $B \subseteq \langle U \rangle$ and hence that $\langle U \rangle = \langle B \rangle$. To complete, we invoke the Principle of Finite Induction. $\square$

Theorem 3.2.13

Suppose V is a vector space over a field F. Then all Hamel bases of V have the same cardinality.

Proof

Let B and U be bases for V with $|B| \leq |U|$. If, on the one hand, B is finite, then we let S be a subset of U with $|S| = |B|$; by 3.2.12, $\langle S \rangle = \langle B \rangle = V$ and it follows that $S = U$, whence $|U| = |B|$. If, on the other hand, B is infinite, then for each $b \in B$, we let Z_b be the smallest finite subset of U such that $b \in \langle Z_b \rangle$. Since B spans V, so does $\bigcup (Z_b)_{b \in B}$. But this union is a subset of U, so must equal U, because U is a minimal spanning set for V. Since U is infinite and each Z_b is finite, we have $|U| \leq |B|$ by 2.4.5, and hence $|U| = |B|$. $\square$

Definition 3.2.14

The cardinality of any Hamel basis of a vector space V will be called the ALGEBRAIC DIMENSION of V and will be denoted by $\dim(V)$. V is said to be FINITE DIMENSIONAL if and only if $\dim(V)$ is finite; otherwise, V is said to be INFINITE DIMENSIONAL.

Quotient Spaces

Suppose V is a vector space over a field F and M is a subspace of V. Then $\{(x,y) \mid x - y \in M\}$ is an equivalence relation. There is a natural way to define addition and scalar multiplication (3.2.15) in order to make the resulting quotient into a vector space. The validity of these definitions depends on the simple fact that, for $a, b, x, y \in V$ and $\alpha \in F$, whenever $a - x \in M$ and $b - y \in M$, we have also $(a + b) - (x + y) \in M$ and $\alpha(a - x) \in M$.

Definition 3.2.15

Suppose V is a vector space over a field F and M is a subspace of V. We shall denote by V/M the quotient of V by the relation $\{(x,y) \mid x - y \in M\}$. For each $x \in V$, the notation x/M will be used for the equivalence class $\{x\} + M$ to which x belongs. We define addition and scalar multiplication for V/M by the equations $x/M + y/M = (x + y)/M$ and $\alpha(x/M) = (\alpha x)/M$ for each $x, y \in V$ and $\alpha \in F$. V/M with these operations is a vector space over F; it is called the QUOTIENT SPACE of V by M.

Example 3.2.16

Suppose V is a vector space and M is a subspace of V. Suppose B is a basis for M and $B \cup C$ is a basis for V, where $B \cap C = \varnothing$. Then $\{c/M \mid c \in C\}$ is a basis for V/M. Note that the c/M are all distinct, so that $\dim(V/M) = |C|$; and therefore, by Q 3.1.1, $\dim(V/M) + \dim(M) = \dim(V)$.

Product Spaces

Definition 3.2.17

Suppose $(V_i)_{i \in I}$ is a non-trivial family of vector spaces over a field F. The product $P = \prod_i V_i$ is a vector space over F when addition and scalar multiplication are defined POINTWISE: for each $f, g \in P$, $\alpha \in F$ and $j \in I$, $(\alpha f)_j = \alpha f_j$ and $(f + g)_j = f_j + g_j$. There may be other ways in which $\prod_i V_i$ can be made into a vector space over F, but unless different operations are specified or understood, we shall assume that such a product is endowed with pointwise operations.

Example 3.2.18

For each vector space V, the product $\prod_{v \in V} V$ is V^V. The pointwise operations which make V^V into a vector space are as follows: if $f: V \to V$ and $g: V \to V$, then $(f + g)(v) = f(v) + g(v)$ and $(\alpha f)(v) = \alpha f(v)$ for all $v \in V$ and $\alpha \in F$.

Example 3.2.19

If V is a vector space over a field F, then, for each $m, n \in \infty \backslash \{0\}$, $\mathcal{M}_{m \times n}(V)$ is the product $V^{m \times n}$ (2.3.16). It is an Abelian group (3.1.23) and scalar multiplication defined pointwise makes it into a vector space over F.

Example 3.2.20

Suppose F is a field and S is a non-empty set. Then F is a vector space over itself and F^S is a vector space with pointwise operations. The members of F^S are called SCALAR FUNCTIONS defined on S, the field F of scalars being understood from the context.

Example 3.2.21

Suppose $(A_i)_{i \in I}$ is a non-empty family of subspaces of V which is LINEARLY INDEPENDENT in the sense that, for every $x \in \prod_i (A_i \backslash \{0\})$, the family (x_i) is linearly independent. Then the subspace $\langle \bigcup(A_i) \rangle$ of V is called the ALGEBRAIC DIRECT SUM of these subspaces. We regard it as a generalization of that of 3.2.3 and write it $\oplus_{i \in I} A_i$ or simply $\oplus_i A_i$; alternatively, if I is finite, the co-ordinate spaces may be listed in any order separated by the symbol $\oplus$, as in $A_1 \oplus A_3 \oplus A_2$. It can be checked that, if, for each $j \in I$, B_j is a basis for A_j, then $\bigcup(B_i)$ is a basis for $\oplus_i A_i$, and that, trivial cases excepted, not all bases of the direct sum are realized in this way. If $V = \oplus_{i \in I} A_i$, this is called a DECOMPOSITION of V.

Example 3.2.22

Suppose $(V_i)_{i \in I}$ is a non-empty family of vector spaces over a field F. Then we define the ALGEBRAIC DIRECT SUM $\oplus_{i \in I} V_i$, or simply $\oplus_i V_i$, to be the subspace of $\prod_i V_i$ whose members are those $(v_i)_{i \in I}$ for which $v_j = 0$ for all except a finite number of values of $j \in I$. If I is finite, this is precisely $\prod_i V_i$. If, for each $i, j \in I$, we set $U_{i,i} = V_i$ and $U_{i,j} = \{0\}$ for $i \neq j$, then each co-ordinate space V_i is identical in form (see 3.2.3) to the subspace $\prod_j U_{i,j}$ of $\prod_i V_i$; the identification of V_i with $\prod_j U_{i,j}$ justifies the notation $\oplus_i V_i$, since this is precisely $\oplus_i \prod_j U_{i,j}$ as defined in 3.2.21. When the indexing set is not mentioned, we shall take it to be of the form $\{k \in \infty \mid 1 \leq k \leq n\}$ for some $n \in \infty \backslash \{0\}$; when we talk, for example, of the direct sum $V \oplus V \oplus V$ of three copies of a single space V, we shall mean the product $V^{\{1,2,3\}}$ with pointwise operations.

EXERCISES

Q 3.2.1 Suppose V is a vector space over a field F and $\infty \leq |F| < |V|$. Show that $\dim(V) = |V|$.

Q 3.2.2 Suppose V is a vector space over a field F and S and T are subspaces of V. How does $\dim(S + T)$ relate to $\dim(S)$ and $\dim(T)$?

Q 3.2.3 Suppose R is a ring. A map $f \colon R \to R$ which satisfies the equations $f(a + b) = f(a) + f(b)$ and $f(ab) = f(a)b + af(b)$ for all $a, b \in R$ is called a DERIVATION on R. Suppose that f and g are derivations on R and show that $(f \circ g) - (g \circ f)$ is also a derivation on R.

Q 3.2.4 Suppose F is a field and $n \in \infty \backslash \{0, 1\}$. Show that $\mathcal{M}_{n \times n}(F)$ is a simple unital ring which is not a division ring (compare Q 3.1.4).

3.3 Algebras

Definition 3.3.1

Suppose that $(A, +, \times)$ is a ring and that $(A, +, \cdot)$ is a vector space over a field F. Then $(A, +, \times, \cdot)$, or more simply A, is called an ALGEBRA over F if and only if $\alpha(xy) = (\alpha x)y = x(\alpha y)$ for all $x, y \in A$ and $\alpha \in F$, where both $\times$ and $\cdot$ are denoted simply by juxtaposition. An algebra A is said to be COMMUTATIVE if and only if $\times$ is commutative, and UNITAL if and only if the ring A is unital. In a unital algebra A, we shall in general use the same notation for scalar multiples of the identity as we use for scalars; so, if α is a scalar, then we shall often write the element $\alpha 1$ of A simply as α.

Example 3.3.2

Every field is an algebra, scalar multiplication and multiplication coinciding.

Example 3.3.3

If V is a vector space over a field F, then the vector space V^V with pointwise operations in general fails to be a ring when composition is designated as the multiplication, because composition does not distribute over addition. But we shall see in 3.4.10 that V^V has an important subspace in which composition does distribute over addition.

Example 3.3.4

Suppose A is a non-unital algebra over a field F. Then the direct sum $\tilde{A} = A \oplus F$ is a vector space (3.2.3). It is made into an algebra by defining multiplication by $(a, \lambda)(b, \mu) = (ab + \lambda b + \mu a, \lambda \mu)$ for each $a, b \in A$ and $\lambda, \mu \in F$. This algebra $\tilde{A}$ is unital, the identity being $(0, 1)$. It is called the UNITIZATION of A. Note that A is a maximal ideal of its unitization.

Spectra

Definition 3.3.5

Suppose A is a unital algebra over a field F and $a \in A$. We define the SPECTRUM $\sigma(a)$ of a to be the set $\{\lambda \in F \mid \lambda 1 - a \notin \mathrm{inv}(A)\}$.

Example 3.3.6

Suppose A is a unital algebra and B is a unital subalgebra of A. If $b \in B$, then b has spectrum $\sigma_B(b)$ as a member of B and spectrum $\sigma_A(b)$ as a member of A. There is certainly inclusion $\sigma_A(b) \subseteq \sigma_B(b)$, but this inclusion may be proper. Suppose that S is a set and F is a field. The vector space F^S is a unital algebra when multiplication is defined pointwise: for $f, g \in F^S$ and $s \in S$, $(fg)(s) = f(s)g(s)$. The constant function $s \mapsto 1$ is the unity. Suppose A is a proper unital subalgebra of F^S. For $f \in A$ and $\lambda \in \mathrm{ran}(f)$, we have $0 \in \mathrm{ran}(\lambda 1 - f)$, yielding $\lambda 1 - f \notin \mathrm{inv}(A)$. Therefore $\mathrm{ran}(f) \subseteq \sigma(f)$; this inclusion may be proper, but if $A = F^S$, then we have $\mathrm{ran}(f) = \sigma(f)$ because every function which does not take the value 0 clearly has an inverse in F^S.

Example 3.3.7

Spectra are defined also in non-unital algebras: if A is a non-unital algebra and $a \in A$, then $\sigma(a)$ is the spectrum of a as a member of the standard unitization of A (3.3.4).

Subalgebras, Algebra Ideals and Quotient Algebras

Definition 3.3.8

Suppose $(A, +, \times, \cdot)$ is an algebra over a field F and $B \subseteq A$. Then $(B, +, \times, \cdot)$, or simply B, is called a SUBALGEBRA of $(A, +, \times, \cdot)$ if and only if $(B, +, \times, \cdot)$ is itself an algebra over F; evidently, B is a subalgebra of A if and only if $B \neq \varnothing$ and B is closed under the three operations of A. If also A is unital and the unity of A is a member of B, then B is called a UNITAL SUBALGEBRA of A. A subalgebra J of A which is also an ideal of A of any type (left, right, two-sided) is called an ALGEBRA IDEAL of A of the same type. If A is a unital algebra, then each left, right or two-sided ring ideal J is certainly an algebra and is therefore an algebra ideal of the appropriate type.

Suppose B is a subalgebra of an algebra A. Then A/B is a vector space. But the natural definition for multiplication in A/B, $((x/B), (y/B)) \mapsto (xy/B)$, is not sound unless $(x - u \in B$ and $y - v \in B) \Rightarrow xy - uv \in B$ for all $x, y, u, v \in A$. In short, we need B to be an algebra ideal of A.

Definition 3.3.9

Suppose that A is an algebra over a field F and that J is a two-sided algebra ideal of A. We define multiplication on the quotient space A/J by setting $(x/J)(y/J) = (xy)/J$ for each $x, y \in A$. Endowed with this multiplication, the vector space A/J becomes an algebra; we call it the QUOTIENT ALGEBRA of A by J.

Product Algebras

Products of algebras are vector spaces and can be made into algebras by defining multiplication pointwise: suppose (A_i) is a family of algebras over a field F and set $(uv)_j = u_j v_j$ for each $u, v \in \prod_i A_i$ and $j \in I$. However, for many important products or subsets of products, the multiplication we use to make them into algebras is not defined pointwise.

Example 3.3.10

If A is an algebra over a field F and S is non-empty, then the product A^S with pointwise operations is an algebra over F. In particular A^A is such an algebra; it is unital if and only if A is unital; then the unity is the constant function $a \mapsto 1$—not the identity function ι_A.

Example 3.3.11

Suppose A is a unital algebra. Denote the identity function ι_A by z. For each $n \in \infty$, the power function z^n is a member of the algebra A^A with operations defined pointwise. Linear combinations of power functions are called POLYNOMIAL FUNCTIONS. The vector space $\text{poly}(A) = \langle \{z^n \mid n \in \infty\} \rangle$ of polynomial functions on A is closed under multiplication; indeed, it is easily checked that the product $(\sum_{i=0}^m \alpha_i z^i)(\sum_{j=0}^n \beta_j z^j)$ of arbitrary polynomial functions is $\sum_{k=0}^{m+n}(\sum_{i=0}^k \alpha_i \beta_{k-i})z^k$. So $\text{poly}(A)$ is a unital subalgebra of A^A. If $p \in \text{poly}(A)$, a member a of A is called a ZERO of p if and only if $p(a) = 0$. If $p \neq 0$, the DEGREE $\deg(p)$ of p is defined to be the smallest $n \in \infty$ such that $p \in \langle \{z^k \mid 0 \leq k \leq n\} \rangle$; we also define $\deg(0) = -\infty$. The polynomial functions of degree 0 are the non-zero constant functions; note, however, that a polynomial function $\sum_{i=0}^n \alpha_i z^i$ with $\alpha_n \neq 0$ need not have degree n, because this representation may not be unique. For each $n \in \infty$, we define $\text{poly}_n(A)$ to be the vector subspace of $\text{poly}(A)$ consisting of polynomial functions on A which have degree not exceeding n; it is not usually an algebra.

Example 3.3.12

Suppose A is a unital algebra which, regarded as a vector space, is finite dimensional. Then every non-zero member of A is either a zero divisor or invertible; even more interesting is that, if $a \in \text{inv}(A)$, then $a^{-1} = p(a)$ for some $p \in \text{poly}(A)$. The proof is similar to that used in 3.1.32. The space $\langle \{a^n \mid n \in \infty\} \rangle$ is finite dimensional; so there exists a minimum member m of $\infty \backslash \{0\}$ such that $\sum_{i=0}^{m} \mu_i a^i = 0$ for some family (μ_i) of scalars with $\mu_m \neq 0$. If $\mu_0 = 0$, then $a(\sum_{i=0}^{m-1} \mu_{i+1} a^i) = 0$, so that a is a zero divisor; if $\mu_0 \neq 0$, then $-\mu_0^{-1} \sum_{i=1}^{m} \mu_i a^{i-1}$ is inverse to a.

Example 3.3.13

If A is an algebra and $n \in \infty \backslash \{0\}$, then $\mathcal{M}_{n \times n}(A)$, being a ring (3.1.28) and a vector space (3.2.19), is an algebra; unless $n = 1$, multiplication is not pointwise. If A is unital, then so is $\mathcal{M}_{n \times n}(A)$. Also, if A is finite dimensional as a vector space, then, despite the fact that multiplication is not in general commutative, we do have $ST = I \Leftrightarrow TS = I$ for $S, T \in \mathcal{M}_{n \times n}(A)$ from 3.3.12, because $\mathcal{M}_{n \times n}(A)$ clearly has dimension $n^2 \dim(A)$.

EXERCISES

Q 3.3.1 Describe operations which make $\{0, 1, 2, 3\}$ into a ring which is not an integral domain.

Q 3.3.2 Suppose A is a unital algebra over a field F and $a, b \in A$. Show that $\sigma(ab) \backslash \{0\} = \sigma(ba) \backslash \{0\}$.

3.4 Preservation of Algebraic Structure

Definition 3.4.1

Suppose S is a set, $\star$ is a binary operation on S, ϕ is a function defined on S and $\circ$ is a binary operation on $\text{ran}(\phi)$. Then ϕ will be said to PRESERVE the operation $\star$ if and only if $\phi(a \star b) = \phi(a) \circ \phi(b)$ for all $a, b \in S$. If the two operations are both interpreted additively or multiplicatively, we shall say simply that ϕ preserves addition or multiplication, as appropriate.

If ϕ preserves the operation $\star$, then simple calculations show that, if $\star$ is associative or commutative, then so is the corresponding operation $\circ$; that if

e is an idempotent or an identity with respect to $\star$, then $\phi(e)$ is likewise an idempotent or an identity with respect to $\circ$; and that if b is a left or right inverse of a with respect to $\star$, then $\phi(b)$ is a left or right inverse of $\phi(a)$ with respect to $\circ$. Moreover, it is easy to show by induction that, for each finite sequence $(a_i)_{i \in I}$ in S and each permutation v of $I \backslash \max I$, we have $\phi(\star_{i \in I}^{v}[a_i]) = \circ_{i \in I}^{v}[\phi(a_i)]$, where the notation is as in 3.1.5.

Definition 3.4.2

Suppose S is a set, F is a field, and S is endowed with a scalar multiplication by members of F—a binary operation from $F \times S$ to S. Suppose that ϕ is a map defined on S and suppose that ran(ϕ) is also endowed with a scalar multiplication by members of F. Then ϕ will be said to preserve the scalar multiplication of S if and only if $\phi(\alpha x) = \alpha\phi(x)$ for all $x \in S$ and $\alpha \in F$.

Homomorphisms and Isomorphisms

Definition 3.4.3

Suppose X is a group, ring, vector space or algebra. A map defined on X which preserves the operations of X is called a HOMOMORPHISM, qualified, if necessary, as a GROUP HOMOMORPHISM, RING HOMOMORPHISM, ALGEBRA HOMOMORPHISM or otherwise, as appropriate. A homomorphism is called an ISOMORPHISM onto its range if and only if it is injective.

By straightforward calculation, the range of a group homomorphism is a group, that of a ring homomorphism is a ring, that of a vector space homomorphism is a vector space, and that of an algebra homomorphism is an algebra, each with the appropriate operations. Moreover, the image under a ring homomorphism of a unital ring is either trivial or unital; a non-zero ring homomorphism defined on a field is necessarily an isomorphism, and its image is a field; and the image of an integral domain under a non-injective ring homomorphism may or may not be an integral domain (see Q 3.4.1). It is easily shown that the inverse of an isomorphism is also an isomorphism.

Definition 3.4.4

Two structures of the same type (groups, vector spaces, algebras etc.) are said to be ISOMORPHIC to each other if and only if there exists an isomorphism from one onto the other.

In practice, mathematicians often regard isomorphic structures as being identical, being careful when making this identification not to consider structure other than that which is preserved under an isomorphism. If, for example, A and B are isomorphic groups, then there is no way of distinguishing between them purely as groups, and, as a group, B may be regarded simply as a relabelling of A; everything concerning group structure which can be proved about A holds also for B. There may of course be other structure which distinguishes A from B: A may be a ring and B not, or A may be a relation and B not, or B may be a group of functions and A not. In 3.2.3, we made scant distinction between a vector space V and an isomorphic copy $V \times \{0\}$; the fact that the members of the second space are ordered pairs is not relevant to the vector space structure. We acted likewise in 3.2.22.

Example 3.4.5

Suppose V is a vector space over a field F and S is a subspace of V. The quotient map $\pi\colon V \to V/S$ given by $x \mapsto x/S$ is a vector space homomorphism. Also, the quotient map of an algebra by an ideal is an algebra homomorphism. In a certain sense (see 3.4.13), quotient maps are typical of all homomorphisms.

Example 3.4.6

Suppose (V_i) is a family of vector spaces over F and the product is endowed with pointwise operations. For each $j \in I$, the natural projection $\pi_j\colon \prod_i V_i \to V_j$ given by $x \mapsto x_j$ is a vector space homomorphism. If (V_i) is a family of algebras with pointwise operations, then the natural projections are all algebra homomorphisms.

Example 3.4.7

Suppose X and Y are vector spaces over a field F. Suppose ϕ is a vector space homomorphism from X to Y, A is a non-empty finite subset of X and $(\mu_a)_{a \in A}$ is a family in F; then, as noted above for $\star$ and $\circ$, $\phi(\sum_a \mu_a a) = \sum_a \mu_a \phi(a)$. On the other hand, homomorphisms can be defined in terms of such sums. Suppose B is a basis for X and $\psi\colon B \to Y$; extend ψ to X as follows: for each $x \in X$, set $\psi(x) = \sum_a \mu_a \psi(a)$, where A is the unique finite subset of B and $(\mu_a)_{a \in A}$ is the unique family indexed by A such that $x = \sum_a \mu_a a$. It is easy to show that this LINEAR EXTENSION of ψ is a vector space homomorphism, and that it is indeed the only such homomorphism from X to Y whose restriction to B is the original map ψ. It is common practice to define vector space homomorphisms by specifying their values only on a basis; the map intended will always be the extension considered here.

Example 3.4.8

An injective function defined on an algebraic structure induces algebraic operations on its range and thus produces an isomorphic copy of its domain; the
induced operations may or may not coincide with any operations which are
already associated with that range. Suppose, for example, that A is an algebra
over a field F and that X is a set which is equinumerous with A. Suppose
$f: A \to X$ is a bijective map. Then operations of addition, scalar multiplication and multiplication are defined on X by $u + v = f(f^{-1}(u) + f^{-1}(v))$,
$\lambda u = f(\lambda f^{-1}(u))$, and $uv = f(f^{-1}(u)f^{-1}(v))$ for all $u, v \in X$ and $\lambda \in F$. With
this structure, X is an algebra and f is an isomorphism from A onto X.

Example 3.4.9

Collections of homomorphisms typically form new algebraic structures. Suppose, for example, that $(G, +)$ and $(H, +)$ are Abelian groups. Then the collection of group homomorphisms from G to H is itself an Abelian group under
pointwise addition; if, moreover, $G = H$, then this group becomes a unital ring
when multiplication is defined to be composition.

Example 3.4.10

Suppose X and Y are vector spaces over a field F. The set of vector space
homomorphisms from X into Y is a non-empty subset of Y^X; calculation con-
firms that it is closed under the pointwise operations of Y^X and is thus a vector
subspace of Y^X. We denote it by $\mathcal{L}(X, Y)$, or by $\mathcal{L}(X)$ if $Y = X$. We usually
denote members of $\mathcal{L}(X, Y)$ by upper case letters and, for $T \in \mathcal{L}(X, Y)$, we
write Tx rather than $T(x)$ for the value of T at x. If X, W and Z are vector
spaces over a field F and Y is a subspace of W, then, for each $T, U \in \mathcal{L}(X, Y)$
and $R, S \in \mathcal{L}(W, Z)$, it is easy to verify that $S \circ T \in \mathcal{L}(X, Z)$ and that
$S \circ (T + U) = (S \circ T) + (S \circ U)$ and $(R + S) \circ T = (R \circ T) + (S \circ T)$.
Composition is associative; we usually call it MULTIPLICATION and denote it
by juxtaposition. Then $\mathcal{L}(X)$ is a unital algebra over F; the unity $x \mapsto x$ is
denoted by I. If $T \in \mathcal{L}(X)$ is bijective, it is easily checked that $T^{-1} \in \mathcal{L}(X)$.

Example 3.4.11

Suppose V is a vector space over a field F. Then, provided $\dim(V) > 1$, the
unital algebra $\mathcal{L}(V)$ has subalgebras with different unities. To verify this, let
$x \in V \backslash \{0\}$ and let B be a basis for V with $x \in B$; define P on B by setting
$Px = x$ and $Py = 0$ for each $y \in B \backslash \{x\}$ and extend P linearly to V as in 3.4.7.
Now $P^2 = P$; so $\mathcal{S} = \{PTP \in \mathcal{L}(V) \mid T \in \mathcal{L}(V)\}$ is a subalgebra of $\mathcal{L}(V)$.
Observe that $I \notin \mathcal{S}$ but that $\mathcal{S}$ is a unital algebra with identity P.

Kernels

Preservation of an operation has a particular interest when it is *faithful*, in the sense that it is effected in a one-to-one manner. Unlike most functions, a homomorphism needs to be tested for injectivity at only one point.

Definition 3.4.12

Suppose f is a function whose range is a subset of an additive group. The subset $f^{-1}\{0\}$ of $\mathrm{dom}(f)$ is called the KERNEL of f; it is denoted by $\ker(f)$.

A group homomorphism maps 0 to 0, so its kernel is not empty. It is easy to show that the kernel of a group or vector space homomorphism is a subgroup or subspace, respectively, of the domain and that the kernel of a ring or algebra homomorphism is a ring or algebra ideal, respectively, of the domain.

Theorem 3.4.13

Suppose that X is an additive group, a ring, a vector space or an algebra and that ϕ is a homomorphism defined on X. Then $X/\ker(\phi)$ is isomorphic to $\phi(X)$. Also, ϕ is injective if and only if $\ker(\phi) = \{0\}$.

Proof

For $a, b \in X$, we have $a/\ker(\phi) = b/\ker(\phi) \Leftrightarrow \phi(a - b) = 0 \Leftrightarrow \phi(a) = \phi(b)$; so the map $x/\ker(\phi) \mapsto \phi(x)$ is well defined on $X/\ker(\phi)$ and is an isomorphism onto $\phi(X)$. If ϕ is injective, certainly $\ker(\phi) = \{0\}$; conversely, if $\ker(\phi) = \{0\}$, then $\phi(a) = \phi(b) \Rightarrow \phi(a - b) = 0 \Rightarrow a = b$.   □

Lemma 3.4.14

Suppose V is a vector space over a field F and $f \in \mathcal{L}(V, F) \setminus \{0\}$. Then $\ker(f)$ is a maximal subspace of V.

Proof

Let $a \in V \setminus \ker(f)$. Suppose $v \in V$ and let $\lambda = f(v)/f(a)$. Then $f(v - \lambda a) = 0$, so that $v - \lambda a \in \ker(f)$ and $v \in \ker(f) + Fa$. Therefore $\ker(f) \oplus Fa = V$.   □

Theorem 3.4.15

Suppose F is a field. If V is a vector space over F, then, for each finite linearly independent sequence $(f_i)_{i \in I}$ in $\mathcal{L}(V, F)$, there exists a linearly independent sequence $(b_i)_{i \in I}$ in V for which $\bigcap \{\ker(f_i) \mid i \in I\} \oplus \langle \{b_i \mid i \in I\} \rangle = V$ and which, for all $i, j \in I$, satisfies the equations $f_i(b_i) = 1$ and $f_i(b_j) = 0$ if $i \neq j$.

Proof

The result for all appropriate sequences of length 1 follows from 3.4.14. We suppose inductively that $k \in \infty \backslash \{0\}$ and that the proposition holds for all sequences of the specified type with length k. Let V be a vector space over F. Suppose $(f_i)_{i \in I}$ is a linearly independent sequence in $\mathcal{L}(V, F)$ of length $k+1$; let $m = \max I$ and $\underline{I} = I \backslash \{m\}$; set $K = \ker(f_m)$. Now $(f_i|_K)_{i \in \underline{I}}$ is a linearly independent sequence in $\mathcal{L}(K, F)$ because, if $(\gamma_i)_{i \in \underline{I}}$ is a sequence of scalars with $\sum_{i \in \underline{I}} \gamma_i f_i|_K = 0$ and if $z \in V \backslash K$, then $\sum_{i \in \underline{I}} \gamma_i f_i = (\sum_{i \in \underline{I}} \gamma_i f_i(z)/f_m(z)) f_m$, and linear independence of $(f_i)_{i \in I}$ yields $\gamma_i = 0$ for all $i \in \underline{I}$. By the inductive hypothesis, there are linearly independent sequences $(c_i)_{i \in \underline{I}}$ in V and $(b_i)_{i \in \underline{I}}$ in K such that, for all $i, j \in \underline{I}$, $f_i(c_j) = f_i(b_j) = 0$ if $i \neq j$ and $f_i(c_i) = f_i(b_i) = 1$, which satisfy $\bigcap \{\ker(f_i) \mid i \in \underline{I}\} \oplus \langle \{c_i \mid i \in \underline{I}\} \rangle = V$ and $\bigcap \{\ker(f_i|_K) \mid i \in \underline{I}\} \oplus \langle \{b_i \mid i \in \underline{I}\} \rangle = K$ respectively. The first decomposition yields the proper inclusion $\bigcap \{\ker(f_i) \mid i \in I\} \subset \bigcap \{\ker(f_i) \mid i \in \underline{I}\}$, because otherwise, for each $v \in V$, we should have $f_m(v) = \sum_{i \in \underline{I}} f_m(c_i) f_i(v)$, contradicting linear independence of $(f_i)_{i \in I}$. So there exists $b_m \in \bigcap \{\ker(f_i) \mid i \in \underline{I}\}$ with $f_m(b_m) = 1$. Since $\bigcap \{\ker(f_i|_K) \mid i \in \underline{I}\} = \bigcap \{\ker(f_i) \mid i \in I\}$, the second decomposition extends to $\bigcap \{\ker(f_i) \mid i \in I\} \oplus \langle \{b_i \mid i \in I\} \rangle = V$. For all $i, j \in I$, we have $f_i(b_j) = 0$ if $i \neq j$ and $f_i(b_i) = 1$. So the result holds for sequences of length $k+1$. Now we invoke the Principle of Finite Induction. $\square$

Corollary 3.4.16

Suppose V is a vector space over a field F and $\mathcal{S}$ is a non-empty finite linearly independent subset of $\mathcal{L}(V, F)$. Then $\dim(V/ \bigcap \{\ker(f) \mid f \in \mathcal{S}\}) = |\mathcal{S}|$.

Corollary 3.4.17

Suppose V is a vector space over a field F, $(f_i)_{i \in I}$ is a finite linearly independent sequence in $\mathcal{L}(V, F)$ and $(\alpha_i)_{i \in I}$ is a corresponding sequence in F. Then there exists $v \in V$ such that $f_i(v) = \alpha_i$ for all $i \in I$.

Proof

By 3.4.15, there exists a finite sequence $(b_i)_{i \in I}$ in V such that, for all $i, j \in I$, $f_i(b_j) = 0$ if $i \neq j$ and $f_i(b_i) = 1$. Let $v = \sum_{i \in I} \alpha_i b_i$. $\square$

Invariant Subspaces

Definition 3.4.18

Suppose V is a vector space over a field F and $T \in \mathcal{L}(V)$. A vector subspace M of V is called an INVARIANT SUBSPACE for T if and only if $T(M) \subseteq M$.

Example 3.4.19

Suppose V is a vector space and $T \in \mathcal{L}(V)$. Then $\ker(T)$ and $\mathrm{ran}(T)$ are invariant subspaces for T. So also are the GENERALIZED RANGE $\bigcap\{\mathrm{ran}(T^n) \mid n \in \infty\}$ of T and, since the collection $\{\ker(T^n) \mid n \in \infty\}$ is nested (see 3.2.4), the GENERALIZED KERNEL $\bigcup\{\ker(T^n) \mid n \in \infty\}$ of T.

Definition 3.4.20

Suppose X is a vector space over a field F and $X = \oplus_{i \in I} A_i$ is a decomposition of X into linearly independent subspaces of X, each of which is invariant under T. Then we say that this decomposition REDUCES T. In this case we may write $T = \oplus_i T_i$ where, for each $j \in I$, the map T_j is the compression of T to A_j.

Vector Space Isomorphisms and Dimension

Theorem 3.4.21

Suppose X and Y are vector spaces, $B \subseteq X$ and $T \in \mathcal{L}(X, Y)$.

- If T is injective and B is linearly independent, then $T(B)$ is a linearly independent subset of Y;
- If T is surjective and B spans X, then $T(B)$ spans Y;
- If T is bijective and B is a basis for X, then $T(B)$ is a basis for Y.

Proof

For the first part, suppose T is injective and B is linearly independent. Suppose A is a finite subset of B, $(\mu_a)_{a \in A}$ is a family in F and $\sum_{a \in A} \mu_a Ta = 0$; then $T(\sum_{a \in A} \mu_a a) = 0$ and, since T is injective, $\sum_{a \in A} \mu_a a = 0$; then linear independence of B yields $\mu_a = 0$ for all $a \in A$. For the second part, suppose instead that B spans X and that T is surjective. Let $y \in Y$. Then there exists $x \in X$ with $Tx = y$. Let A be a finite subset of B and $(\mu_a)_{a \in A}$ be a family in $F \backslash \{0\}$ such that $x = \sum_a \mu_a a$. Then $y = Tx = \sum_a \mu_a Ta$, which is a linear combination of members of $T(B)$. The third part follows from the first two. $\square$

Corollary 3.4.22

Suppose X and Y are vector spaces over a field F. Then X and Y are isomorphic if and only if they have the same dimension.

Proof

Firstly, suppose X and Y are isomorphic and let $S \in \mathcal{L}(X, Y)$ be an isomorphism. If B is a basis for X, then $S(B)$ is a basis for Y, by 3.4.21. Since S is bijective, $\dim(Y) = |S(B)| = |B| = \dim(X)$. Towards the converse, suppose that $\dim(X) = \dim(Y)$ and that B and C are bases for X and Y respectively. Since $|B| = |C|$, there exists a bijective map $f : B \to C$. Let T be the linear extension of f to X (3.4.7). Suppose A is a finite subset of B and $(\alpha_b)_{b \in A}$ is a family in F; then $T(\sum_{b \in A} \alpha_b b) = \sum_{b \in A} \alpha_b f(b)$ which, being a linear combination of distinct members of C, is not 0 unless $\alpha_b = 0$ for all $b \in A$. So $\ker(T) = \{0\}$ and T is injective by 3.4.13. Also, since $C \subseteq T(X)$, we have $Y = \langle C \rangle \subseteq T(X)$; so T is surjective and is an isomorphism onto Y.   □

Homomorphisms on Finite Dimensional Spaces

Definition 3.4.23
Suppose X and Y are vector spaces over a field F and $T \in \mathcal{L}(X, Y)$. The dimension of the subspace $T(X)$ of Y is called the RANK of T and is denoted by $\mathrm{rank}(T)$; the dimension of the subspace $\ker(T)$ of X is called the NULLITY of T and is denoted by $\mathrm{nul}(T)$.

Theorem 3.4.24   RANK-NULLITY THEOREM
Suppose X and Y are vector spaces over a field F and $T \in \mathcal{L}(X, Y)$. Then $\dim(X) = \mathrm{rank}(T) + \mathrm{nul}(T)$.

Proof
By 3.4.13, $X/\ker(T)$ is isomorphic to $\mathrm{ran}(T)$; so $\mathrm{rank}(T) = \dim(X/\ker(T))$ by 3.4.22. The equation then follows from 3.2.16.   □

Corollary 3.4.25
Suppose V is a finite dimensional vector space and $T \in \mathcal{L}(V)$. Then T is injective if and only if T is surjective.

Example 3.4.26
Suppose X and Y are finite dimensional vector spaces over a field F, with $\dim(X) = n$ and $\dim(Y) = m$. Then the vector spaces $\mathcal{L}(X, Y)$ and $\mathcal{M}_{m \times n}(F)$ are isomorphic. This can be established simply by checking that each of these spaces has dimension mn, but more is learnt by specifying an isomorphism as follows. Let $\{x_j \mid 1 \le j \le n\}$ and $\{y_i \mid 1 \le i \le m\}$ be any indexed bases for X and Y respectively. For each $T \in \mathcal{L}(X, Y)$, there exist unique scalars $\tau_{i,j}$ such that, for $1 \le j \le n$, $Tx_j = \sum_{i=1}^{m} \tau_{i,j} y_i$. It is easy to verify that the map $T \mapsto (\tau_{i,j})$ defined on $\mathcal{L}(X, Y)$ is a vector space isomorphism and that, where

appropriate, composition of homomorphisms is preserved by vector space isomorphisms of this type as multiplication of matrices. It follows, in particular,
that, when $X = Y$, $\mathcal{L}(X)$ and $\mathcal{M}_{n \times n}(F)$ are isomorphic algebras. The matrix
$(\tau_{i,j})$ is said to REPRESENT the homomorphism T with respect to the given
bases; we thus identify the algebra of homomorphisms between finite dimensional vector spaces with the algebra of matrices of the appropriate size (but
any change to either of the given bases, even one of order, will change the
representative matrices). Homomorphisms between arbitrary vector spaces can
thus be regarded as generalizations of matrices; but many results from matrix
theory do not carry over to the more general theory (see, for example, 3.4.29).

Eigenvalues

Invertibility has two aspects, injectivity and surjectivity; and a homomorphism fails to have an inverse homomorphism when it lacks either one of these
attributes. But *eigenvalues* depend on considering only non-injectivity.

Definition 3.4.27

Suppose V is a vector space over a field F and $T \in \mathcal{L}(V)$. A scalar λ is
called an EIGENVALUE of T if and only if $\ker(\lambda - T)$ is non-trivial. The space
$\ker(\lambda - T)$ is called the EIGENSPACE of T associated with λ. Since the collection
$\{\ker((\lambda - T)^n) \mid n \in \infty\}$ is nested, its union is also a subspace of V; it is called
the GENERALIZED EIGENSPACE of T associated with λ; its dimension is called
the ALGEBRAIC MULTIPLICITY of the eigenvalue λ for T.

Theorem 3.4.28

Suppose V is a vector space over F and $T \in \mathcal{L}(V)$. Then every eigenvalue of T
is a member of $\sigma(T)$. The converse also is true if V is finite dimensional.

Proof

Let $\lambda \in F$. If λ is an eigenvalue of T then $\lambda - T$ is not injective, so is not
invertible. It follows that every eigenvalue of T is in $\sigma(T)$. If $\dim(V) < \infty$ and
$\lambda \in \sigma(T)$, then $\lambda - T$ is not invertible, therefore not injective by 3.4.25. $\square$

Example 3.4.29

The finite stipulation in 3.4.28 is needed. Suppose B is a basis for a vector
space V and $|B| \geq \infty$. Well order B by the Well Ordering Principle. Define T
on B by $Tb = b_+$ and extend linearly to V. T is injective because B is linearly
independent; so 0 is not an eigenvalue of T. But $\min B \notin T(V)$; so $0 \in \sigma(T)$.

Example 3.4.30

Let V be a vector space over a field F, $T \in \mathcal{L}(V)$ and Λ the set of eigenvalues of T. For each $\lambda \in \Lambda$, let E_λ be the corresponding generalized eigenspace—it is clearly invariant under T; let T_λ be the compression of T to E_λ. We show that $(E_\lambda)_{\lambda \in \Lambda}$ is a linearly independent family of subspaces of V (3.2.21) and that $\sigma(T_\lambda) = \{\lambda\}$ for each $\lambda \in \Lambda$. If $\Lambda = \varnothing$, this is trivial; we suppose otherwise.

Firstly, we show that $\{E_\lambda \backslash \{0\} \mid \lambda \in \Lambda\}$ is disjointed. If $\lambda, \mu \in \Lambda$ and $E_\lambda \cap E_\mu \neq \{0\}$, then there exist $p, q \in \infty \backslash \{0\}$ with minimum sum $p + q$, for which there exists a non-zero vector $z \in \ker(\lambda - T)^p \cap \ker(\mu - T)^q$. But then it follows that $(\lambda - T)z \in \ker(\lambda - T)^{p-1} \cap \ker(\mu - T)^q$, whence $(\lambda - T)z = 0$; similarly $(\mu - T)z = 0$, so that $(\lambda - \mu)z = 0$ and, since $z \neq 0$, $\lambda = \mu$.

Secondly, we show that (E_λ) is a linearly independent family. Suppose otherwise and let $\sum_{i=1}^n v_i = 0$, the v_i being non-zero vectors in corresponding distinct generalized eigenspaces E_{μ_i} where the μ_i are distinct members of Λ, and $n \in \infty$ being the smallest possible for any such combination. Certainly $n > 1$. There exists $k \in \infty \backslash \{0\}$ with $v_n \in \ker(\mu_n - T)^k$, whence $0 = (\mu_n - T)^k \sum_{i=1}^n v_i = \sum_{i=1}^{n-1} w_i$ where, for $1 \leq i \leq n-1$, we have $w_i = (\mu_n - T)^k v_i \in E_{\mu_i}$ and $w_i \neq 0$ because $E_{\mu_i} \cap E_{\mu_n} = \{0\}$. This contradicts the minimality of n and proves our claim.

Thirdly, we show that, for each $\lambda \in \Lambda$ and $\alpha \in F \backslash \{\lambda\}$, the compression $\alpha - T_\lambda$ is bijective. It is injective because either $\alpha \notin \Lambda$ or $E_\lambda \cap E_\alpha = \{0\}$. To show surjectivity of $\alpha - T_\lambda$, we use induction. Certainly $\ker(\lambda - T)^0 \subseteq (\alpha - T)(E_\lambda)$. Suppose $k \in \infty \backslash \{0\}$ and $\ker(\lambda - T)^k \subseteq (\alpha - T)(E_\lambda)$. If $z \in \ker(\lambda - T)^{k+1}$, then $(\alpha - T)z \in (\alpha - T)(E_\lambda)$ and $(\lambda - T)z \in \ker(\lambda - T)^k \subseteq (\alpha - T)(E_\lambda)$, so that $(\alpha - \lambda)z \in (\alpha - T)(E_\lambda)$. Since $\lambda \neq \alpha$, we have $z \in (\alpha - T)(E_\lambda)$. Then the claim is justified by induction.

Involutions

Definition 3.4.31

Suppose that R and S are rings and that $\psi \colon R \to S$ is a group homomorphism which REVERSES PRODUCTS, in the sense that $\psi(ab) = \psi(b)\psi(a)$ for all $a, b \in R$. Then ψ is called a ring ANTI-HOMOMORPHISM. An anti-homomorphism from R to R which has the property that $\psi^2 = \iota_R$ is called an INVOLUTION on R.

Involutions are often denoted by $*$; the image of an element a under $*$ is then denoted by a^*. An involution $*$ on a ring R is bijective because its square ι_R is bijective; and, if R is unital, then $1^* = 1$. If R is an algebra over a field F, and F has a natural involution $\lambda \mapsto \overline{\lambda}$, then $*$ will be called an INVOLUTION on the algebra R only if we also have $(\lambda a)^* = \overline{\lambda} a^*$ for all $\lambda \in F$ and $a \in R$.

Example 3.4.32

Involutions on a commutative ring are simply isomorphisms whose square is the identity map; in particular, the identity map itself is an involution. But there are more interesting examples on non-commutative rings. Suppose R is a ring and $m, n \in \infty \backslash \{0\}$; then for each matrix $A = (a_{i,j}) \in \mathcal{M}_{m \times n}(R)$, the $n \times m$ matrix $(a_{j,i})$ is called the TRANSPOSE of A and is denoted by A^t. The map $A \mapsto A^t$ is an involution on the ring $\mathcal{M}_{n \times n}(R)$ of square matrices.

Definition 3.4.33

Suppose X and Y are rings or algebras with involutions, each denoted by *, and suppose that $\phi \colon X \to Y$ is a ring or algebra homomorphism which also preserves involution in the sense that $(\phi(a))^* = \phi(a^*)$ for all $a \in X$. Then ϕ is called a *-HOMOMORPHISM. If ϕ is injective then ϕ is called a *-ISOMORPHISM onto its range; it is easy to verify that in this case ϕ^{-1} also preserves involution. If there exists a *-isomorphism from X onto Y, then X and Y are said to be *-ISOMORPHIC rings or algebras.

EXERCISES

Q 3.4.1 Find a ring homomorphism from $-\infty \cup \infty$ onto $\{0, 1, 2, 3\}$ which is not an integral domain homomorphism.

Q 3.4.2 Let A be an arbitrary unital algebra. Show that there is an isomorphism from A onto a subalgebra of $\mathcal{L}(A)$.

Q 3.4.3 Let X be a vector space. Show that the only members of $\mathcal{L}(X)$ which commute with every member of $\mathcal{L}(X)$ are the scalar multiples of the identity mapping.

Q 3.4.4 Suppose X is a vector space and $T \in \mathcal{L}(X)$. Show that T is left invertible in $\mathcal{L}(X)$ if and only if T is injective. Show that T is right invertible in $\mathcal{L}(X)$ if and only if T is surjective.

Q 3.4.5 Suppose V is a vector space, $T \in \mathcal{L}(V)$ and $p \in \mathrm{poly}(\mathcal{L}(V))$. Show that $\ker(p(T))$ and $\mathrm{ran}(p(T))$ are subspaces of V invariant under T.

Q 3.4.6 Suppose X is a vector space and $T \in \mathcal{L}(X)$. Let $\mathcal{R}(T)$ denote the generalized range $\bigcap\{T^n(X) \mid n \in \infty\}$ of T. Then $T(\mathcal{R}(T)) \subseteq \mathcal{R}(T)$. Show that, if $\mathrm{nul}(T) < \infty$, then $T(\mathcal{R}(T)) = \mathcal{R}(T)$.

Q 3.4.7 Suppose X is a vector space over a field F and $T \in \mathcal{L}(X)$. Show that there exists an idempotent $P \in \mathcal{L}(X)$ such that $T - P \in \mathrm{inv}(\mathcal{L}(X))$.

Analytic Structure

The several chairs of order look you scour
With juice of balm and every precious flower:
Each fair instalment, coat, and several crest,
With loyal blazon, evermore be blest!
And nightly, meadow-fairies, look you sing,
Like to the Garter's compass, in a ring.   *Merry Wives of Windsor, V,v.*

It was hardly to be expected that the Axiom of Infinity would propel us inexorably from a discrete universe into a *continuous* one. But that is precisely what it does. And whether or not continuity is an intrinsic feature of creation, scientific enquiry into the mysteries of our world is greatly aided by a mathematics of continuity. This mathematics begins with a marriage of algebraic structure and order structure, from which arises a completely ordered field—essentially unique—which we call the *Real Number System*.

4.1 Ordered Algebraic Structure

Ordered Groups

Definition 4.1.1

An ordered triple $(G, P, +)$ is called an ORDERED GROUP if and only if $(G, +)$ is an Abelian group, $\{P, -P, \{0\}\}$ is a partition of G and $(P, +)$ is a semigroup. In that case, the members of P are said to be POSITIVE and the members of $-P$ NEGATIVE; P will be denoted by G^+ and $-P$ by G^-. We usually say simply that G is an ordered group (with operation $+$ and positive subset G^+). The set $\{0\} \cup G^-$ of NON-POSITIVE members of G will be denoted by $G^\ominus$ and the set $\{0\} \cup G^+$ of NON-NEGATIVE members by $G^\oplus$. An ordered group $(S, S^+, +)$ will be called an ORDERED SUBGROUP of an ordered group $(G, G^+, +)$ if and only if $(S, +)$ is a subgroup of $(G, +)$ and $S^+ = S \cap G^+$.

Theorem 4.1.2

Suppose $(G, G^+, +)$ is an ordered group. The relation $\{(a, b) \mid b - a \in G^+\}$ on G, denoted hereafter by $<$, is a total ordering on G. Then $G^+ = \{g \in G \mid 0 < g\}$ and $G^- = \{g \in G \mid g < 0\}$, and, for each $a, b \in G$, we have $a < b \Leftrightarrow -b < -a$.

Proof

Suppose $a, b, c \in G$. Then $a - a = 0 \notin G^+$; so $<$ is anti-reflexive. If $a < b$ and $b < c$, then $b - a \in G^+$ and $c - b \in G^+$, and, as G^+ is closed under addition, $c - a = (c + 0) - a = (c + (-b + b)) - a = ((c - b) + b) - a = (c - b) + (b - a) \in G^+$, so that $a < c$. Moreover, $b - a$ and $a - b$ are inverse to one another; so, provided $b \neq a$, one of them is in G^+ and the other in G^-, yielding either $a < b$ or $b < a$. Therefore $<$ is a total ordering on G. The rest follows easily. $\square$

Example 4.1.3

If the induced total ordering (4.1.2) on an ordered group is a complete ordering (1.3.10), we use the term COMPLETE ORDERED GROUP to describe the structure. The ordering on the Abelian group $-\infty \cup \infty$ (3.1.24), with the set of positive elements defined to be $\infty \backslash \{0\}$ is an extension of the ordering of the counting numbers; it is the standard ordering and clearly differs from that of 2.3.10. It is evidently enumerative and complete but not dense.

Example 4.1.4

No non-trivial ordered group is well ordered; indeed it can have neither maximal nor minimal elements: suppose G is a non-trivial ordered group; there exists $x \in G^+$; then $-x < 0 < x$, so that 0 is neither maximal nor minimal. And for each $g \in G \backslash \{0\}$, either $0 < g < g + g$ or $g + g < g < 0$.

Ordered Rings and Fields

Definition 4.1.5

An ordered quadruple $(F, P, +, \times)$ is called an

- ORDERED RING if and only if $(F, +, \times)$ is a ring, $(F, P, +)$ is an ordered group and P is closed under $\times$;
- ORDERED DOMAIN if and only if it is an ordered ring and $(F, +, \times)$ is an integral domain;
- ORDERED FIELD if and only if it is an ordered ring and $(F, +, \times)$ is a field.

Usually we say simply that F is an ordered ring, domain or field (with the specified structure); if the ordering is complete, then that word may be included in the appellation. A subring, subdomain or subfield S of F is called an ordered subring, subdomain or subfield, as appropriate, if and only if $S^+ = S \cap F^+$.

Suppose R is an ordered ring and $a, b \in R$. Since $(-a)b = -(ab) = a(-b)$ and $(-a)(-b) = ab$ (3.1.26), it follows that ab is positive if and only if either a and b are both positive or a and b are both negative; in like vein, ab is negative if and only if one of a and b is positive and the other negative. Then, if $a, b, c, d \in R$ and $0 < a < b$ and $0 < c < d$, we have $0 < ac < bc < bd$. Also, if R is unital, then, since $1 \cdot 1 = 1$, we have $1 \in R^+$; and, if $a, b \in \mathrm{inv}(R)$ and $0 < a < b$, then $0 < b^{-1} < a^{-1}$. These facts we shall henceforth use without comment.

Example 4.1.6
By 3.1.31, $-\infty \cup \infty$ is an integral domain; and by 4.1.3, it is an ordered group. Since $\infty \backslash \{0\}$ is closed under multiplication, $-\infty \cup \infty$ is an ordered domain.

Example 4.1.7
Every ordered field is densely ordered. To show this, suppose $(F, F^+, +, \times)$ is an ordered field. The ordering is total by 4.1.2. Also $2 = 1+1 \in F^+$ is invertible and $2^{-1} \in F^+$. Suppose $a, b \in F$ and $a < b$. Then $2a < a+b < 2b$ and it follows that $a < 2^{-1}(a + b) < b$.

Order Isomorphisms

We show in this subsection that every ordered group includes an isomorphic copy of $-\infty \cup \infty$ in which the order is preserved. Then every ordered unital ring includes exactly one such copy which identifies the unities; and the relevant isomorphism also preserves multiplication. These facts imply that no complete ordered field has a complete ordered proper subfield and move us a step nearer to showing that the *Real Number System* is essentially unique.

Definition 4.1.8
A strictly increasing homomorphism between ordered groups, rings, integral domains or fields is called an ORDER HOMOMORPHISM; the strictly increasing nature of such a map implies its injectivity and it is therefore called also an ORDER ISOMORPHISM onto its range. The inverse of an order isomorphism is clearly also an order isomorphism. Structures compared by a surjective order isomorphism are said to be ORDER ISOMORPHIC to one another. If it is necessary to avoid ambiguity, terms such as GROUP ORDER ISOMORPHISM may be used.

Theorem 4.1.9

Suppose G is a non-trivial ordered group and $g \in G^+$. Then there exists a unique order isomorphism ϕ from the ordered group $-\infty \cup \infty$ onto an ordered subgroup of G such that $\phi(1) = g$. Moreover

- if the order on G is complete, then, for each $x \in G$, there exists a unique $n \in -\infty \cup \infty$ such that $\phi(n-1) \leq x < \phi(n)$;
- if G is a complete ordered field, then $\inf_G\{(\phi(n))^{-1} \mid n \in \infty\backslash\{0\}\} = 0$.

Proof

Consider the translation $x \mapsto x + g$ defined on G. By the Finite Recursion Theorem II (Q 2.3.4), there exists a unique function $\phi: -\infty \cup \infty \to G$ which satisfies $\phi(0) = 0$ and $\phi(n) = \phi(n-1) + g$ for all $n \in -\infty \cup \infty$. Induction shows that this is an order isomorphism; its uniqueness with the stated properties implies that it is the only one which maps 1 to g. Now suppose that the order on G is complete. Then $\text{ran}(\phi)$ is not bounded above; indeed, if it were then there would exist $z = \sup \text{ran}(\phi) \in G$ and, since $z - g < z$, there would exist $n \in -\infty \cup \infty$ such that $z - g < \phi(n)$, yielding the contradiction $z < \phi(n+1)$. Similarly, $\text{ran}(\phi)$ is not bounded below. So, for each $x \in G$, the set $\phi^{-1}(\hat{x})$ is a non-empty subset of $-\infty \cup \infty$ which is bounded below in $-\infty \cup \infty$. Let n be its infimum; then $\phi(n-1) \leq x < \phi(n)$. Lastly, suppose G is a complete ordered field and $z \in G^+$. Then $0 < z^{-1}$ and there exists $m \in \infty\backslash\{0\}$ such that $z^{-1} < \phi(m)$; so $0 < \phi(m)^{-1} < z$ and $0 \leq \inf_G\{(\phi(n))^{-1} \mid n \in \infty\backslash\{0\}\} \leq \inf G^+$. But $\inf G^+ = 0$ because the ordering is dense by 4.1.7. $\square$

Corollary 4.1.10 EUCLIDEAN ALGORITHM

Suppose $a \in -\infty \cup \infty$ and $b \in \infty\backslash\{0\}$. Then there exist $q, r \in -\infty \cup \infty$ such that $a = bq + r$ and $0 \leq r < b$.

Proof

$-\infty \cup \infty$ is a complete ordered group and b is positive. Let ϕ be the order isomorphism of 4.1.9 with $\phi(1) = b$. By 4.1.9, there exists $q \in -\infty \cup \infty$ such that $\phi(q) \leq a < \phi(q + 1)$. Set $r = a - bq$ and the result follows. $\square$

If G is a unital ring, then an inductive argument shows that the order isomorphism of 4.1.9 which identifies the unity of G with the counting number 1 preserves multiplication. It follows that every ordered field F includes a unique ring order isomorphic copy of $-\infty \cup \infty$ in which the unities are identified with each other. It is customary to use the symbols $\ldots, -3, -2, -1, 0, 1, 2, 3, \ldots$ for the images in F of the corresponding members of $-\infty \cup \infty$.

Theorem 4.1.11

Suppose F is a complete ordered field and S is a complete ordered subfield of F. Then $S = F$.

Proof

Let $n \mapsto \underline{n}$ be the unique ring order isomorphism from $-\infty \cup \infty$ to F with $\underline{1} = 1$; then $\underline{n} \in S$ for each $n \in -\infty \cup \infty$ by 4.1.9. Let $z \in F$. Since the ordering of F is complete, 4.1.9 implies that $A = \{x \in S \mid x < z\}$ is non-empty and bounded above in S and that $B = \{x \in S \mid z < x\}$ is non-empty and bounded below in S. Let $a = \sup_S A$ and $b = \inf_S B$. We claim that $a = z = b$. Indeed, if we had $a < z$, then, again by 4.1.9, there would exist $n \in \infty \backslash \{0\}$ such that $\underline{n}^{-1} < z - a$ and then $m \in -\infty \cup \infty$ such that $\underline{m} - \underline{1} = \underline{m-1} \leq \underline{n}a < \underline{m}$, which together imply that $a < \underline{n}^{-1}\underline{m} < z$, contradicting the definition of a because $\underline{n}^{-1}\underline{m} \in S$; therefore $z \leq a$; similarly $b \leq z$; since $a \leq b$, we have $a = z = b$ and therefore $z \in S$.   □

It is important to note that the more conditions we try to impose upon a set, the more likely it becomes that no set or very few sets can be endowed with them. Conditions we impose on sets are usually designed to reflect structure which occurs naturally either in mathematics or in another discipline where mathematics is used. Even so, we should always ensure that sets of the type we are discussing exist. At present, we do not know that there is any complete ordered field. We certainly hope that there is, because the structure is precisely that which we expect the *Real Number System* to enjoy. If there were no such field, then there would be no structured set corresponding exactly to our notion of the *Real Line*. Happily, ZF ensures the existence of such a set, as we shall see in the next section.

EXERCISES

Q 4.1.1 Define a total ordering $<'$ on $-\infty \cup \infty$ as follows: for $m, n \in \infty$, set $m <' n$ if and only if $m < n$; for $m \in -\infty \backslash \{0\}$ and $n \in \infty$, set $m <' n$; for $m, n \in -\infty \backslash \{0\}$, set $m <' n$ if and only if $n < m$. Show that $-\infty \cup \infty$ with this ordering is not similar to $-\infty \cup \infty$ with the usual ordering of 4.1.3, nor to $-\infty \cup \infty$ with the ordering of 2.3.10.

Q 4.1.2 Suppose R is an ordered ring. Show that R has no zero divisors; deduce that every commutative ordered ring is an ordered domain.

4.2 Number Systems

The Real Number System

We have described the properties of a complete ordered field, anticipating that
there exists a set of real numbers with such structure. In this section, we define
real numbers; we indicate briefly, omitting details, the steps that might be
taken, using 4.2.1 below freely, to prove that they form a complete ordered
field and to show that this structure is essentially unique, in the sense that
every other complete ordered field is order isomorphic to our Real Number
System.

For the purposes of this section, we denote the set $(-\infty \cup \infty) \times (\infty\backslash\{0\})$ by
Ω. We define a relation $<$ on Ω by specifying that, for each (a, b) and (c, d) in Ω,
$(a, b) < (c, d)$ if and only if $ad < bc$. If $(a, b), (c, d), (e, f) \in \Omega$ and $(a, b) < (c, d)$
and $(c, d) < (e, f)$, then $ad < bc$ and $cf < de$, whence, applying the facts
stated after 4.1.5 to the ordered domain $-\infty \cup \infty$, we have $adf < bcf < bde$
and therefore $af < be$, giving $(a, b) < (e, f)$. So $<$ is a partial order relation on
Ω; it is not a total order relation, because, for example, $(0, 1)$ and $(0, 2)$ are not
comparable. For each $z = (a, b) \in \Omega$ we shall use alternative notation $\mathrm{seg}(a, b)$
for the lower segment $\grave{z}$ determined by (a, b). The partial ordering of Ω has the
special properties of 4.2.1 below.

Lemma 4.2.1
Ω has no maximal and no minimal elements. Moreover, for each $x, y \in \Omega$ with
$x < y$, we have $\acute{x} \cap \grave{y} \neq \varnothing$ and $\acute{x} \cup \grave{y} = \Omega$.

Proof
If $(r, s) \in \Omega$ then $(r - 1, s) < (r, s) < (r + 1, s)$, so that (r, s) is neither maximal
nor minimal. Suppose $(a, b), (c, d) \in \Omega$ and $(a, b) < (c, d)$. Then $ad < bc$; it
follows that $a(b + d) < b(a + c)$ and $(a + c)d < (b + d)c$ and hence that
$(a, b) < (a + c, b + d) < (c, d)$. And, if $(e, f) \in \Omega$ and $(e, f) \not< (c, d)$, then
$fc \leq ed$; but $ad < bc$; so $adf < bcf \leq bed$, then $af < be$ and $(a, b) < (e, f)$. $\square$

Definition 4.2.2
A non-empty proper subset r of Ω will be called

- a REAL NUMBER if and only if r is a union of lower segments of Ω;
- a RATIONAL NUMBER if and only if r is a lower segment of Ω;
- an INTEGER if and only if $r = \mathrm{seg}(n, 1)$ for some $n \in -\infty \cup \infty$;
- a NATURAL NUMBER if and only if $r = \mathrm{seg}(n, 1)$ for some $n \in \infty\backslash\{0\}$.

The classes of numbers thus defined are subsets of $\mathcal{P}(\Omega)$. We shall denote the set of real numbers by $\mathbb{R}$, the set of rational numbers by $\mathbb{Q}$, the set of integers by $\mathbb{Z}$ and the set of natural numbers by $\mathbb{N}$. Evidently $\mathbb{N} \subseteq \mathbb{Z} \subseteq \mathbb{Q} \subseteq \mathbb{R}$. Easy calculations show that $\operatorname{seg}(0,1) \in \mathbb{Z}\backslash\mathbb{N}$ and that $\operatorname{seg}(1,2) \in \mathbb{Q}\backslash\mathbb{Z}$; so the first two inclusions are proper; the third inclusion is also proper (see 4.2.7).

For $r, s \in \mathbb{R}$, define $r + s$ to be $\bigcup\{\operatorname{seg}(ad + bc, bd) \mid (a,b) \in r, (c,d) \in s\}$; it is easy to show that $r + s \neq \Omega$, which ensures that $r + s \in \mathbb{R}$. Then it must be shown that $+$ is associative and that $(\mathbb{R}, +)$ is an Abelian group, the zero (which we shall denote by 0) being $\operatorname{seg}(0,1)$ and the inverse of $r \in \mathbb{R}$ being $\bigcup\{\operatorname{seg}(a,b) \mid (-a,b) \in \Omega\backslash r\}$ (the Euclidean Algorithm, 4.1.10, is crucial in the proof); this last is a member of $\mathbb{Q}$ or $\mathbb{Z}$ precisely when r is, so that $\mathbb{Q}$ and $\mathbb{Z}$ are subgroups of $\mathbb{R}$. Then $\mathbb{R}$ is an ordered group when its set of positive elements is defined to be $\{r \in \mathbb{R} \mid (0,1) \in r\}$, and the order thus induced is precisely that of proper inclusion. Moreover, if $\mathcal{A}$ is any non-empty set of real numbers which is bounded above in $\mathbb{R}$, then $\bigcup\mathcal{A} \neq \Omega$, whence $\bigcup\mathcal{A}$ is the supremum of $\mathcal{A}$ in $\mathbb{R}$, so that $\mathbb{R}$ is completely ordered. This complete ordering will henceforth be denoted, as is usual, by $<$. The set of real numbers with this ordering is referred to as the REAL LINE or as the CONTINUUM.

For $r \in \mathbb{R}$, we define $0r = 0 = r0$; then for each $r, s \in \mathbb{R}^+$, we define rs and $(-r)(-s)$ both to be $\bigcup\{\operatorname{seg}(ac, bd) \mid a, c \in \infty, (a,b) \in r, (c,d) \in s\}$ and $(-r)s$ and $r(-s)$ both to be $-(rs)$. These are real numbers and multiplication thus defined distributes over addition, making the group $\mathbb{R}$ into a unital ring; it is an ordered ring because $\mathbb{R}^+$ is closed under multiplication. The unity $\operatorname{seg}(1,1)$ of $\mathbb{R}$ is denoted by 1. For each $r \in \mathbb{R}^+$, $(a,b) \in \Omega\backslash r \Rightarrow a \in \infty\backslash\{0\}$; then r has multiplicative inverse $r^{-1} = \bigcup\{\operatorname{seg}(b,a) \mid (a,b) \in \Omega\backslash r\}$ and $-r$ has inverse $-(r^{-1})$, these inverses being in $\mathbb{Q}$ whenever $r \in \mathbb{Q}$. So $\mathbb{R}$ is a complete ordered field and $\mathbb{Q}$ is an ordered subfield of $\mathbb{R}$.

The ring order isomorphic copy of $-\infty \cup \infty$ in $\mathbb{R}$ is $\mathbb{Z}$, the isomorphism being the map $n \mapsto \operatorname{seg}(n,1)$. In view of this, for each $n \in -\infty \cup \infty$, we shall denote by n the integer $\operatorname{seg}(n,1)$ and blur the distinction between $-\infty \cup \infty$ and $\mathbb{Z}$. The set of positive members of the ordered domain $\mathbb{Z}$ is $\mathbb{N}$; and, since $\mathbb{N}$ is an order isomorphic copy of $\infty\backslash\{0\}$, it is well ordered. The subset $\mathbb{N} \cup \{0\}$ of $\mathbb{Z}$ is an order isomorphic copy of ∞.

Each rational number $r = \operatorname{seg}(m,n)$ determined by $(m,n) \in \Omega$ is equal to m/n, where m and n this time denote the corresponding members of $\mathbb{Z}$ and $\mathbb{N}$ respectively (see 3.1). There are, of course, many such ways of representing r as a quotient; but, since $\mathbb{N}$ is well ordered, there exists a minimum value for n in such a quotient; denoting this by q, it can be shown that there exists a unique $p \in \mathbb{Z}$ such that the representations of r as a quotient of an integer by a natural number are precisely those which can be written kp/kq where $k \in \mathbb{N}$. We say that the rational number r is given in STANDARD FORM by the representation

p/q. Moreover, the map $p/q \mapsto (p,q)$ from $\mathbb{Q}^{\oplus}$ to $\infty \times \infty$, where each member of $\mathbb{Q}^{\oplus}$ is presented in standard form as p/q, is injective; so $\mathbb{Q}^{\oplus}$ is countable by 2.4.4. Similarly $\mathbb{Q}^{\ominus}$ is countable; therefore $\mathbb{Q} = \mathbb{Q}^{\ominus} \cup \mathbb{Q}^{\oplus}$ is countable, and so are its subsets $\mathbb{Z}$ and $\mathbb{N}$. Since $\mathbb{N}$ is in one-to-one correspondence with $\infty \backslash \{0\}$, we have $|\mathbb{N}| = |\mathbb{Z}| = |\mathbb{Q}| = \infty$. We shall show later (4.3.16) that $|\mathbb{R}| = |2^{\infty}|$; establishing that, in a particular sense, most real numbers are not rational.

The identification of $-\infty \cup \infty$ with $\mathbb{Z}$ by the order isomorphism given above yields the opportunity to make the following slight shift in notation: if X is a set and $n \in \mathbb{N}$, then X^n will be used to denote the set of functions from $\{m \in \mathbb{N} \mid 1 \le m \le n\}$ into X; its members are finite sequences indexed in the generally accepted way starting with 1 rather than 0; regarded as being generalizations of ordered pairs, triples and so on, they are called ORDERED ENTUPLES and may be denoted by expressions like $(x_1, \ldots, x_n)$. We shall also write X^2 for $X \times X$, disregarding the difference between these sets (see 1.2.24).

In future, unless we state otherwise, we shall generally assume that the indexing set for a well ordered sequence is $\mathbb{N}$, or $\{m \in \mathbb{N} \mid 1 \le m \le n\}$ if it is a finite sequence of length n. To indicate that the indexing set is necessarily one of these, we shall sometimes denote it by $\mathbb{I}$. Similarly, the indexing set for an $m \times n$ matrix will be assumed to be $\{i \in \mathbb{N} \mid 1 \le i \le m\} \times \{i \in \mathbb{N} \mid 1 \le i \le n\}$ unless we state otherwise. In the same vein, when we denote a power z^n of a member z of a multiplicative semigroup (see 3.1.18), we shall generally assume that n is an integer rather than a member of $-\infty \cup \infty$.

Density

Every ordered field is densely ordered by 4.1.7. But $\mathbb{R}$ has the property that between every two real numbers, there is not simply another real number, but a rational number—and, in consequence, an infinite set of rational numbers.

Theorem 4.2.3 DENSITY THEOREM
Suppose $a, b \in \mathbb{R}$ and $a < b$. Then there exists $p \in \mathbb{Q}$ such that $a < p < b$. Also, for each $r \in \mathbb{R}$, we have $\sup_{\mathbb{R}} \{q \in \mathbb{Q} \mid q < r\} = r = \inf_{\mathbb{R}} \{q \in \mathbb{Q} \mid r < q\}$.

Proof
Let $x \in \mathbb{R}$ satisfy $a < x < b$. Let $(m, n) \in b \backslash x$. Then $a < x \le \text{seg}(m, n) < b$. Set $p = \text{seg}(m, n)$ to prove the first assertion; the second follows immediately.   □

Theorem 4.2.4
Suppose F is a complete ordered field. Then F is order isomorphic to $\mathbb{R}$.

Proof

F includes an order isomorphic copy of $-\infty \cup \infty$. So there exists an order isomorphism ϕ from $\mathbb{Z}$ onto this copy with $\phi(1) = 1$. The extension of ϕ to $\mathbb{Q}$ given by $q \mapsto \phi(m)/\phi(n)$, where m/n is q in standard form, is evidently also an order isomorphism. Using 4.2.3, we infer that the map defined on $\mathbb{R}$ by $r \mapsto \sup\{\phi(q) \mid q \in \mathbb{Q},\ q < r\}$ is an extension of ϕ and is an order isomorphism; so its range is a complete ordered subfield of F. By 4.1.11, this extension is surjective. $\square$

Irrational Numbers

We establish in this subsection that $\mathbb{Q} \neq \mathbb{R}$. In fact, most real numbers are not rational, in the sense that $|\mathbb{R}\backslash\mathbb{Q}| = |\mathbb{R}| = |2^\infty|$, whereas $|\mathbb{Q}| = \infty$.

Definition 4.2.5

For $r \in \mathbb{R}^+$, we define $0^r = 0$. For each $z \in \mathbb{R}^+$, POWERS z^r have been defined when $r \in \mathbb{Z}$; we extend this definition to $r \in \mathbb{R}$ as follows: for $n \in \mathbb{N}$, we define $z^{1/n}$ to be the unique solution $x \in \mathbb{R}^+$ of the equation $x^n = z$; this is readily shown to be $\sup\{y \in \mathbb{R}^+ \mid y^n \leq z\}$; then, for each other $q \in \mathbb{Q}$, we define z^q to be $(z^m)^{1/n}$, where $q = m/n$ in standard form; and for each $r \in \mathbb{R}\backslash\mathbb{Q}$, if $1 \leq z$ we define $z^r = \sup\{z^q \mid q \in \mathbb{Q} : q < r\}$, which also equals $\inf\{z^q \mid q \in \mathbb{Q} : r < q\}$; and if $0 < z < 1$, we define $z^r = (1/z)^{-r}$, which is easily checked to be equal to both $\sup\{z^q \mid q \in \mathbb{Q} : r < q\}$ and $\inf\{z^q \mid q \in \mathbb{Q} : q < r\}$. For $z \in \mathbb{R}^\oplus$, we adopt the usual convention of denoting the POSITIVE SQUARE ROOT $z^{1/2}$ of z by $\sqrt{z}$. Evidently, if $r \neq 0$, then $z \mapsto z^r$ maps $\mathbb{R}^+$ onto $\mathbb{R}^+$. It is just a little harder to show that, if $z \in \mathbb{R}^+\backslash\{1\}$, then $r \mapsto z^r$ maps $\mathbb{R}$ onto $\mathbb{R}^+$—it is strictly increasing if $z > 1$ and strictly decreasing if $z < 1$.

Theorem 4.2.6

Suppose $m \in \mathbb{N}$ and $\sqrt{m} \in \mathbb{Q}$. Then $\sqrt{m} \in \mathbb{N}$.

Proof

Suppose $\sqrt{m} = p/q$ in standard form. Then $mq/p = p/q$; so there exists $k \in \mathbb{Z}\backslash\{0\}$ such that $p = kq$; whence $\sqrt{m} = p/q = kq/q = k \in \mathbb{Z}$. Since $\sqrt{m} > 0$ by definition, we have $\sqrt{m} \in \mathbb{N}$. $\square$

Since $1 < 2 < 4$, we have $1 < \sqrt{2} < 2$ and then, because $2 = 1_+$ in $\mathbb{Z}$, $\sqrt{2}$ is not an integer. So 4.2.6 ensures that $\sqrt{2}$ is not rational, and we have shown that $\mathbb{Q} \neq \mathbb{R}$. Members of $\mathbb{R}\backslash\mathbb{Q}$ are called IRRATIONAL NUMBERS.

Theorem 4.2.7 SECOND DENSITY THEOREM

Suppose $r, s \in \mathbb{R}$ and $r < s$. Then there exists $t \in \mathbb{R}\backslash\mathbb{Q}$ such that $r < t < s$.

Proof

By 4.2.3, there exist $p, q \in \mathbb{Q}$ such that $r < p < q < s$. Using 4.1.9, there exists $n \in \mathbb{N}$ such that $2/(q - p) < n$. Now $p + (\sqrt{2}/n) \notin \mathbb{Q}$ because $\mathbb{Q}$ is a field and $\sqrt{2} \notin \mathbb{Q}$. And $r < p < p + n^{-1} < p + n^{-1}\sqrt{2} < p + 2n^{-1} \leq q < s$.   □

Extended Number Systems

Non-empty subsets of $\mathbb{R}$ may or may not be bounded above or below because $\mathbb{R}$ has neither a maximum nor a minimum member. For some purposes, it is easier to consider $\mathbb{R}$ as a totally ordered subset of a set $\tilde{\mathbb{R}}$ which does have such members. Every non-empty subset of $\mathbb{R}$ is order-bounded in $\tilde{\mathbb{R}}$.

Definition 4.2.8

We define the set $\tilde{\mathbb{R}}$ of EXTENDED REAL NUMBERS to be $\mathbb{R} \cup \{-\infty, \infty\}$ and extend the total ordering of $\mathbb{R}$ to $\tilde{\mathbb{R}}$ by specifying that $-\infty < \infty$ and that $-\infty < r < \infty$ for all $r \in \mathbb{R}$. Since $\mathbb{R}$ is completely ordered, every non-empty subset A of $\tilde{\mathbb{R}}$ has both an infimum and a supremum in $\tilde{\mathbb{R}}$; we shall habitually use the notation $\sup A$ and $\inf A$ to refer to the supremum and infimum, respectively, of A in $\tilde{\mathbb{R}}$. We extend addition—but not to the whole of $\tilde{\mathbb{R}}$—by specifying that $\infty + r = \infty = r + \infty$ and $-\infty + r = -\infty = r + (-\infty)$ for all $r \in \mathbb{R}$; and we extend multiplication by setting $\infty r = (-\infty)(-r) = \infty = (-r)(-\infty) = r\infty$ if $r \in \mathbb{R}^+$ and $\infty r = (-\infty)(-r) = -\infty = (-r)(-\infty) = r\infty$ if $r \in \mathbb{R}^-$. We define $\tilde{\mathbb{N}} = \mathbb{N} \cup \{\infty\}$, $\tilde{\mathbb{Z}} = \mathbb{Z} \cup \{-\infty, \infty\}$ and $\tilde{\mathbb{Q}} = \mathbb{Q} \cup \{-\infty, \infty\}$; we endow these sets with the ordering and operations induced from $\tilde{\mathbb{R}}$.

Intervals

Suppose $a, b \in \tilde{\mathbb{R}}$ and $a \leq b$. We adopt the notation $[a, b]$, (a, b), $(a, b]$ and $[a, b)$ for the sets $\{x \in \tilde{\mathbb{R}} \mid a \leq x \leq b\}$, $\{x \in \tilde{\mathbb{R}} \mid a < x < b\}$, $\{x \in \tilde{\mathbb{R}} \mid a < x \leq b\}$, and $\{x \in \tilde{\mathbb{R}} \mid a \leq x < b\}$ respectively. It should always be clear from the context whether (a, b) refers to such a set or to an ordered pair.

Definition 4.2.9

A non-empty subset I of $\tilde{\mathbb{R}}$ is called an INTERVAL if and only if, for each $x \in \mathbb{R}$, $\inf I < x < \sup I \Rightarrow x \in I$. An interval which is a subset of $\mathbb{R}$ is called a REAL INTERVAL. An interval I is called an OPEN INTERVAL if and only if $\inf I \notin I \backslash \{-\infty\}$ and $\sup I \notin I \backslash \{\infty\}$; it is called a CLOSED INTERVAL if and only if either it is real and $\inf I \in I \cup \{-\infty\}$ and $\sup I \in I \cup \{\infty\}$ or it is not real and both $\inf I \in I$ and $\sup I \in I$; other intervals are said to be HALF OPEN or HALF CLOSED. A DEGENERATE INTERVAL is an interval which consists of a single point; such an interval is closed. A BOUNDED INTERVAL is one which has real infimum and supremum; and a subset of $\mathbb{R}$ is said to be BOUNDED in $\mathbb{R}$ if and only if it is included in a bounded interval. Evidently, a non-empty subset of $\mathbb{R}$ is bounded in $\mathbb{R}$ if and only if it is order-bounded in $\mathbb{R}$. It is clear that a subset I of $\tilde{\mathbb{R}}$ is an interval if and only if I is not empty and I is of the form $[a, b]$, (a, b), $(a, b]$ or $[a, b)$, where $a = \inf I$ and $b = \sup I$. Moreover, it follows easily from the denseness of the ordering that a real interval I is open if and only if, for each $u \in I$, there exists $\epsilon \in \mathbb{R}^+$ such that $(u - \epsilon, u + \epsilon) \subseteq I$.

The Complex Number System

Complex numbers are often presented as ordered pairs of real numbers. We modify the presentation so that the set of complex numbers is a superset of $\mathbb{R}$.

Definition 4.2.10

Let $\mathbb{C} = \mathbb{R} \cup (\mathbb{R} \times \mathbb{R} \backslash \{0\})$, and extend addition and multiplication of real numbers to $\mathbb{C}$ by setting, for each a, c, r in $\mathbb{R}$ and $b, d, s \in \mathbb{R} \backslash \{0\}$:

- $(a, b) + (c, d) = (a + c, b + d)$ if $b + d \neq 0$ and $(a, b) + (c, d) = a + c$ otherwise;
- $(a, b)(c, d) = (ac - bd, bc + ad)$ if $bc + ad \neq 0$ and $(a, b)(c, d) = ac - bd$ otherwise;
- $r + (c, d)$ and $(c, d) + r$ both to be $(r + c, d)$;
- $s(c, d)$ and $(c, d)s$ both to be (sc, sd); and $0(c, d)$ and $(c, d)0$ both to be 0.

It is easily verified that $\mathbb{C}$ with these operations is a field which has $\mathbb{R}$ as a proper subfield. We call it the field of COMPLEX NUMBERS or the Complex Plane and denote it simply by $\mathbb{C}$. The complex number $(0, 1)$ is denoted by i; note that $i^2 = -1$. Each $z \in \mathbb{C}$ can be written uniquely in STANDARD FORM as $a + ib$ where $a, b \in \mathbb{R}$; a is called the REAL PART of z and is denoted by $\Re(z)$; b is called the IMAGINARY PART of z and is denoted by $\Im(z)$. If $z = a + ib$ in standard form, we denote the COMPLEX CONJUGATE $a - ib$ of z by $\bar{z}$ and call the map $z \mapsto \bar{z}$ CONJUGATION; this is clearly an involution on $\mathbb{C}$. We denote the MODULUS, $\sqrt{a^2 + b^2}$, of z by $|z|$. If $z \in \mathbb{R}$, then $|z|$ is often called the ABSOLUTE VALUE of z; in this case, $|z| = z$ if $z \geq 0$ and $|z| = -z$ otherwise. An inductive argument establishes that, if S is a finite subset of $\mathbb{C}$, then $\left| \sum_{z \in S} z \right| \leq \sum_{z \in S} |z|$ and $\left| \prod_{z \in S} z \right| = \prod_{z \in S} |z|$.

Definition 4.2.11

Any set D which satisfies $\{z \in \mathbb{C} \mid |z - a| < r\} \subseteq D \subseteq \{z \in \mathbb{C} \mid |z - a| \leq r\}$ for some $a \in \mathbb{C}$ and $r \in \mathbb{R}^+$ is called a DISC. The set $D[a\,;r] = \{z \in \mathbb{C} \mid |z - a| < r\}$ is called the OPEN DISC of RADIUS r CENTRED at a and the corresponding set $D[a\,;r] = \{z \in \mathbb{C} \mid |z - a| \leq r\}$ is called the CLOSED DISC of RADIUS r CENTRED at a. A subset S of $\mathbb{C}$ is said to be BOUNDED in $\mathbb{C}$ if and only if it is included in a disc; clearly, if $S \subseteq \mathbb{R}$ then S is bounded in $\mathbb{C}$ if and only if S is bounded in $\mathbb{R}$. We shall adopt special notation for two bounded subsets of $\mathbb{C}$: the UNIT CIRCLE of $\mathbb{C}$ is $\mathbb{T} = \{z \in \mathbb{C} \mid |z| = 1\}$ and the CLOSED UNIT DISC is $\mathbb{D} = \{z \in \mathbb{C} \mid |z| \leq 1\}$.

EXERCISES

Q 4.2.1 Show that $\mathbb{Q}$ is the smallest ordered subfield of $\mathbb{R}$.

Q 4.2.2 Show that the complete ordering of $\mathbb{R}$ can not be extended to a complete ordering on $\mathbb{C}$.

4.3 Real and Complex Functions

Historically, number systems are extended in order to help in the solving of problems. There is no such thing as a negative sheep; yet negative numbers facilitate the solution of problems relating to fluctuations in the size of a flock due to barter, slaughter or natural death. Similarly, half a live racehorse is a purely abstract concept; but it is one which might bring joy and distraction to a person of moderate wealth. Fractions themselves display an inadequacy for calculating naturally occurring lengths; the ratio of the length of the diagonal of a square to its side is the irrational number $\sqrt{2}$, for example. So the progression from natural numbers to integers to rational numbers to real numbers has a practical purpose. It is not yet clear from our exposition what purpose complex numbers have, apart from providing a solution to the equation $x^2 = -1$. The set $\mathbb{R}$, being a complete ordered field, has the most extensive algebraic and order structure we have so far devised, whereas its extension $\mathbb{C}$ does not (see Q 4.2.2). Why then extend? Even at this early stage, we should indicate that there is a better justification than the desire to widen our playground. In fact, complex numbers are invaluable in many disciplines and much of their power derives from the simple but important fact that, in $\mathbb{C}$, there are solutions for all polynomial equations with real or complex coefficients. This *Fundamental Theorem of Algebra* (4.3.28), is sufficient justification for introducing $\mathbb{C}$.

Definition 4.3.1

Suppose X is a set. Members of $\mathbb{C}^X$ are called COMPLEX FUNCTIONS and members of $\mathbb{R}^X$ are called REAL FUNCTIONS, the domain X, if not stated, being understood from the context. When $X = \mathbb{N}$, these functions are, of course, sequences. Sequences whose terms are real or complex are called REAL SEQUENCES and COMPLEX SEQUENCES respectively. For $f \in \mathbb{C}^X$, the maps $x \mapsto \Re(f(x))$, $x \mapsto \Im(f(x))$ and $x \mapsto |f(x)|$ defined on X, and denoted by $\Re f$, $\Im f$ and $|f|$ respectively, are called the REAL PART, the IMAGINARY PART and the MODULUS of f. A real function is said to be BOUNDED ABOVE or BOUNDED BELOW if and only if its range is bounded above or below respectively. A complex function is said to be BOUNDED if and only if its range is a bounded set.

Inferior and Superior Limits of Sequences

Definition 4.3.2

Suppose (x_n) is a sequence in $\tilde{\mathbb{R}}$. Then we define the LIMIT SUPERIOR of (x_n) to be $\limsup x_n = \inf\{\sup t_k \mid k \in \mathbb{N}\}$ and the LIMIT INFERIOR of (x_n) to be $\liminf x_n = \sup\{\inf t_k \mid k \in \mathbb{N}\}$, where $t_k = \{x_m \mid k \leq m\}$ is the kth tail of (x_n). Certainly, $\liminf x_n \leq \limsup x_n$. If $\liminf x_n = \limsup x_n$, then we define the LIMIT of (x_n) to be their value and denote it by $\lim x_n$; also, for $z \in \tilde{\mathbb{R}}$, we may write $x_n \to z$ to indicate that $\lim x_n = z$.

Example 4.3.3

If (x_n) is the sequence $1, 1, 1, 1/2, 1, 1/3, 1, 1/4, \ldots$, then $\limsup x_n = 1$ and $\liminf x_n = 0$.

Example 4.3.4

A bounded sequence in $\mathbb{R}$ clearly has real limit superior and limit inferior. But this observation does not split into two parts; a sequence of real numbers might be bounded above or bounded below yet have non-real limit superior and limit inferior. An example is the sequence $(-n)_{n \in \infty}$ which has limit $-\infty$.

Example 4.3.5

Every increasing sequence in $\tilde{\mathbb{R}}$ has a limit in $\tilde{\mathbb{R}}$. Suppose (x_n) such a sequence. For each $k \in \mathbb{N}$, let t_k denote the kth tail of (x_n); then $\sup t_k = \sup\{x_n \mid n \in \mathbb{N}\}$ and $\inf t_k = x_k$. It follows that $\limsup x_n = \sup\{x_n \mid n \in \mathbb{N}\}$ and also that $\liminf x_n = \sup\{x_n \mid n \in \mathbb{N}\}$. So $\lim x_n = \sup\{x_n \mid n \in \mathbb{N}\}$, which is real if (x_n) is bounded above in $\mathbb{R}$. Similar statements hold for decreasing sequences.

Theorem 4.3.6

Suppose (x_n) is a sequence in $\tilde{\mathbb{R}}$ and I is an open interval of $\tilde{\mathbb{R}}$. If either $\limsup x_n \in I$ or $\liminf x_n \in I$, then no tail of (x_n) is included in $\tilde{\mathbb{R}}\backslash I$.

Proof

For each $k \in \mathbb{N}$, let t_k denote the kth tail of (x_n). Let $z \in I$. If $z \in \mathbb{R}$, then, since I is open, there is $\epsilon \in \mathbb{R}^+$ with $[z - \epsilon, z + \epsilon] \subseteq I$. Suppose $m \in \mathbb{N}$ and $t_m \subseteq \tilde{\mathbb{R}}\backslash I$. For all $k \in \mathbb{N}$ with $m \leq k$, we have $t_k \subseteq \tilde{\mathbb{R}}\backslash I$ and so $\sup t_k \notin (z - \epsilon, z + \epsilon)$ and $\inf t_k \notin (z - \epsilon, z + \epsilon)$. So $\inf\{\sup t_n \mid n \in \mathbb{N}\} = \inf\{\sup t_k \mid m \leq k\} \neq z$ and $\sup\{\inf t_n \mid n \in \mathbb{N}\} = \sup\{\inf t_k \mid m \leq k\} \neq z$. If $z \in \{-\infty, \infty\}$, a modification of the argument gives the same result. So $\limsup x_n \notin I$ and $\liminf x_n \notin I$. $\quad\square$

Theorem 4.3.7

Suppose (x_n) is a sequence in $\tilde{\mathbb{R}}$ and $w \in \tilde{\mathbb{R}}$. Then $x_n \to w$ if and only if every open interval I of $\tilde{\mathbb{R}}$ with $w \in I$ includes a tail of (x_n).

Proof

Let $l = \liminf x_n$ and $u = \limsup x_n$ and, for each $k \in \mathbb{N}$, let t_k denote the kth tail of (x_n). For the forward implication, suppose I is an open interval of $\tilde{\mathbb{R}}$ and $w \in I$ and $x_n \to w$. Then $w = \inf\{\sup t_n \mid n \in \mathbb{N}\} = \sup\{\inf t_n \mid n \in \mathbb{N}\}$. Since $w \in I$ and I is open, we have either $w \in \{-\infty, \infty\}$ or $\inf I < w < \sup I$; in either case there exist $k, m \in \mathbb{N}$ such that $\sup t_k \in I$ and $\inf t_m \in I$; so I includes the tail $t_{\max\{m,k\}}$. For the converse, suppose the condition is satisfied and $a \in \tilde{\mathbb{R}}$. If $a < w$, there exists $z \in \mathbb{R}$ with $a < z < w$, and $\tilde{\mathbb{R}}\backslash[-\infty, z)$ includes a tail of (x_n); so $a \neq \limsup x_n$ and $a \neq \liminf x_n$ by 4.3.6. If $w < a$, a similar argument yields the same conclusion. So $\liminf x_n = w = \limsup x_n$. $\quad\square$

Theorem 4.3.8 Bolzano–Weierstrass theorem I

Suppose (x_n) is a sequence in $\tilde{\mathbb{R}}$ and $\limsup x_n = u$ and $\liminf x_n = l$. Then there exist subsequences (x_{k_n}) and (x_{j_n}) of (x_n) such that $x_{k_n} \to u$ and $x_{j_n} \to l$.

Proof

We consider only the case for u; the other is similar. For each $n \in \mathbb{N}$, define an open interval I_n as follows: if $u \in \mathbb{R}$, set $I_n = (u - n^{-1}, u + n^{-1})$; if $u = \infty$, set $I_n = (n, \infty]$; if $u = -\infty$, set $I_n = [-\infty, -n)$. In any case, (I_n) is a nest with $\bigcap(I_n)_{n\in\mathbb{N}} = \{u\}$. Moreover, $|\{m \in \mathbb{N} \mid x_m \in I_n\}| = \infty$ for each $n \in \mathbb{N}$, by 4.3.6. Let $k_0 = 0$ and, for each $n \in \mathbb{N}$, recursively define k_n to be the least integer greater than k_{n-1} for which $x_{k_n} \in I_n$. Every open interval which contains u includes some I_n and hence a tail of (x_{k_n}). So $x_{k_n} \to u$ by 4.3.7. $\quad\square$

Convergence

The variant of the celebrated Bolzano–Weierstrass Theorem presented in 4.3.8 has far reaching consequences in analysis. Our interest is chiefly in real and complex sequences; the purpose of this subsection is to define *convergence* for such sequences and to adapt 4.3.8 to them.

Definition 4.3.9

Suppose (z_n) is a complex sequence and $w \in \mathbb{C}$. We say that (z_n) CONVERGES to w in $\mathbb{C}$ if and only if every open disc which contains w includes a tail of (z_n). A complex sequence which is not CONVERGENT is said to be DIVERGENT.

It is clear that, if (z_n) is a real sequence, then real open intervals can replace open discs in 4.3.9. It is easy to show that a complex sequence (z_n) converges to $w \in \mathbb{C}$ if and only if both $(\Re z_n)$ converges to $\Re w$ and $(\Im z_n)$ converges to $\Im w$; then 4.3.7 implies that w is uniquely determined as $\lim \Re z_n + i \lim \Im z_n$ and is real if (z_n) is a real sequence. Extending notation, we write $z_n \to w$ and call w the LIMIT of z_n, which we denote by $\lim z_n$. Every convergent sequence is bounded and every subsequence of a convergent sequence converges to the same limit (Q 4.3.4). Moreover, it is easy to check that if (z_n) and (w_n) are convergent complex sequences and $\lambda \in \mathbb{C}$, then $(z_n + w_n)$ converges to $\lim z_n + \lim w_n$ and (λz_n) converges to $\lambda \lim z_n$.

Theorem 4.3.10 BOLZANO–WEIERSTRASS THEOREM II
Every bounded sequence in $\mathbb{R}$ or $\mathbb{C}$ has a convergent subsequence.

Proof

The result for $\mathbb{R}$ follows immediately from 4.3.8. Suppose (z_n) is a bounded complex sequence. Then there exists an open disc $D[0\,;r]$ which contains every term of (z_n). Set $x_n = \Re z_n$ and $y_n = \Im z_n$ for each $n \in \mathbb{N}$. Then $|x_n| \le r$ for all $n \in \mathbb{N}$, so that, by 4.3.8, (x_n) has a convergent subsequence (x_{j_n}) with limit $a \in \mathbb{R}$. Also $|y_{j_n}| \le r$ for all $n \in \mathbb{N}$, so that, also by 4.3.8, (y_{j_n}) has a convergent subsequence y_{k_n}, with limit $b \in \mathbb{R}$. Now note that (x_{k_n}) is a subsequence of (x_{j_n}), so converges to the same limit a. So (z_{k_n}) converges to $a + ib$. □

Convergent Series

If it is defined by 4.3.2, the limit of a real or complex series $\sum_{n \in \mathbb{N}} z_n$ is called its SUM and is denoted by $\sum_{n=1}^{\infty} z_n$. Clearly, a series $\sum_{n \in \mathbb{N}} z_n$ cannot converge unless the sequence (z_n) of its terms converges to 0; but the converse of this is not true (4.3.18). Some other tests for convergence of series are presented in this subsection. As for other sequences, if $\sum_{n\in\mathbb{N}} z_n$ and $\sum_{n\in\mathbb{N}} w_n$ converge and $\lambda \in \mathbb{C}$, then $\sum_{n\in\mathbb{N}}(z_n + w_n)$ converges to $\sum_{n=1}^{\infty} z_n + \sum_{n=1}^{\infty} w_n$ and $\sum_{n\in\mathbb{N}} \lambda z_n$ converges to $\lambda \sum_{n=1}^{\infty} z_n$.

Example 4.3.11

A GEOMETRIC SERIES $\sum_{n\in\mathbb{N}} ar^{n-1}$, with RATIO $r \in \mathbb{C}$, where $a \in \mathbb{C}\backslash\{0\}$, converges if and only if $|r| < 1$. In fact, the terms do not converge to 0 if $|r| \geq 1$, and, if $|r| < 1$, it is easy to show by induction that the nth partial sum is $a(1 - r^n)/(1 - r)$ and then that the series converges to $a/(1 - r)$.

Test 4.3.12 COMPARISON TEST

Suppose $\sum_{n\in\mathbb{N}} a_n$ is a series whose terms are all in $\mathbb{R}^+$. If there exists a series $\sum_{n\in\mathbb{N}} b_n$ of positive terms which converges and whose terms satisfy $a_n \leq kb_n$ for some $k \in \mathbb{R}^+$ and all $n \in \mathbb{N}$, then $\sum_{n\in\mathbb{N}} a_n$ is an increasing real sequence bounded above by $k\sum_{n=1}^{\infty} b_n$, and therefore converges (4.3.5).

Test 4.3.13 ROOT TEST

Suppose $\sum_{n\in\mathbb{N}} a_n$ is a series whose terms are all in $\mathbb{R}^+$. If $\limsup a_n^{1/n} < 1$, then there exist $r \in (0,1)$ and $k \in \mathbb{N}$ such that $a_n < r^n$ for all $n \in \mathbb{N}$ with $k \leq n$. Since the geometric series $\sum_{n\in\mathbb{N}} r^{n+k}$ converges, a simple argument using the comparison test establishes convergence of $\sum_{n\in\mathbb{N}} a_n$. Conversely, if $\limsup a_n^{1/n} \geq 1$, then (a_n) does not converge to 0 and the series diverges.

Test 4.3.14 RATIO TEST

Suppose $\sum_{n\in\mathbb{N}} a_n$ is a series whose terms are all in $\mathbb{R}^+$. If $\limsup a_{n+1}/a_n < 1$, then there exist $r \in (0,1)$ and $k \in \mathbb{N}$ such that $a_{n+1} < ra_n$ for all $n \in \mathbb{N}$ with $k \leq n$. By induction, $a_{n+k} \leq a_k r^n$ for all $n \in \mathbb{N}$, so that, since the geometric series $\sum_{n\in\mathbb{N}} a_k r^n$ converges, the comparison test establishes that $\sum_{n\in\mathbb{N}} a_{n+k}$ converges and it follows that $\sum_{n\in\mathbb{N}} a_n$ converges. Conversely, if $\limsup a_{n+1}/a_n \geq 1$, then (a_n) does not converge to 0 and the series diverges.

Test 4.3.15 CONDENSATION TEST

Suppose (a_n) is a decreasing sequence in $\mathbb{R}^+$. An inductive argument shows that $\sum_{i=1}^{n} 2^{i-1}a_{2^i} \leq \sum_{i=1}^{2^n-1} a_i \leq \sum_{i=0}^{n-1} 2^i a_{2^i}$ for each $n \in \mathbb{N}$. So the partial sums of $\sum_{n\in\mathbb{N}} a_n$ are bounded if and only if the partial sums of $\sum_{n\in\mathbb{N}} 2^n a_{2^n}$ are bounded. Therefore (4.3.5) the series either both converge or both diverge.

Example 4.3.16

For each $r \in [0,1)$ we define a sequence recursively as follows: let $x_1 = 0$ if $r < 1/2$ and $x_1 = 1$ otherwise; then, for each $n \in \mathbb{N}$, let $x_{n+1} = 0$ if $r - \sum_{i=1}^{n} x_i/2^i < 1/2^{n+1}$ and $x_{n+1} = 1$ otherwise. Then $\sum_{n \in \mathbb{N}} x_n/2^n$ is an increasing sequence of non-negative terms bounded above by r; it is easy to verify that r is its supremum, so that $r = \sum_{n=1}^{\infty} x_n/2^n$. The series $\sum_{n \in \mathbb{N}} x_n/2^n$ is called the BINARY EXPANSION of r, and it is a consequence of 4.2.3 that no two members of $[0,1)$ have the same binary expansion. By easy calculation, binary expansions of numbers in $[0,1)$ are precisely all series of the type $\sum_{n \in \mathbb{N}} x_n/2^n$, where each x_n is either 0 or 1 and the set $\{n \in \mathbb{N} \mid x_n \neq 1\}$ is infinite. Ternary and other expansions can be defined similarly; in particular, the DECIMAL EXPANSIONS of numbers in $[0,1)$ are all the series of the form $\sum_{n \in \mathbb{N}} x_n/10^n$ where each x_n is an integer with $0 \leq x_n \leq 9$ and the set $\{n \in \mathbb{N} \mid x_n \neq 9\}$ is infinite.

Example 4.3.17

It is now easy to show that $|\mathbb{R}| = |2^\infty|$. The function $(x_n)_{n \in \mathbb{N}} \mapsto \sum_{n \in \mathbb{N}} x_n/3^n$ defined on $\{0,1\}^{\mathbb{N}}$ is injective into $\mathbb{R}$, so that $|2^\infty| \leq |\mathbb{R}|$. For the reverse inequality, let $\Omega = (-\infty \cup \infty) \times (\infty \backslash \{0\})$; then $|\Omega| = \infty$ by 2.4.5; and $|\mathcal{P}(\Omega)| = |2^\infty|$ by 2.2.8; so, since $\mathbb{R} \subseteq \mathcal{P}(\Omega)$, we have $|\mathbb{R}| \leq |2^\infty|$ by 2.2.6. Hence $|\mathbb{R}| = |2^\infty|$.

Example 4.3.18

If $p \in \mathbb{R}^+$, then $\sum_{n \in \mathbb{N}} 1/n^p$ converges if and only if $p > 1$: the sequence $(1/n^p)$ is decreasing; by the condensation test, $\sum_{n \in \mathbb{N}} 1/n^p$ converges if and only if $\sum_{n \in \mathbb{N}} 2^n/2^{np}$ converges; this is a geometric series with ratio 2^{1-p}, so converges if and only if $p > 1$. Note that the HARMONIC SERIES $\sum_{n \in \mathbb{N}} 1/n$ diverges.

Absolute Convergence

Definition 4.3.19

A complex series $\sum_{n \in \mathbb{N}} z_n$ is said to be ABSOLUTELY CONVERGENT if and only if the series $\sum_{n \in \mathbb{N}} |z_n|$ converges.

Theorem 4.3.20

Suppose $\sum_{n \in \mathbb{N}} z_n$ is a complex series which is absolutely convergent. Then $\sum_{n \in \mathbb{N}} z_n$ converges.

Proof

Let $s = \sum_{n=1}^{\infty}|z_n|$. Then, for all $m \in \mathbb{N}$, $|\sum_{i=1}^{m} z_i| \leq \sum_{i=1}^{m}|z_i| \leq s$. So $\sum_{n\in\mathbb{N}} z_n$ is a bounded sequence and, by 4.3.10, has a subsequence which converges to some $w \in \mathbb{C}$. Let $\epsilon \in \mathbb{R}^+$. There exists $k \in \mathbb{N}$ such that both $|w - \sum_{i=1}^{k} z_i| < \epsilon$ and the kth tail of $\sum_{n\in\mathbb{N}}|z_n|$ is included in $(s - \epsilon, s]$. So, for each $m \in \mathbb{N}$ with $k < m$, we have $|w - \sum_{i=1}^{m} z_i| \leq |w - \sum_{i=1}^{k} z_i| + \sum_{i=1}^{m}|z_i| - \sum_{i=1}^{k}|z_i| < 2\epsilon$. □

Example 4.3.21

For each $z \in \mathbb{C}$, the series $\sum_{n\in\mathbb{N}} z^n/n!$ converges absolutely by the ratio test, so converges. The function $z \mapsto \sum_{n=1}^{\infty} z^n/n!$ on $\mathbb{C}$ is the EXPONENTIAL FUNCTION.

Example 4.3.22

The series $\sum_{n\in\mathbb{N}}(-1)^{n+1}/n$ does not converge absolutely (4.3.18), but it does converge: for each $k \in \mathbb{N}$, we have $0 < \sum_{n=1}^{2k}(-1)^{n+1}/n \leq \sum_{n=1}^{k} 1/n^2$; the series $\sum_{n\in\mathbb{N}} 1/n^2$ converges (4.3.18) and the result follows easily.

Open and Closed Subsets of the Complex Plane

Definition 4.3.23

Suppose $S \subseteq \mathbb{C}$. Then S will be called an OPEN subset of $\mathbb{C}$ if and only if S can be expressed as a union of open discs of $\mathbb{C}$. S will be called a CLOSED subset of $\mathbb{C}$ if and only if $\mathbb{C}\backslash S$ is an open subset of $\mathbb{C}$.

Example 4.3.24

$\varnothing$, being the trivial union, is open in $\mathbb{C}$; and $\mathbb{C}$, being the union of all its open discs, is also open. It follows that $\mathbb{C}$, being the complement of $\varnothing$, and $\varnothing$, being the complement of $\mathbb{C}$, are both closed in $\mathbb{C}$. No other subset of $\mathbb{C}$ has the property of being both open and closed in $\mathbb{C}$ (Q 4.3.5).

Example 4.3.25

Certainly every open disc of $\mathbb{C}$ is open in $\mathbb{C}$; indeed, it is clear that a subset S of $\mathbb{C}$ is open in $\mathbb{C}$ if and only if, for each $u \in S$, there is an open disc D of $\mathbb{C}$ such that $u \in D \subseteq S$. And every closed disc is closed: consider a closed disc $D = D[a\,;r]$; for each $z \in \mathbb{C}\backslash D$, we have $|z - a| > r$, so that $D[z\,;|z - a| - r) \subseteq \mathbb{C}\backslash D$, whence $\mathbb{C}\backslash D = \bigcup\{D[z\,;|z - a| - r) \mid z \in \mathbb{C}\backslash D\}$ is open in $\mathbb{C}$ and D is therefore closed in $\mathbb{C}$.

The Fundamental Theorem of Algebra

Here we prove by induction that every non-constant complex polynomial function has a zero in $\mathbb{C}$ (4.3.28). Part of the work is to establish that, for every such function p, the function $|p|$ attains a minimum value; to do this, we use the two basic properties of polynomial functions given in 4.3.26—the first is that such functions are *continuous*—together with the Bolzano-Weierstrass Theorem and the Axiom of Choice. An alternative argument is available in ZF using 4.3.26 and 11.1.8—but we must not run before our horse to market.

Theorem 4.3.26

Suppose $p \in \mathrm{poly}(\mathbb{C})$ is not constant and $D = D[a\,;r]$ is an open disc of $\mathbb{C}$. Then $p^{-1}(D)$ is open and bounded.

Proof

Let $p = \sum_{k=0}^{n} \beta_k z^k$ where $n \in \mathbb{N}$ and $\beta_n \neq 0$. Let $c = \sum_{k=0}^{n} |\beta_k|$ and let $m = 2(c + r + |a|)/|\beta_n|$. Then $m > 2$. Also, $p^{-1}(D) \subseteq D[0\,;m/2]$; indeed, if $v \in \mathbb{C}$ and $|v| \geq m/2$, then $|v| > 1$ and $p(v) = v^{n-1}(\beta_n v + \sum_{k=0}^{n-1} \beta_k/v^{n-1-k})$, from which we get $|p(v) - a| \geq |\beta_n| |v| - c - |a| \geq r$ and therefore $p(v) \notin D$. In particular, $p^{-1}(D)$ is bounded. Now suppose $s \in p^{-1}(D)$. Then, from above, $|s| < m/2$; and, since D is open and $p(s) \in D$, there exists $\epsilon \in \mathbb{R}^+$ such that $D[p(s)\,;\epsilon) \subseteq D$. Let $\delta = \min\{m/2, \epsilon/n^2 m^n c\}$. Our claim now is that $D[s\,;\delta) \subseteq p^{-1}(D)$; indeed, if $u \in D[s\,;\delta)$, then $|u| < |s| + m/2 < m$ and, using the easily checked equations $u^k - s^k = (u - s)\sum_{j=0}^{k-1} u^j s^{k-j-1}$, we have

$$|p(u) - p(s)| = \left| \sum_{k=1}^{n} \beta_k(u^k - s^k) \right| \leq c|u - s| \sum_{k=1}^{n} \sum_{j=0}^{k-1} |u|^j |s|^{k-j-1} < c\delta n^2 m^n \leq \epsilon.$$

So $p(u) \in D[p(s)\,;\epsilon) \subseteq D$. Arbitrariness of u in $D[s\,;\delta)$ yields $D[s\,;\delta) \subseteq p^{-1}(D)$; and, since s is arbitrary in $p^{-1}(D)$, it follows that $p^{-1}(D)$ is open in $\mathbb{C}$. □

Lemma 4.3.27

Suppose $p \in \mathbb{N}$ and $\alpha \in \mathbb{T}$. Let $\omega = (1 + i)/\sqrt{2}$. Then $\left|1 + \alpha\omega^{2q(2p-1)}\right| < 1$ for some $q \in \mathbb{N}$.

Proof

$\omega^8 = 1$, and $\{\omega^{2q(2p-1)} \mid q \in \mathbb{N}\} = \{1, i, -1, -i\}$. If $\alpha = a + ib$ in standard form, then $a^2 + b^2 = 1$ and the only moduli of the form $\left|1 + \alpha\omega^{2q(2p-1)}\right|$ are the non-negative square roots of $2 + 2a$, $2 - 2a$, $2 + 2b$ and $2 - 2b$; the equation $a^2 + b^2 = 1$ ensures that either $|a| \geq 1/\sqrt{2}$ or $|b| \geq 1/\sqrt{2}$ and hence that at least one of the stated moduli does not exceed than $2 - \sqrt{2}$. □

Theorem 4.3.28 FUNDAMENTAL THEOREM OF ALGEBRA

Every member of poly($\mathbb{C}$) of positive degree has a zero in $\mathbb{C}$.

Proof

The result is certainly true for polynomial functions of degree 1. Let $n \in \mathbb{N} \backslash \{1\}$ and suppose that the assertion is true for all polynomial functions of degree less than n. Let p be a polynomial function of degree n and let $s = \inf \mathrm{ran}|p|$. For each $n \in \mathbb{N}$, the set $\{z \in \mathbb{C} \mid |p(z)| - s < 1/n\}$ is non-empty; so, by the Product Theorem, there exists a sequence (w_n) in $\mathbb{C}$ such that $|p(w_n)| - s < 1/n$ for each $n \in \mathbb{N}$. Then $(p(w_n))$ is a bounded sequence and, by 4.3.26, so is (w_n). By 4.3.10, there exists $a \in \mathbb{C}$ such that $w_{k_n} \to a$ for some subsequence (w_{k_n}) of (w_n). For each $m \in \mathbb{N}$, $p^{-1}(D[p(a)\,;1/m))$ is open (4.3.26); it contains a and therefore includes a tail of (w_{k_n}); so $D[p(a)\,;1/m)$ includes a tail of $(p(w_{k_n}))$, whence $|p(a)| - s < 2/m$; since m is arbitrary in $\mathbb{N}$, it follows that $|p(a)| = s$. We suppose $s \neq 0$ and achieve a contradiction. The function $z \mapsto p(a + z)/p(a)$ is clearly a polynomial function of the form $\sum_{k=0}^{n} \beta_k z^k$ whose values all lie outside the disc $D[0\,;1)$; evidently also, $\beta_0 = 1$ and $\beta_n \neq 0$. Let $m = \min\{k \mid 1 \le k \le n; \beta_k \neq 0\}$. Then $m < n$, for, if n is composite, two applications of the inductive hypothesis yield some $\zeta \in \mathbb{C}$ for which $1 + \beta_n \zeta^n = 0$; and, if n is prime, 4.3.27 yields some $k \in \mathbb{N}$ such that $|1 + \beta_n z^n| < 1$ when $z = \omega^k/|\beta_n|^{1/n}$ and $\omega = (1+i)/\sqrt{2}$. Let $c = \sum_{k=0}^{n}|\beta_k|$ and $\gamma = \min\{1/2, |\beta_m|^{m+1}/(2^{m+1}c^m)\}$. Then, by the inductive hypothesis, there exists $\mu \in \mathbb{C}$ such that $\beta_m \mu^m = -2\gamma$. Note that $|\mu| \le |\beta_m|/2c < 1$, whence

$$\left| \sum_{k=m+1}^{n} \beta_k \mu^k \right| \le c|\mu^m|\,|\mu| \le c \frac{2\gamma}{|\beta_m|} \frac{|\beta_m|}{2c} = \gamma.$$

Then $p(a + \mu)/p(a) = 1 + \sum_{k=m}^{n} \beta_k \mu^k = 1 - 2\gamma + \sum_{k=m+1}^{n} \beta_k \mu^k$, whence $|p(a + \mu)/p(a)| \le |1 - 2\gamma| + \gamma = 1 - \gamma$, and we have our contradiction. We infer that $s = 0$, and the Principle of Induction yields the result. $\square$

It follows easily from 4.3.28 that every non-constant complex polynomial function is surjective onto $\mathbb{C}$ and can be written as a product $c \prod_{i=1}^{n}(z - \zeta_i)$, for some $c \in \mathbb{C} \backslash \{0\}$, where n is the degree of the polynomial and the ζ_j are all the zeroes, which cannot therefore be more than n in number; the number of times each occurs is called the MULTIPLICITY of the zero. Another easy consequence is that the representation of a non-zero complex polynomial function as $\sum_{i=0}^{n} \alpha_i z^i$ with $\alpha_n \neq 0$ is unique.

Roots of Unity

It is a consequence of the Fundamental Theorem of Algebra that, for each $n \in \mathbb{N}$, the equation $z^n = 1$ has at most n solutions or ROOTS in $\mathbb{C}$. In fact there are exactly n roots and they are called the nth ROOTS OF UNITY; moreover, for each $n \in \mathbb{N}$, there is at least one n^{th} root of unity ω for which $\{\omega^k \mid k \in \mathbb{N}\}$ comprises all n roots (Q 4.3.7); such a root is called a PRIMITIVE n^{th} ROOT of unity. The roots of unity yield the following rather nice average.

Theorem 4.3.29 AVERAGE INVERSE THEOREM
Suppose A is a complex unital algebra and $a \in A$. Let $n \in \mathbb{N}$ and let ω be a primitive n^{th} root of unity. Then $1 - \omega^k a \in \text{inv}(A)$ for all $k \in \{j \in \mathbb{N} \mid 1 \le j \le n\}$ if and only if $1 - a^n \in \text{inv}(A)$ and, in that case, $(1 - a^n)^{-1} = \frac{1}{n} \sum_{k=1}^{n}(1 - \omega^k a)^{-1}$.

Proof

Since $\omega^n = 1$, we have $(1 - a^n) = (1 - \omega^k a)\sum_{j=0}^{n-1} \omega^{kj} a^j$, for each k with $1 \le k \le n$, so that, because this product commutes, invertibility of $1 - a^n$ certainly implies invertibility of $1 - \omega^k a$ for all such k. Now suppose that $1 - \omega^k a$ is invertible for each such k and rewrite the equations given above as $(1 - a^n)(1 - \omega^k a)^{-1} = \sum_{j=0}^{n-1} \omega^{kj} a^j$. For each $j \in \mathbb{N}$ with $0 < j < n$, we have $(1 - \omega^j)\sum_{k=1}^{n} \omega^{kj} = 0$; so, since $\omega^j \ne 1$, we have $\sum_{k=1}^{n} \omega^{kj} = 0$. Then addition of the n equations above yields, by reversing summation in accord with 3.1.20,

$$(1 - a^n)\sum_{k=1}^{n}(1 - \omega^k a)^{-1} = \sum_{k=1}^{n}\sum_{j=0}^{n-1} \omega^{kj} a^j = \sum_{j=0}^{n-1}\sum_{k=1}^{n} \omega^{kj} a^j = n$$

from which the result follows.   □

EXERCISES

Q 4.3.1 Show that $|\mathbb{C}| = |2^\infty|$.

Q 4.3.2 Suppose I is a non-degenerate real interval. Show that $|I| = |2^\infty|$. Show also that $|\mathbb{T}| = |\mathbb{D}| = |2^\infty|$.

Q 4.3.3 Show that $\mathbb{R}^{[0,1]}$ has cardinality $\left|2^{2^\infty}\right|$.

Q 4.3.4 Suppose (x_n) is a complex sequence which converges to $z \in \mathbb{C}$. Show that (x_n) is bounded and that every subsequence of (x_n) converges to z. Give an example of a sequence of real numbers which is bounded below in $\mathbb{R}$ but has no convergent subsequence.

Q 4.3.5 Show that $\mathbb{C}$ has no proper non-empty open subset which is closed.

Q 4.3.6 Suppose that $f\colon \mathbb{C} \to \mathbb{C}$ and that, for every open disc D of $\mathbb{C}$, $f^{-1}(D)$ is bounded and open in $\mathbb{C}$. Show that, for every closed subset S of $\mathbb{C}$, $f(S)$ is also closed in $\mathbb{C}$.

Q 4.3.7 Let $n \in \mathbb{N}$. Show that there are exactly n n^{th} roots of unity and that at least one of them is primitive.

4.4 Inequalities

Inequalities are grist to the mill in analysis. In this subsection, we shall prove a few rather neat ones, culminating in that of Minkowski which will enable us to present some interesting examples of vector spaces in Chapter 5. Firstly, 4.1.9 ensures that $\mathbb{N}$ is not bounded above in $\mathbb{R}$ and allows us, with the enumerative well ordering of $\mathbb{N}$, to make Definition 4.4.1.

Definition 4.4.1

For each $r \in \mathbb{R}$, we define the INTEGER PART or FLOOR $\lfloor r \rfloor$ of r to be the largest integer not greater than r; its counterpart, the CEILING of r, denoted by $\lceil r \rceil$ is the smallest integer not smaller than r. The NON-INTEGER PART of r is $r - \lfloor r \rfloor$; this is a non-negative real number less than 1.

Lemma 4.4.2

Suppose $r, x \in \mathbb{R}$ with $1 \le r$ and $-1 \le x$. Then $1 + rx \le (1 + x)^r$.

Proof

Suppose $z \in \mathbb{R}^{\oplus}$ and $q \in \mathbb{Q}$ with $q > 1$. Let $q = m/n$ in standard form. Then $n < m$ and, for each $k \in \mathbb{N} \cup \{0\}$, we have $z^{\lfloor k/n \rfloor} \le z^{\lfloor k/m \rfloor}$ if $0 \le z \le 1$ and, conversely, $z^{\lfloor k/m \rfloor} \le z^{\lfloor k/n \rfloor}$ if $1 < z$. So

$$n(1 - z^m) = (1 - z) \sum_{k=0}^{mn-1} z^{\lfloor k/n \rfloor} \le (1 - z) \sum_{k=0}^{mn-1} z^{\lfloor k/m \rfloor} = m(1 - z^n).$$

Since $x \ge -1$, this inequality holds when $z = (1 + x)^{1/n}$; in that case it yields $n(1 - (1+x)^{m/n}) \le -mx$, and therefore $1 + qx \le (1+x)^q$. If $0 \le x$, we then have $1 + rx = \inf\{1 + qx \mid r < q\} \le \inf\{(1 + x)^q \mid r < q\} = (1 + x)^r$; if $-1 \le x < 0$, then $1 + rx = \sup\{1 + qx \mid r < q\} \le \sup\{(1 + x)^q \mid r < q\} = (1 + x)^r$. $\square$

Lemma 4.4.3 YOUNG'S INEQUALITY

Suppose $p, q \in \mathbb{R}^+$ and $1/p + 1/q = 1$. Then, for each $a, b \in \mathbb{R}^\oplus$, we have $a^{1/p} b^{1/q} \leq a/p + b/q$.

Proof

Certainly the inequality holds if $b = 0$; so we assume $b \neq 0$. Clearly $p > 1$. It follows that $(a - b)/pb \geq -1/p > -1$, so that, by 4.4.2, we have

$$\frac{a}{b} = 1 + \frac{a - b}{b} \leq \left(1 + \frac{a - b}{pb} \right)^p.$$

Multiplying both sides of this by b^p and noting that $p - 1 = p/q$, we get

$$ab^{p/q} \leq \left(\frac{a}{p} + \frac{b}{q} \right)^p.$$

The result follows by raising each side of the inequality to the power of $1/p$. □

Definition 4.4.4

Suppose $p \in \mathbb{R}^+$ and $a = (a_i)_{i \in \mathbb{I}}$ is a sequence or finite sequence in $\mathbb{C}$. If the series $\sum_{i \in \mathbb{I}} |a_i|^p$ converges in $\mathbb{R}$, we define the p-NORM $\|a\|_p$ of a to be $\left(\sum_{i \in \mathbb{I}} |a_i|^p \right)^{1/p}$; otherwise, we define $\|a\|_p$ to be ∞. We also define $\|a\|_\infty$ to be $\sup\{|a_i| \mid i \in \mathbb{I}\}$; if it is finite, it is called the SUPREMUM NORM of a.

Theorem 4.4.5 HÖLDER'S INEQUALITY

Suppose a and b are sequences or finite sequences in $\mathbb{C}$ with the same indexing set $\mathbb{I}$. Let ab denote the pointwise product $(a_i b_i)_{i \in \mathbb{I}}$. Suppose $p, q \in \mathbb{R}^+ \cup \{\infty\}$ with either $1/p + 1/q = 1$ or ($p = 1$ and $q = \infty$). Then $\|ab\|_1 \leq \|a\|_p \|b\|_q$.

Proof

The case when $\{p, q\} = \{1, \infty\}$ is straightforward, so we assume that p and q are real. We assume also that $\|a\|_p$ and $\|b\|_q$ are finite and that neither a nor b is the zero sequence; other cases are trivial. Then $\|a\|_p > 0$ and $\|b\|_q > 0$, and, by Young's inequality, we have

$$\frac{|a_i|}{\|a\|_p} \frac{|b_i|}{\|b\|_q} \leq \frac{|a_i|^p}{p\|a\|_p^p} + \frac{|b_i|^q}{q\|b\|_q^q}$$

for each $i \in \mathbb{I}$. Then, for each $n \in \mathbb{I}$, we have, by addition,

$$\sum_{i=1}^n \frac{|a_i b_i|}{\|a\|_p \|b\|_q} \leq \frac{1}{p} + \frac{1}{q} = 1,$$

giving $\sum_{i=1}^n |a_i b_i| \leq \|a\|_p \|b\|_q$. So $\|ab\|_1 = \lim \sum_{i=1}^n |a_i b_i| \leq \|a\|_p \|b\|_q$. □

Corollary 4.4.6   CAUCHY'S INEQUALITY

Suppose a and b are sequences or finite sequences in $\mathbb{C}$ with the same indexing set. Then $(\sum |a_n b_n|)^2 \le (\sum |a_n|^2)(\sum |b_n|^2)$.

Theorem 4.4.7   MINKOWSKI'S INEQUALITY

Suppose that a and b are sequences or finite sequences in $\mathbb{C}$ with the same indexing set $\mathbb{I}$. Suppose that $p \in [1, \infty]$. Then $\|a + b\|_p \le \|a\|_p + \|b\|_p$.

Proof

When $a + b$ is the zero sequence or either of $\|a\|_p$ or $\|b\|_p$ is infinite, the result is trivial; the inequality follows easily from the elementary properties of the modulus if $p = 1$; and the case when $p = \infty$ is straightforward. So we assume that $a + b \ne 0$, that $p \in (1, \infty)$ and that $\|a\|_p$ and $\|b\|_p$ are real. Set $q = p/(p-1)$. Then $1/p + 1/q = 1$. For each $n \in \mathbb{I}$, using Hölder's inequality,

$$
\begin{aligned}
\sum_{i=1}^{n} |a_i + b_i|^p &= \sum_{i=1}^{n} |a_i + b_i|\, |a_i + b_i|^{p/q} \\
&\le \sum_{i=1}^{n} |a_i|\, |a_i + b_i|^{p/q} + \sum_{i=1}^{n} |b_i|\, |a_i + b_i|^{p/q} \\
&\le \left(\left(\sum_{i=1}^{n} |a_i|^p \right)^{1/p} + \left(\sum_{i=1}^{n} |b_i|^p \right)^{1/p} \right) \left(\sum_{i=1}^{n} |a_i + b_i|^p \right)^{1/q} \\
&\le (\|a\|_p + \|b\|_p)\|a + b\|_p^{p/q} .
\end{aligned}
$$

So $\|a + b\|_p^p \le (\|a\|_p + \|b\|_p)\|a + b\|_p^{p/q}$. The result follows because $a + b \ne 0$. $\quad\square$

Example 4.4.8

The condition that $p \ge 1$ in 4.4.7 is necessary. Suppose p is real and $0 < p < 1$. Then $\|(1,0)\|_p = 1 = \|(0,1)\|_p$, but $\|(1,1)\|_p = 2^{1/p} > 2$.

EXERCISES

Q 4.4.1   Suppose $n \in \mathbb{N}$ and $a \in \mathbb{K}^n$. Show that $\|a\|_\infty = \inf\{\|a\|_p \mid p \in \mathbb{R}^+\}$.

Q 4.4.2   Suppose A is a finite subset of $\mathbb{R}^3$. Set $S_x = \{(y,z) \mid (x,y,z) \in A\}$; $S_y = \{(z,x) \in \mathbb{R}^2 \mid (x,y,z) \in A\}$; $S_z = \{(x,y) \in \mathbb{R}^2 \mid (x,y,z) \in A\}$. Show that $|A|^2 \le |S_x|\,|S_y|\,|S_z|$.

Q 4.4.3   (ARITHMETIC-GEOMETRIC MEAN INEQUALITY) Suppose A is a nonempty finite subset of $\mathbb{R}^\oplus$. Show that $\left(\prod_{a \in A} a\right)^{1/|A|} \le \left(\sum_{a \in A} a\right)/|A|$.

5

Linear Structure

Vector spaces over $\mathbb{R}$ and over $\mathbb{C}$ have special properties which derive from the fields which underlie them. Since each such space is a union of its *lines*, we shall call them LINEAR SPACES, the former REAL LINEAR SPACES and the latter COMPLEX LINEAR SPACES. A vector subspace S of a linear space X will be called a LINEAR SUBSPACE of X and X will be called a LINEAR SUPERSPACE of S. Vector space homomorphisms between real linear spaces or between complex linear spaces will be called LINEAR MAPS or LINEAR OPERATORS. Algebras over $\mathbb{R}$ will be called REAL ALGEBRAS and algebras over $\mathbb{C}$ COMPLEX ALGEBRAS. Many of the statements we shall make regarding linear spaces are valid where either $\mathbb{R}$ or $\mathbb{C}$ is the field involved. We shall therefore use $\mathbb{K}$ to identify either $\mathbb{R}$ or $\mathbb{C}$, and the reader may assume that, in any context where it appears, the symbol $\mathbb{K}$ may be replaced consistently by either $\mathbb{R}$ or $\mathbb{C}$; where no field is mentioned, it is to be understood that either $\mathbb{R}$ or $\mathbb{C}$ is intended.

5.1 Linear Spaces and Algebras

In this section, we merely present some examples of linear spaces and algebras.

Example 5.1.1
Each of $\mathbb{Q}$, $\mathbb{R}$ and $\mathbb{C}$ is an algebra over itself. Moreover, $\mathbb{C}$ is a real algebra and a rational algebra; and $\mathbb{R}$ is a rational algebra. But differences occur in dimension: $\mathbb{C}$ has dimension 1 as a vector space over itself, dimension 2 as a vector space over $\mathbb{R}$, and dimension $|2^{\infty}|$ as a vector space over $\mathbb{Q}$ (see Q 3.2.1).

Example 5.1.2

For any non-empty set X, the set $\mathbb{R}^X$ of all real functions defined on X is a real algebra with pointwise operations. Similarly, the set $\mathbb{C}^X$ of all complex functions defined on X is a complex algebra with pointwise operations. These are the usual operations on these algebras; both are unital, the constant function $x \mapsto 1$ being the unity; and a function $g \in \mathbb{K}^X$ is invertible in $\mathbb{K}^X$ if and only if $0 \notin \operatorname{ran}(g)$. CONJUGATION $f \mapsto \overline{f}$, where $\overline{f}(x) = \overline{f(x)}$ for all $x \in X$, is the standard non-trivial involution on $\mathbb{C}^X$. Also, for each $f \in \mathbb{C}^X$, $\Re f$, $\Im f$ and $|f|$ are members of $\mathbb{R}^X$ and $f = \Re f + i\Im f$, where the operations are those of $\mathbb{C}^X$.

Example 5.1.3

$\mathbb{R}^{\mathbb{N}}$ and $\mathbb{C}^{\mathbb{N}}$ are the linear spaces of real and complex sequences respectively. For each $n \in \mathbb{N}$, $\mathbb{K}^n$ is isomorphic to the subspace $\{a \in \mathbb{K}^{\mathbb{N}} \mid a_m = 0 \text{ for all } m > n\}$ of $\mathbb{K}^{\mathbb{N}}$. If a and b are sequences in $\mathbb{K}$ which converge to r and s respectively, then $a + b$ converges to $r + s$ and λa converges to λr for all $\lambda \in \mathbb{K}$ (4.3.9). It follows that the convergent sequences in $\mathbb{K}$ form a linear subspace of $\mathbb{K}^{\mathbb{N}}$. The sequences which converge to 0 form a smaller subspace.

Example 5.1.4

There is a whole raft of SEQUENCE SPACES determined by Minkowski's inequality (4.4.7): for each $p \in [1, \infty]$, Minkowski's inequality ensures that the set $\{x \in \mathbb{K}^{\mathbb{N}} \mid \|x\|_p < \infty\}$ is closed under addition; it is easily seen to be closed under scalar multiplication and so to be a linear subspace of $\mathbb{K}^{\mathbb{N}}$.

Example 5.1.5

It is a consequence of the Fundamental Theorem of Algebra that the sequence $(z^n)_{n \in \mathbb{N} \cup \{0\}}$ of power functions on $\mathbb{K}$ is linearly independent; indeed, each linear combination of power functions is a product $c \prod_{i=0}^{n}(z - \zeta_i)$ which, if $c \neq 0$, has non-zero values on the set $\mathbb{K} \backslash \{\zeta_i \mid 0 \leq i \leq n\}$. Then $\{z^n \mid n \in \mathbb{N} \cup \{0\}\}$ is the STANDARD BASIS for $\operatorname{poly}(\mathbb{K})$. There are some distinguished operators on $\operatorname{poly}(\mathbb{R})$: $D: \operatorname{poly}(\mathbb{R}) \to \operatorname{poly}(\mathbb{R})$ defined on the basis by $z^n \mapsto nz^{n-1}$, and extended linearly, is the linear map known as DIFFERENTIATION; and $\int_0^1: \operatorname{poly}(\mathbb{R}) \to \mathbb{R}$ defined on the basis by $z^n \mapsto 1/(n+1)$, and extended linearly, is known as INTEGRATION on $[0, 1]$. These linear maps can be extended to larger spaces of functions; but this theory is beyond the scope of our work.

Example 5.1.6

A polynomial function on a unital algebra is said to be REDUCIBLE if and only if it can be expressed as a product of polynomial functions of lesser degree.

Reducibility is a property which depends on the underlying field. If A is a unital algebra over $\mathbb{K}$, then, as in 5.1.5, the power functions z^n are all distinct and $\{z^n \mid n \in \mathbb{N} \cup \{0\}\}$ is a basis for $\mathrm{poly}(A)$; if B is any other unital algebra over $\mathbb{K}$, then the function which maps the power functions on A to the corresponding ones on B extends linearly to an algebra isomorphism between $\mathrm{poly}(A)$ and $\mathrm{poly}(B)$. This isomorphism identifies reducible members of $\mathrm{poly}(A)$ with those of $\mathrm{poly}(B)$. By the Fundamental Theorem of Algebra, every member of $\mathrm{poly}(\mathbb{C})$ with degree greater than 1 is reducible; it follows, therefore, that the same is true in any complex unital algebra. Real unital algebras admit polynomial functions which are IRREDUCIBLE, the primary example being $z^2 + 1$.

Example 5.1.7

Suppose X is a complex linear space; then $\mathcal{L}(X)$ is a complex algebra. The fact that the polynomial functions of degree greater than 1 on $\mathcal{L}(X)$ are reducible (5.1.6) leads to partial reductions of X by its linear maps. Suppose $T \in \mathcal{L}(X)$. As in 3.4.30, let Λ (which may, of course, be empty) denote the set of eigenvalues of T and, for each $\lambda \in \Lambda$, let E_λ denote the corresponding generalized eigenspace; let $E_\mu = \{0\}$ if $\mu \in \mathbb{C} \backslash \Lambda$. Recall from 3.4.30 that these spaces are invariant under T and linearly independent, and also that $\sigma(T_\lambda) = \{\lambda\}$ for each $\lambda \in \Lambda$, where T_λ is the compression of T to E_λ. Set $E_\Lambda = \oplus_{\lambda \in \Lambda} E_\lambda$. The partial reduction result is that either X/E_Λ is infinite dimensional or $X = E_\Lambda$. To prove this, suppose $X \neq E_\Lambda$ and X/E_Λ has finite dimension. Let $v \in X \backslash E_\Lambda$; then the sequence $(T^n v / E_\Lambda)_{n \in \mathbb{N} \cup \{0\}}$ in X/E_Λ is linearly dependent. So there exists a minimum $m \in \mathbb{N}$ for which there exists $p \in \mathrm{poly}_m(\mathcal{L}(X)) \backslash \{0\}$ such that $p(T)v \in E_\Lambda$. Since X is complex, we have $p(T) = (\xi - T)q(T)$ for some scalar ξ and some $q \in \mathrm{poly}_{m-1}(\mathcal{L}(X)) \backslash \{0\}$. But the compression of $\xi - T$ to $\oplus_{\lambda \in \Lambda \backslash \{\xi\}} E_\lambda$ is surjective because $\sigma(T_\lambda) = \{\lambda\}$ for each $\lambda \in \Lambda$, so that $(\xi - T)q(T)v = p(T)v = z + (\xi - T)x$ for some $z \in E_\xi$ and $x \in \oplus_{\lambda \in \Lambda \backslash \{\xi\}} E_\lambda$. It follows that $(\xi - T)(q(T)v - x) \in E_\xi$ and hence that $q(T)v - x \in E_\xi$. So $q(T)v \in E_\Lambda$. Minimality of m forces $\deg(q) = 0$ and hence $v \in E_\Lambda$, which is a contradiction. The implications of this for non-trivial finite dimensional complex spaces are immediate—that $X = \oplus_{\lambda \in \Lambda} E_\lambda$, that $\Lambda \neq \varnothing$ and that the sum of the algebraic multiplicities of the eigenvalues of T is the dimension of X.

EXERCISES

Q 5.1.1 Suppose $n \in \mathbb{N}$ and $A \in \mathcal{M}_{n \times n}(\mathbb{C})$. Show that A has at least one eigenvalue. Show that not every real square matrix has a real eigenvalue.

Q 5.1.2 Suppose $A \in \mathcal{M}_{n \times n}(\mathbb{C})$. Then $\sigma(A)$ is the set of eigenvalues of A, by 3.4.28. If the entries of A are all real, it is clear that its spectrum as an element of $\mathcal{M}_{n \times n}(\mathbb{R})$ may be smaller, though certainly not larger, than its spectrum as an element of $\mathcal{M}_{n \times n}(\mathbb{C})$. Does this contradict 3.3.7?

Q 5.1.3 Suppose A is a complex algebra. What is the cardinality of $\mathrm{poly}(A)$?

5.2 Linear Shapes

Each non-trivial vector space over $\mathbb{R}$ or $\mathbb{C}$ may be regarded as a union of *lines*. Each line has distinguished subsets known as *line segments*. These line segments determine a rich assortment of *convex shapes* which play an important rôle in the analysis of linear spaces.

Lines

Definition 5.2.1
Suppose that X is a linear space, that $a \in X$ and that $v \in X \backslash \{0\}$. We define the LINE THROUGH a IN THE DIRECTION v to be the subset $\{a\} + \mathbb{R}v$ of X. It is easily checked that, given distinct points $a, b \in X$, the line $\{a\} + \mathbb{R}(b - a)$ is the unique line which includes $\{a, b\}$; it will be denoted by $L_{a,b}$. We define the CLOSED HALF-LINE FROM a IN THE DIRECTION v to be the subset $\{a\} + \mathbb{R}^{\oplus} v$ of X and the OPEN HALF-LINE FROM a IN THE DIRECTION v to be the subset $\{a\} + \mathbb{R}^{+} v$ of X; in each case, a is called the ENDPOINT of the half-line.

In a linear space X, the lines through the origin are precisely the one dimensional subspaces $\mathbb{R}v$ of X considered as a real linear space, where $v \in X \backslash \{0\}$. The lines of X are all the translates of these special lines of X. Each of them has two natural complete orderings, one the reverse of the other, and there are many different similarity mappings which identify a given line L as a copy of $\mathbb{R}$. Indeed, we can choose any pair of distinct points of L, one to associate with the origin by translation and the other to associate with 1 on the Real Line, and use these choices to determine such a mapping; whether $a \leq b$ or $b \leq a$ then depends on which of the two half-lines is associated with $\mathbb{R}^{\oplus}$. Specifically, consider a line $L = \{a\} + \mathbb{R}v$ of a linear space X and an arbitrary point $x = a + rv$ of L. Then $L - \{x\} = \mathbb{R}v$ which is an order isomorphic copy of the linear space $\mathbb{R}$ under any of the isomorphisms $t \mapsto \beta t v$ for $\beta \in \mathbb{R} \backslash \{0\}$, and each such isomorphism imposes one of the two natural complete orderings on the line L.

Definition 5.2.2

Suppose X is a linear space and $a, b \in X$. We define the CLOSED LINE SEGMENT $[a, b]$ of X to be the subset $\{a + t(b - a) \mid t \in [0, 1]\}$ of X. We define (a, b) to be $\{a + t(b - a) \mid t \in (0, 1)\}$, $(a, b]$ to be $\{a + t(b - a) \mid t \in (0, 1]\}$, and $[a, b)$ to be $\{a + t(b - a) \mid t \in [0, 1)\}$; if $a \neq b$, the first of these is called an OPEN LINE SEGMENT of X and the other two HALF-OPEN LINE SEGMENTS of X. The points a and b are called the ENDPOINTS of these line segments.

Line segments are generalizations of bounded intervals of the Real Line. There is, however, a slight difference, in that, when we write $[a, b]$ for a closed interval of the Real Line, it is usually understood that $a \leq b$; when we write $[a, b]$ for a closed line segment of an arbitrary linear space, we might just as well write $[b, a]$ for the same line segment.

Singleton sets are degenerate closed line segments. Every other line segment has exactly two endpoints uniquely determined; similarly, the single endpoint of a half-line is uniquely determined. The direction of a half-line or line, however, can be multiplied by any member of $\mathbb{R}^+$ or $\mathbb{R} \backslash \{0\}$ respectively without effecting any alteration of the set.

Convex Sets

Convexity, defined below, relates to real line segments even if the space is complex. Consequently, whether a complex linear space is regarded as complex or real has no relevance in deciding on the convexity or otherwise of its subsets.

Definition 5.2.3

Suppose X is a linear space and $S \subseteq X$. Then S is said to be CONVEX if and only if, for each $a, b \in S$, the line segment $[a, b]$ is included in S.

Example 5.2.4

The unit disc of $\mathbb{C}$ is convex; the unit circle is not. All lines, half-lines and line segments of a linear space are convex. All subspaces of a linear space are convex. All translates of convex sets are convex; and it is easy to check that sums of scalar multiples of convex sets are convex.

Example 5.2.5

Suppose X is a linear space and $S \subseteq X$. We define the CONVEX HULL of S, denoted by $co(S)$, to be the intersection of all convex supersets of S in X. For example $co(\mathbb{T}) = \mathbb{D}$. It is clear that every intersection of convex subsets of X

is convex, and, since X is convex, it follows that $S \subseteq \mathrm{co}(S)$ and that $\mathrm{co}(S)$ is the smallest convex subset of X which includes S. If $S \neq \varnothing$, it is easily shown by induction that $\mathrm{co}(S)$ is the collection of all linear combinations $\sum_{i=1}^{n} \alpha_i x_i$ of members of S where the coefficients are in $\mathbb{R}^+$ and have sum 1; such sums are called CONVEX COMBINATIONS of members of S.

Polygons

For our purposes, a *polygon* is simply a finite union of closed line segments joined end to end.

Definition 5.2.6
Suppose X is a linear space and $P \subseteq X$. Then P is called a POLYGON in X if and only if there exists a finite sequence $(a_i)_{i \in \mathbb{I}}$ in P such that P can be expressed as the union $\bigcup\{[a_{i-1}, a_i] \mid i \in \mathbb{I} \setminus \{1\}\}$.

Theorem 5.2.7
Suppose C is a convex subset of a linear space, $P \subseteq C$, $p \in P$ and $x \in C$. If P is a polygon, then $P \cup [p, x]$ is also a polygon included in C.

Proof
There exists a finite sequence $(a_i)_{i \in \mathbb{I}}$ in P with $P = \bigcup\{[a_{i-1}, a_i] \mid i \in \mathbb{I} \setminus \{1\}\}$. And there exists $k \in \mathbb{I} \setminus \{1\}$ such that $p \in [a_{k-1}, a_k]$. For $i \in \mathbb{N}$ with $1 \leq i < k$, set $b_i = a_i$. Set $b_k = b_{k+2} = p$ and $b_{k+1} = x$; lastly, for $i \in \mathbb{N}$ with $k \leq i \leq \max \mathbb{I}$, set $b_{i+3} = a_i$. Then $P \cup [p, x] = \bigcup\{[b_{i-1}, b_i] \mid 1 < i \leq \max \mathbb{I} + 3\}$ is a polygon, and, since C is convex, it is included in C. □

Definition 5.2.8
Suppose X is a linear space and $S \subseteq X$. Then S is said to be POLYGONALLY CONNECTED if and only if, for each $x, y \in S$, S includes a polygon which includes $\{x, y\}$.

Example 5.2.9
Every interval of $\mathbb{R}$ is polygonally connected; indeed, every convex set is polygonally connected and only line segments are needed to connect its points. The set $\{z \in \mathbb{R}^2 \mid z = 0 \text{ or } 0 < z_2 < |z_1|\}$ is polygonally connected, but is not convex. The set $\{z \in \mathbb{R}^2 \mid z = 0 \text{ or } 0 < z_2 < z_1^2\}$ is not polygonally connected.

Inside Points and Extreme Points

We introduce the concept of an *inside point* of a subset of a linear space; this will play a rôle in some important theorems later in this chapter.

Definition 5.2.10

Suppose X is a linear space and $U \subseteq X$. We shall call a point z of U an INSIDE POINT of U if and only if, for every $x \in \langle U - \{z\} \rangle$, there exists $t \in \mathbb{R}^+$ such that $(-tx, tx) \subseteq U - \{z\}$. The set of inside points of U will be denoted by $\mathfrak{I}(U)$.

Theorem 5.2.11

Suppose X is a linear space and $a, z \in X$, and $U, A \subseteq X$ are non-empty. Then

- $\mathfrak{I}(-U) = -\mathfrak{I}(U)$;
- $(z \in \mathfrak{I}(U)$ and $A - A \subseteq \langle U - \{z\} \rangle) \Rightarrow \{z\} - A \subseteq \mathfrak{I}(U - A)$;
- $\mathfrak{I}(U - \{a\}) = \mathfrak{I}(U) - \{a\}$;
- if X is a real space, then $\mathfrak{I}(U) \subseteq \mathfrak{I}(\mathbb{R}^{\oplus} U)$.

Proof

For the first assertion, simply note that $\langle U - \{z\} \rangle = \langle -U + \{z\} \rangle$ and that $-(-tx, tx) = (-tx, tx)$. For the second, suppose $z \in \mathfrak{I}(U)$, $A - A \subseteq \langle U - \{z\} \rangle$ and $v \in \{z\} - A$. Then $U - A - \{v\} \subseteq (A - A) + (U - \{z\}) \subseteq \langle U - \{z\} \rangle$, so that, since $z \in \mathfrak{I}(U)$, for each $x \in \langle U - A - \{v\} \rangle$, there exists $t \in \mathbb{R}^+$ such that $(-tx, tx) \subseteq U - \{z\} \subseteq U - A - \{v\}$; then $v \in \mathfrak{I}(U - A)$ and the second assertion is proved. Since $\{a\} - \{a\} = \{0\}$, this now implies both $\mathfrak{I}(U) - \{a\} \subseteq \mathfrak{I}(U - \{a\})$ and $\mathfrak{I}(U - \{a\}) + \{a\} \subseteq \mathfrak{I}((U - \{a\}) + \{a\}) = \mathfrak{I}(U)$, yielding the third assertion. For the fourth, we assume X is real and $z \in \mathfrak{I}(U)$. Note that $\langle \mathbb{R}^{\oplus} U - \{z\} \rangle = \langle U - \{z\} \rangle + \mathbb{R}z$. Suppose that $x = a + \beta z$, where $a \in \langle U - \{z\} \rangle$ and $\beta \in \mathbb{R}$. There exists $t \in \mathbb{R}^+$ be such that $t|\beta| < 1$ and $(-ta, ta) \subseteq U - \{z\}$. Set $u = t/(1 + t|\beta|)$. Then, for $r \in (-u, u)$, we have $|r\beta| \leq u|\beta| \leq t|\beta| < 1$ and $|r/(1 + r\beta)| < u/(1 - u|\beta|) = t$; therefore $ra/(1 + r\beta) \in U - \{z\}$, and so $ra \in \mathbb{R}^{\oplus} U - \{z + r\beta z\}$, yielding $rx \in \mathbb{R}^{\oplus} U - \{z\}$. Since r is arbitrary in $(-u, u)$, it follows that $(-ux, ux) \subseteq \mathbb{R}^{\oplus} U - \{z\}$. Since x is arbitrary in $\langle \mathbb{R}^{\oplus} U - \{z\} \rangle$, we conclude that $z \in \mathfrak{I}(\mathbb{R}^{\oplus} U)$. $\square$

Theorem 5.2.12

Suppose X is a linear space and U is a subset of X. Then

- $\mathfrak{I}(U) \subseteq \{w \in U \mid \mathbb{R}^{\oplus}(U - \{w\}) = \langle U - \{w\} \rangle\}$;
- if U is convex, then $\mathfrak{I}(U) = \{w \in U \mid \mathbb{R}^{\oplus}(U - \{w\}) = \langle U - \{w\} \rangle\}$;
- if U is convex and $0 \in U$, then $\mathfrak{I}(U) = \{w \in U \mid \mathbb{R}^{\oplus}(U - \{w\}) = \langle U \rangle\}$.

Proof

If $a \in \mathfrak{I}(U)$, then, for each $x \in \langle U - \{a\}\rangle$, we have $\mathbb{R}^+ x \cap (U - \{a\}) \neq \varnothing$; therefore $x \in \mathbb{R}^{\oplus}(U - \{a\})$, and the first assertion follows. Now suppose U is convex and $z \in U$ and $\mathbb{R}^{\oplus}(U - \{z\}) = \langle U - \{z\}\rangle$; then, for each $x \in \langle U - \{z\}\rangle$, there exist $r, s \in \mathbb{R}^+$ such that $rx \in U - \{z\}$ and $-sx \in U - \{z\}$. Since U is convex, $[z - sx, z + rx] \subseteq U$, and hence $(-tx, tx) \subseteq U - \{z\}$, where $t = \min\{r, s\}$. So $z \in \mathfrak{I}(U)$ and the second assertion follows. Lastly, if U is convex, $0 \in U$ and $z \in U$, then $\langle U - \{z\}\rangle = \langle U\rangle$ (Q 5.2.4), which justifies the third assertion. □

Theorem 5.2.13

Suppose X is a linear space, C is a convex subset of X and $z \in C$. If $z \in \mathfrak{I}(C)$, then, for each $x \in C \backslash \{z\}$, z is not an endpoint of $C \cap L_{z,x}$; moreover, the converse is true if X is real.

Proof

Firstly, if $z \in \mathfrak{I}(C)$ and $x \in C$, then $z - x \in \langle C - \{z\}\rangle$; so there exists $t \in \mathbb{R}^+$ such that $t(z - x) \in C - \{z\}$, whence $z + t(z - x) \in C$, showing that, if $x \neq z$, then z is not an endpoint of $C \cap L_{z,x}$. For the converse, suppose $x \in C \backslash \{z\}$ and z is not an endpoint of $C \cap L_{z,x}$. There exists $t \in \mathbb{R}^+$ such that $z + t(z - x) \in C$; then $z - x \in \mathbb{R}^+(C - \{z\})$, whence $\mathbb{R}(x - z) \subseteq \mathbb{R}^+(C - \{z\})$. If this is so for all $x \in C \backslash \{z\}$ and X is real, then every linear combination of members of $C - \{z\}$ is a positive multiple of a convex combination of members of $C - \{z\}$; since $C - \{z\}$ is convex, this implies that $\langle C - \{z\}\rangle \subseteq \mathbb{R}^+(C - \{z\})$ and, by 5.2.12, that $z \in \mathfrak{I}(C)$. □

Definition 5.2.14

Suppose X is a linear space, C is a convex subset of X and $e \in C$. Then e is called an EXTREME POINT of C if and only if, for each line L of X, either $e \notin L$ or e is an endpoint of $C \cap L$.

Example 5.2.15

Convex sets admit an interesting partial ordering. Suppose X is a linear space and C is a convex subset of X; then for $a, b \in C$, we say $a < b$ if and only if b is an endpoint of $C \cap L_{a,b}$ and a is not. The relation $<$ is certainly anti-reflexive; that it is transitive can be checked as follows. Suppose $a, b, c \in C$ and $a < b$ and $b < c$; then a, b and c are distinct and there exist $r, s \in \mathbb{R}^+$ such that $a + r(a - b) \in C$ and $b + s(b - c) \in C$; it follows from the convexity of C that $a + (rs(a - c)/(1 + r + s)) \in C$, so that a is not an endpoint of $C \cap L_{a,c}$. Also, for $u \in \mathbb{R}^+$, if we had $c + u(c - a) \in C$, then we should have

$b + (us(b - a)/(1 + u + s)) \in C$, contradicting $a < b$; so c is an endpoint of $L_{a,c}$. Therefore $a < c$. Then $<$ is a partial ordering on C as claimed. This partial ordering may or may not admit maximal or minimal elements, but 5.2.13 shows that inside points of C are minimal and that, if the space is real, the two concepts coincide. Moreover, extreme points of C are maximal and, provided C satisfies a fairly mild condition, every maximal point is also extreme (Q 5.2.6). Extreme points play an important practical rôle in analysis which we do not pursue in this book.

Hyperplanes

A PLANE of $\mathbb{R}^3$ is any translate of a maximal subspace of $\mathbb{R}^3$. This property defines its generalization, the *hyperplane*, in an arbitrary linear space. The hyperplanes of $\mathbb{R}$ are its singleton subsets; the hyperplanes of $\mathbb{R}^2$ are its lines, and those of $\mathbb{R}^3$ are its planes.

Definition 5.2.16
Suppose X is a linear space. A subset H of X is called a HYPERPLANE of X if and only if H is a translate of some maximal subspace of X.

Hyperplanes are convex sets (5.2.4). And each hyperplane H is a translate of exactly one maximal subspace of X (Q 5.2.3); we shall denote this subspace below by $\tilde{H}$; note that $H - H = \tilde{H}$.

Half-spaces

A plane in $\mathbb{R}^3$ divides its complement in $\mathbb{R}^3$ into two distinct *half-spaces*; no point in one half can be connected by a line segment to any point in the other without passing through the plane. A similar situation obtains in an arbitrary real linear space when the plane is replaced by a hyperplane.

Definition 5.2.17
Suppose X is a real linear space, H is a hyperplane of X and $a \in X \backslash \tilde{H}$, where $\tilde{H}$ denotes the maximal subspace of X which is a translate of H. Then the set $H + \mathbb{R}^+ a$ will be called a HALF-SPACE of X determined by H.

Theorem 5.2.18
Suppose X is a real linear space and H is a hyperplane of X. Then H determines precisely two half-spaces of X, which can be expressed as $H + \mathbb{R}^+ a$ and

$H + \mathbb{R}^- a$, where a is any member of $X \backslash \tilde{H}$. Moreover, $\{H + \mathbb{R}^- a, H, H + \mathbb{R}^+ a\}$ is a partition of X.

Proof

Let $a \in X \backslash \tilde{H}$ and let $z \in X$ be such that $H = \tilde{H} + \{z\}$. Then we have $(H + \mathbb{R}^- a) \cup H \cup (H + \mathbb{R}^+ a) = H + \mathbb{R}a = \tilde{H} + \mathbb{R}a + \{z\} = X + \{z\} = X$. Moreover $H + \mathbb{R}^- a$ and $H + \mathbb{R}^+ a$ are certainly half-spaces; and, if any two members of $\{H + \mathbb{R}^- a, H, H + \mathbb{R}^+ a\}$ were not disjoint, then we should have $a \in \mathbb{R}^+ (H - H) = \tilde{H}$, which is a contradiction. It remains to show that there is no other half-space determined by H. Let $b \in X \backslash \tilde{H}$; then either $b \in \tilde{H} + \mathbb{R}^+ a$ or $b \in \tilde{H} + \mathbb{R}^- a$. Suppose $b \in \tilde{H} + \mathbb{R}^+ a$. Then $\tilde{H} + \mathbb{R}^+ b \subseteq \tilde{H} + \mathbb{R}^+ a$. Also, there exists $t \in \mathbb{R}^+$ such that $b - ta \in \tilde{H}$, whence $a - t^{-1} b \in \tilde{H}$, yielding $a \in \tilde{H} + \mathbb{R}^+ b$. Therefore $\tilde{H} + \mathbb{R}^+ a \subseteq \tilde{H} + \mathbb{R}^+ b$ and hence $\tilde{H} + \mathbb{R}^+ a = \tilde{H} + \mathbb{R}^+ b$. It follows that $H + \mathbb{R}^+ b = H + \mathbb{R}^+ a$. If we had $b \in \tilde{H} + \mathbb{R}^- a$, a similar calculation would yield $H + \mathbb{R}^+ b = H + \mathbb{R}^- a$. Because b is arbitrary in $X \backslash \tilde{H}$, the half-spaces $H + \mathbb{R}^+ a$ and $H + \mathbb{R}^- a$ are the only ones determined by H.   □

Wedges

The importance of *wedges* for analysis rests in their relationship with scalar linear maps; this will be discussed later in 5.3.4.

Definition 5.2.19

Suppose X is a linear space. A non-empty subset W of X will be called a WEDGE of X if and only if both $W + W \subseteq W$ and $\mathbb{R}^\oplus W \subseteq W$. Every wedge of X other than X itself is said to be PROPER. A wedge W of X will be called a CONE of X if and only if $W \cap (-W) = \{0\}$.

Example 5.2.20

The FIRST QUADRANT, $\{z \in \mathbb{R}^2 \mid 0 < z_1, \, 0 \le z_2\}$, of $\mathbb{R}^2$ is a cone. The wedge $\{z \in \mathbb{R}^3 \mid 0 \le z_1, \, 0 \le z_2\}$ of $\mathbb{R}^3$ is not a cone because it includes the line through the origin in the direction $(0, 0, 1)$.

Example 5.2.21

Suppose X is a linear space and C is a convex subset of X. Then, for $a, b \in C$ and $s, t \in \mathbb{R}^\oplus$ with $s + t \ne 0$, we have $sa/(s + t) + tb/(s + t) \in C$ and hence $sa + tb \in \mathbb{R}^+ C$. It follows that $\mathbb{R}^\oplus C$ is a wedge of X. If $C \cap (-C) \subseteq \{0\}$ then an easy calculation shows that $\mathbb{R}^\oplus C \cap \mathbb{R}^\ominus C = \{0\}$, so that $\mathbb{R}^\oplus C$ is a cone.

Maximal Wedges

In a real linear space, every maximal subspace M determines two half-spaces, each of which, when united with M, gives a *maximal wedge*. The converse (5.2.26) is far less easy to prove than our intuition would lead us to suspect.

Definition 5.2.22

Suppose X is a linear space. A proper wedge W of X is called a MAXIMAL WEDGE of X if and only if no proper wedge of X properly includes W.

Example 5.2.23

The wedge $\{z \in \mathbb{R}^3 \mid 0 \leq z_1\}$ of $\mathbb{R}^3$ is maximal. Note that this wedge properly includes the maximal subspace $\{z \in \mathbb{R}^3 \mid z_1 = 0\}$ of $\mathbb{R}^3$; we shall see below (5.2.26) that every maximal wedge includes a maximal subspace.

Example 5.2.24

If W is a wedge of a linear space X and $W + \mathbb{R}^\oplus z = X$ for some $z \in X$, it does not necessarily follow that W is a maximal wedge of X: let Q denote the first quadrant (5.2.20) of $\mathbb{R}^2$; for any $z \in \mathbb{R}^2 \backslash Q$, the wedge $W = Q + \mathbb{R}^\oplus z$ properly includes Q; but $W = X$ if and only if $z \in \mathbb{R}^- \times \mathbb{R}^-$.

Lemma 5.2.25

Suppose that X is a linear space and that W is a maximal wedge of X. Then $W \cup (-W) = X$. Moreover, W is not a subspace of X.

Proof

Since $W \neq X$, there exists $z \in X \backslash W$. Then $W + \mathbb{R}^\oplus z$ is a wedge of X which properly includes W. By maximality of W, we have $X = W + \mathbb{R}^\oplus z$. In particular $-z \in W + \mathbb{R}^\oplus z$, whence $-z \in W$. Since $z \notin W$, W is not a subspace of X. $\square$

Theorem 5.2.26

Suppose X is a real linear space, W is a maximal wedge of X and $a \in X \backslash W$. Then $M = W \cap (-W)$ is the unique maximal subspace of X included in W. Moreover, $W = M + \mathbb{R}^\ominus a$ and $X = M + \mathbb{R}a = W + \mathbb{R}^\oplus a$.

Proof

Since $W + W \subseteq W$ and $\mathbb{R}^\oplus W \subseteq W$, M is certainly a subspace of X; its inclusion in W is proper by 5.2.25. Suppose $b \in W \backslash M$. Then $-b \in X \backslash W$.

Since $a \notin W$, maximality of W ensures that $X = W + \mathbb{R}^{\oplus} a$ and hence that $-b \in W + \mathbb{R}^+ a$ and therefore that the set $S = \{ r \in \mathbb{R}^+ \mid -b - ra \in W \}$ is not empty. Let $\mu = \inf S$. We claim that $b + \mu a \in M$ and justify this claim in two parts. Firstly, noting that $-b \notin W$, we have $-b - t\mu a/(1+t) \notin W$ for each $t \in \mathbb{R}^{\oplus}$, whence $-b - t(b + \mu a) \notin W$, giving $-b \notin W + \mathbb{R}^{\oplus}(b + \mu a)$, which is not therefore equal to X. So maximality of W implies $b + \mu a \in W$. Secondly, for each $t \in \mathbb{R}^+$, the definition of μ yields some $s \in [\mu, \mu + 1/t)$ such that $-b - sa \in W$. Now $\mathbb{R}^+ a \cap [-b - sa, b + (\mu + 1/t)a] \neq \varnothing$ and $\mathbb{R}^+ a \cap W = \varnothing$; so $[-b - sa, b + (\mu + 1/t)a]$ is not a subset of W. Since W is convex and $-b - sa \in W$, this implies that $b + (\mu + 1/t)a \notin W$ and hence that $a + t(\mu a + b) \notin W$. Noting that also $a \notin W$, we therefore have $a \notin W + \mathbb{R}^{\ominus}(\mu a + b)$, which is not therefore equal to X. Maximality of W gives $-\mu a - b \in W$, and our claim that $b + \mu a \in M$ is justified. Therefore $b \in M + \mathbb{R}^{\ominus} a$. Since b is arbitrary in $W \backslash M$, this gives $W = M + \mathbb{R}^{\ominus} a$. Since $a \notin W$, it follows also that $X = W + \mathbb{R}^{\oplus} a = M + \mathbb{R}a$, so that M is a maximal subspace of X. Finally, since the sum of distinct maximal subspaces of X must equal X, M is uniquely determined. $\qquad \square$

The argument is two dimensional

Inclusion of Wedges in Maximal Wedges

Wedges which are sufficiently small are certainly included in maximal wedges. To be precise, if X is a linear space and W is a wedge of X which is included in a proper subspace of X (if X is real, this simply means that $W - W \neq X$), then that subspace is included in a maximal subspace by 3.2.11 and any such maximal subspace is included in a maximal wedge obtained by adding appropriate half-lines. But we have seen in 5.2.24 that there are non-maximal wedges which are not included in any maximal subspace; such wedges are the primary subject of 5.2.27 below.

Theorem 5.2.27 SPATIAL HAHN–BANACH THEOREM

Suppose X is a real linear space and V is a proper wedge of X with $\Im(V) \neq \varnothing$. Then V is included in a maximal wedge of X. If V is not a subspace of X, then, for every $z \in \Im(V)$, we have $-z \notin V$ and there exists a maximal wedge W of X with $V \subseteq W$ and $W + \mathbb{R}^{\ominus} z = X$.

Proof

Firstly, if V is a subspace of X then it is proper and, by 3.2.11, there exists a maximal subspace M of X with $V \subseteq M$; then, for any $a \in X \backslash M$, $M + \mathbb{R}^{\oplus} a$ is a maximal wedge of X which includes V. For the rest, suppose V is not a subspace of X and $z \in \mathfrak{I}(V)$. Then $\langle V \rangle = \mathbb{R}^{\oplus}(V - \{z\})$ by 5.2.12, whence $V + \mathbb{R}^{\ominus} z = \langle V \rangle$ and $-z \notin V$. Let Y be a subspace of X such that $\langle V \rangle \oplus Y = X$. Let $\mathcal{C}$ be the collection of all proper wedges of X which include $V + Y$. Certainly $V + Y$ is such a wedge, because $-z \in \langle V \rangle \backslash V$. So $\mathcal{C} \neq \varnothing$. Also $-z$ is not a member of any member of $\mathcal{C}$ because $V + Y + \mathbb{R}^{\ominus} z = X$. Suppose $\mathcal{N}$ is any non-trivial nest in $\mathcal{C}$; then $-z \notin \bigcup \mathcal{N}$ and $V + Y \subseteq \bigcup \mathcal{N}$, and it is easy to check that $\bigcup \mathcal{N}$ is a wedge. So $\bigcup \mathcal{N} \in \mathcal{C}$. By Zorn's Lemma, $\mathcal{C}$ has a maximal member W, which is clearly a maximal wedge of X. Since $-z \notin W$, we have $W + \mathbb{R}^{\ominus} z = X$.   □

Example 5.2.28

The condition on z in 5.2.27 cannot in general be weakened to $z \in V \backslash (-V)$: let $V = \{w \in \mathbb{R}^2 \mid w_1 \in \mathbb{R}^+$ or $(w_1 = 0$ and $w_2 \in \mathbb{R}^{\ominus})\}$ and $z = (0, -1)$; then $V + \mathbb{R}^{\ominus} z$ is the only proper wedge of X which properly includes V.

Separation

Most pairs of disjoint convex subsets of $\mathbb{R}^2$ can be *separated* by some line, in the sense that one set is on one *side* of the line and the other on the other side. Usually there are many choices of line to achieve this separation. In some cases, however, we may be forced to allow the line to have non-empty intersection with just one of the sets—for example, for the sets $\{z \in \mathbb{R}^2 \mid z_2 \geq 0\}$ and $\{z \in \mathbb{R}^2 \mid z_2 < 0\}$. There are also cases in which even this sort of separation is not possible: consider the sets $\{z \in \mathbb{R}^2 \mid z_2 > 0$ or $(z_1 > 0$ and $z_2 = 0)\}$ and $\{z \in \mathbb{R}^2 \mid z_2 < 0$ or $(z_1 < 0$ and $z_2 = 0)\}$. After generalizing the idea of separation, we explore below conditions under which disjoint convex subsets of a real linear space can be separated by a hyperplane.

Definition 5.2.29

Suppose X is a real linear space, H is a hyperplane of X, and A and B are subsets of X. We say that H WEAKLY SEPARATES A and B in X if and only if A is disjoint from one half-space determined by H and B is disjoint from the other; if also $A \cap H = \varnothing$, we say that H SEPARATES A from B; if, moreover, $B \cap H = \varnothing$, we say that H STRONGLY SEPARATES A and B.

Example 5.2.30

Although strong and weak separation, as defined above, are symmetrical concepts, our notion of separation itself is not. For example, hyperplanes of $\mathbb{R}$ are singleton sets and the interval $[0,1)$ of $\mathbb{R}$ is separated from the interval $[1,2]$ by $\{1\}$; but there is no hyperplane of $\mathbb{R}$ which separates $[1,2]$ from $[0,1)$.

Theorem 5.2.31 Hahn–Banach separation theorem

Suppose X is a real linear space, A and U are non-empty disjoint subsets of X, and $A-U$ is convex. Suppose there exists $z \in \mathfrak{I}(U)$ such that $A-A \subseteq \langle U - \{z\}\rangle$. Then there is a hyperplane H of X which weakly separates U and A and separates $\mathfrak{I}(U)$ from A. If $\mathfrak{I}(U) = U$, then H separates U from A.

Proof

$0 \notin A-U$; so $\mathbb{R}^\oplus(A-U)$ is a cone. Let $w \in A-\{z\}$. Since $A-A \subseteq \langle U - \{z\}\rangle$, $w \in \mathfrak{I}(\mathbb{R}^\oplus(A-U))$, by 5.2.11. By 5.2.26 and 5.2.27, there is a maximal subspace M of X with $w \notin M$ such that $\mathbb{R}^\oplus(A-U) \subseteq M+\mathbb{R}^\oplus w$ and $M+\mathbb{R}w = X$. Then $P = \{\alpha \in \mathbb{R} \mid A \cap (M + \alpha w) \neq \varnothing\}$ and $Q = \{\beta \in \mathbb{R} \mid U \cap (M + \beta w) \neq \varnothing\}$ are non-empty and $-\infty < \sup Q \leq \inf P < \infty$. Let $\mu = (\sup Q + \inf P)/2$ and set $H = M+\mu w$. Then $A \subseteq H+\mathbb{R}^\oplus w$ and $U \subseteq H + \mathbb{R}^\ominus w$, as required. Finally, suppose $h \in \mathfrak{I}(U) \cap H$. Then, for each $x \in \langle U - \{h\}\rangle$, there exists $r \in \mathbb{R}^+$ such that $rx \in U - \{h\} \subseteq M + \mathbb{R}^\ominus w$; so $\langle U - \{h\}\rangle \subseteq M + \mathbb{R}^\ominus w$, whence $\langle U - \{h\}\rangle \subseteq M$ and $U \subseteq H$ and $U - U \subseteq M$. But $z \in U$; so $A - A \subseteq \langle U - \{z\}\rangle \subseteq M$, so that $A \subseteq M + \{\beta w\}$ for some $\beta \in \mathbb{R}$. Since $h \in U \cap H$, we have $\sup Q = \mu = \inf P$ and so $\beta = \mu$ and $A \subseteq H$. Then $w \in A - U \subseteq H - H = M$, a contradiction. □

Corollary 5.2.32

Suppose X is a real linear space and C is a proper convex subset of X with $\mathfrak{I}(C) \neq \varnothing$. Then, for each $x \in X\backslash C$, there exists a hyperplane H of X which weakly separates $\{x\}$ and C and satisfies $H \cap \mathfrak{I}(C) = \varnothing$.

Corollary 5.2.33

Suppose X is a real linear space and C is a proper convex subset of X and $\mathfrak{I}(C) = C$. Then C is the intersection of all half-spaces of X which include C.

Proof

If $C = \varnothing$, the result is obvious, so we suppose otherwise. Let $z \in C$. Then, for each $x \in X\backslash C$, 5.2.32 gives us a hyperplane H of X which weakly separates C from $\{x\}$; and, because $H \cap C = H \cap \mathfrak{I}(C) = \varnothing$, this is a separation of C from $\{x\}$. Since x is arbitrary in $X\backslash C$, the result follows at once. □

EXERCISES

Q 5.2.1 Suppose X is a linear space and $W \subseteq X$. Show that W is a wedge of X if and only if $W = \bigcup \{ \mathbb{R}^{\oplus} v \mid v \in W \}$ and W is convex.

Q 5.2.2 Suppose X is a linear space and $W \subseteq X$. Show that W is a cone of X if and only if W is a wedge which includes no line through the origin.

Q 5.2.3 Suppose X is a linear space. Show that every half-space of X is a translate of exactly one maximal subspace of X.

Q 5.2.4 Suppose X is a linear space, C is a convex subset of X and $z \in C$. Show that $0 \in C \Rightarrow \langle C - \{z\} \rangle = \langle C \rangle$.

Q 5.2.5 Suppose X is a linear space, C is a convex subset of X and $e \in C$. Show that e is extreme if and only if no open line segment of X is included in C and contains e. Show also that e is extreme if and only if $C \backslash \{e\}$ is convex.

Q 5.2.6 Suppose X is a linear space and C is a convex subset of X. Show that every extreme point of C is maximal with respect to the ordering of 5.2.15 and that this characteristic identifies extreme points if, for every line L of X, $C \cap L$ is either empty or contains an endpoint.

5.3 Linear Functionals

Scalar linear maps defined on linear spaces are called *functionals*. *Integration on an appropriate space of functions is an important example.*

Definition 5.3.1

Suppose X is a linear space. Then each member of $\mathcal{L}(X, \mathbb{K})$ is known as a LINEAR FUNCTIONAL on X. Each member of $\mathcal{L}(X, \mathbb{R})$, where X is regarded as a real linear space, is known as a REAL-LINEAR FUNCTIONAL on X.

When X is a real linear space, the real-linear functionals on X are precisely the linear functionals on X. If X is a complex space, the two concepts differ, but the multiplicative structure of $\mathbb{C}$ imposes a very tight relationship between real and complex functionals which ensures that we can identify the latter by knowing the former. 5.3.2 below implies that the map $f \mapsto \Re f$ is a one-to-one correspondence between the set of linear functionals on a given linear space and the set of real-linear functionals on the same space. But it says even more.

Theorem 5.3.2

Suppose X is a complex linear space.

- If $f: X \to \mathbb{C}$ is a linear map and $u = \Re f$, then f is $x \mapsto u(x) - iu(ix)$.
- If $u: X \to \mathbb{R}$ is a real-linear map, then the map f defined on X to be $x \mapsto u(x) - iu(ix)$ is linear.

Proof

If u is the real part of a linear map $f: X \to \mathbb{C}$, then, for each $x \in X$, we have $f(x) = u(x) + (f(x) - \overline{f(x)})/2 = u(x) - i(f(ix) + \overline{f(ix)})/2 = u(x) - iu(ix)$, and the first part is proved. If u is a real-linear map on X and f is as stated, then, for each $x, y \in X$, we have $f(x + y) = f(x) + f(y)$ and, for each $a, b \in \mathbb{R}$,

$$
\begin{aligned}
f((a + ib)x) &= u((a + ib)x) - iu((-b + ia)x) \\
&= au(x) + bu(ix) + ibu(x) - iau(ix) \\
&= (a + ib)u(x) - i(a + ib)u(ix) \\
&= (a + ib)f(x),
\end{aligned}
$$

which justifies the second assertion.   □

Maximal Wedges and Real-Linear Functionals

The key connection between wedges and functionals is that each functional determines a *positive wedge*. Specifically, if X is a linear space and f is a linear or real-linear functional on X, then $\{x \in X \mid f(x) \in \mathbb{R}^+\} \cup \{0\}$ is clearly a cone of X, and the union of this cone with $\ker(f)$ is the wedge $\{x \in X \mid f(x) \in \mathbb{R}^{\oplus}\}$. This wedge is the one which we shall associate with f. It is proper if f is non-zero. If X is a complex space, then the wedge associated with f is not maximal. But the wedge associated with each non-zero real-linear functional on X is maximal and, moreover, there is a one-to-one correspondence between maximal wedges of X and sets of the type $\{ru \mid r \in \mathbb{R}^+\}$, where u is a real-linear functional on X (5.3.4).

Definition 5.3.3

Suppose X is a linear space and f is a linear or real-linear functional on X. The wedge $\{x \in X \mid f(x) \in \mathbb{R}^{\oplus}\}$ will be called the POSITIVE WEDGE DETERMINED BY f and will be denoted by wedge(f).

Theorem 5.3.4

Suppose X is a real linear space. Then the map $\mathbb{R}^{\oplus}u \mapsto \text{wedge}(u)$ is a one-to-one correspondence between the closed half-lines from the origin of $\mathcal{L}(X, \mathbb{R})$ and the maximal wedges of X.

Proof

Firstly, if $u \in \mathcal{L}(X, \mathbb{R}) \setminus \{0\}$, then $\ker(u)$ is a maximal subspace of X (3.4.14), and it follows that wedge (u) is a maximal wedge of X. Moreover, it is clear that wedge $(ru) = \text{wedge}(u)$ for all $r \in \mathbb{R}^{+}$. Now suppose W is a maximal wedge of X. Invoking 5.2.26, let $M = W \cap (-W)$ be the maximal subspace of X included in W and let $a \in X \setminus W$; then $X = M + \mathbb{R}a$ and $W = M + \mathbb{R}^{\ominus}a$. Define f on X by setting $f(m - ta) = t$ for each $m \in M$ and $t \in \mathbb{R}$. Clearly f is a functional and $f(W) \subseteq \mathbb{R}^{\oplus}$. Moreover, $X \setminus W = M + \mathbb{R}^{+}a$, so that $f(X \setminus W) \subseteq \mathbb{R}^{-}$. Therefore $W = \text{wedge}(f)$. Now suppose that $g \in \mathcal{L}(X, \mathbb{R})$ and wedge $(g) = W$. Then $\ker(g)$ is a maximal subspace of X included in W; since this is unique by 5.2.26, we have $\ker(g) = M$. Also $-a \in \text{wedge}(g) \setminus \ker(g)$, so that $g(-a) \in \mathbb{R}^{+}$. For each $m \in M$ and $t \in \mathbb{R}$, we now have $g(m - ta) = tg(-a) = g(-a)f(m - ta)$. Hence $g = g(-a)f \in \mathbb{R}^{+}f$. □

The Algebraic Hahn–Banach Theorem

A linear functional defined on a subspace of a real linear space X can be extended to a linear functional on X. This is easy to prove. But we generally require an extension to exhibit some important property. The particular property which we consider below is domination by a given *sublinear* map. Such maps are natural generalizations of the modulus function; the most important examples are *norms*, which we shall discuss in Chapter 6.

Definition 5.3.5

Suppose X is a linear space. A function $p: X \to \mathbb{R}$ is called a SUBLINEAR FUNCTIONAL on X if and only if $p(a + b) \leq p(a) + p(b)$ and $p(ta) = tp(a)$ for all $a, b \in X$ and $t \in \mathbb{R}^{\oplus}$.

Theorem 5.3.6 ALGEBRAIC HAHN–BANACH THEOREM

Suppose X is a real linear space, M is a proper subspace of X, p is a sublinear functional on X and f is a linear functional on M dominated by $p|_M$. Then there exists a linear extension of f to X which is dominated by p.

Proof

Let N be any subspace of X such that $M \oplus N = X$. Define q on X by setting $q(m+n) = p(m+n) - f(m)$ for each $m \in M$ and $n \in N$; q is certainly sublinear and $q(M) \subseteq \mathbb{R}^\oplus$. Let $V = \{a \in X \mid q(a) < 0\} \cup \{0\}$; since q is sublinear, V is a cone of X. Then $V \cap (M - V) = \{0\}$. For every $v \in V \backslash \{0\}$ and $x \in X$, there exists $t \in \mathbb{R}^+$ such that $tq(x) < -q(v)$, whence $x + t^{-1}v \in V$, yielding $x \in V - \mathbb{R}^+ v = \mathbb{R}^\oplus (V - \{v\})$ and then $\mathbb{R}^\oplus (V - \{v\}) = X$. So $\mathfrak{I}(V) = V \backslash \{0\}$ by 5.2.12. By the Separation Theorem, there exists a hyperplane K of X which separates $V \backslash \{0\}$ from $M - V$; certainly $0 \in K$, because $0 \in V \cap (M - V)$; so K is a maximal subspace of X and $M \subseteq K$. Then there exists $z \in N \backslash K$. Note that $X = M \oplus (K \cap N) \oplus \mathbb{R}z$ and define g to be the unique functional on X for which $g|_M = f$ and $g(K \cap N) = \{0\}$ and $g(z) = \inf\{q(z + k) \mid k \in K\}$. Suppose $x \in X$ and $x = m + a + tz$, where $m \in M$, $a \in K \cap N$ and $t \in \mathbb{R}$. If $t \in \mathbb{R}^\oplus$, then $g(x) = f(m) + \inf\{q(tz + k) \mid k \in K\} \leq f(m) + q(x) = p(x)$; and, if $t \in \mathbb{R}^-$, then $g(x) = f(m) - \inf\{q(-tz + k) \mid k \in K\} \leq f(m) + q(m + a + tz) = p(x)$, the last inequality holding because $q(m + a + tz) + q(-tz + k) \geq q(m + a + k) \geq 0$ for all $k \in K$. Since x is arbitrary in X, we have $g \leq p$ as required.   □

EXERCISES

Q 5.3.1 A function f between additive groups is said to be ADDITIVE if and only if $f(x + y) = f(x) + f(y)$ for all $x, y \in \text{dom}(f)$. Does there exist an additive non-linear map from $\mathbb{R}$ to $\mathbb{R}$?

Q 5.3.2 Find a proper wedge which is not included in any maximal wedge.

Q 5.3.3 Find two disjoint convex subsets A and U of $\mathbb{R}^2$ for which $\mathfrak{I}(A) \backslash A$ and $\mathfrak{I}(U) \backslash U$ are singleton sets, and for which there is no line in $\mathbb{R}^2$ which separates either A from U or U from A.

Q 5.3.4 Find two disjoint convex subsets A and U of $\mathbb{R}^2$ for which $\mathfrak{I}(U) = U$ and $\mathfrak{I}(A) = A$, and for which there is no line in $\mathbb{R}^2$ which separates U from A.

6

Geometric Structure

The desire to quantify is at the heart of primitive mathematics. Its first realization is in the art of counting, which develops naturally into a science of discrete mathematics. But both precision, pursued as an ideal, and the scientific need for fine measurement propel us beyond the bounds of the discrete environment into the world of *geometry*—literally *earth measurement*—in which the Real Number System is the appropriate tool for measurement.

6.1 Semimetrics and Metrics

The only geometry inherent in an arbitrary set is that of coincidence—members are either distinct or they are not. On the Real Line and in the Complex Plane, there is a well developed geometry in which the distance between two numbers is the modulus of their difference. These distances are the values of the function $d: \mathbb{C} \times \mathbb{C} \to \mathbb{R}$ given by $(z, w) \mapsto |z - w|$. Our task in this section is to impose on arbitrary sets a geometry which generalizes that of $\mathbb{R}$ and $\mathbb{C}$. We therefore examine the fundamental properties which the function d possesses. They are, firstly, that d distinguishes between points; secondly, that d is symmetric; and thirdly, that the distance between two points is never shortened by making a detour through a third point. If we identify complex numbers with points in a plane, the intuitive realization of this third property is that no side of a triangle is longer than the sum of the lengths of the other two sides; it is therefore known as the *triangle inequality*. These three properties inform our definition of a distance function, which we call a *metric*.

Definition 6.1.1
Suppose X is a set. A function $d: X \times X \to \mathbb{R}$ will be called a SEMIMETRIC on X if and only if, for each $a, b, c \in X$,

- $d(a, a) = 0$;
- $d(a, b) = d(b, a)$;
- (TRIANGLE INEQUALITY) $d(a, c) \leq d(a, b) + d(b, c)$.

A semimetric d is called a METRIC on X if and only if, for all $a, b \in X$, it satisfies the implication $d(a, b) = 0 \Rightarrow a = b$. It can be easily established by induction that, if $(a_i)_{i \in \mathbb{I}}$ is a finite sequence in X of length $n \in \mathbb{N}$, then $d(a_1, a_n) \leq \sum_{i=1}^{n-1} d(a_i, a_{i+1})$.

Example 6.1.2
The function $d: \mathbb{C} \times \mathbb{C} \to \mathbb{R}$ given by $d(z, w) = |z - w|$ for each $z, w \in \mathbb{C}$ is called the USUAL or EUCLIDEAN METRIC on $\mathbb{C}$; its restriction to $\mathbb{R} \times \mathbb{R}$ is the USUAL metric on $\mathbb{R}$.

Example 6.1.3
The empty function is the only metric on $\varnothing$. Suppose X is any non-empty set. The zero map is the simplest semimetric on X. The simplest metric on X is the DISCRETE METRIC, defined by setting $d(a, b) = 1$ for each $a, b \in X$ with $a \neq b$ and $d(a, a) = 0$ for each $a \in X$. No new geometry is imposed on a set by the discrete metric; it simply captures the inherent geometry of coincidence.

Example 6.1.4
Semimetrics have non-negative values: $0 = d(a, a) \leq d(a, b) + d(b, a) = 2d(a, b)$. On the other hand, the range of a semimetric may or may not be bounded above; the range of the Euclidean metric on $\mathbb{R}$ is certainly not. However, given any metric, it is always possible to define a related metric on the same set whose range is bounded: let X be a non-empty set and d a metric on X; a little calculation shows that $(a, b) \mapsto d(a, b)/(1 + d(a, b))$ is a metric on X whose range is included in the interval $[0, 1)$.

Example 6.1.5
Suppose that X is a non-empty set and that $d: X \times X \to \mathbb{R}$ is non-negative, symmetric, zero only on the set $\{(x, x) \mid x \in X\}$, and satisfies the inequality $d(a, c) \leq \max\{d(a, b), d(b, c)\}$ for all $a, b, c \in X$. Then d is called an ULTRA-METRIC on X. Ultrametrics are, of course, metrics.

Definition 6.1.6

An ordered pair (X, d) is called a SEMIMETRIC SPACE if and only if d is a semimetric on X; it is called a METRIC SPACE if and only if d is a metric. When it is clear from the context which function is intended, we generally say simply that X is a semimetric space or metric space. If Y is a subset of X, then the restriction of d to $Y \times Y$, which we denote also by d, is clearly a semimetric on Y; (Y, d), or simply Y, is called a SEMIMETRIC SUBSPACE or METRIC SUBSPACE of (X, d), as appropriate, and (X, d) is called a a SEMIMETRIC SUPERSPACE or METRIC SUPERSPACE of (Y, d).

Example 6.1.7

There are several ways of making $\mathbb{R}^2$ into a metric space, any of which might be regarded as natural depending on the intended application. The following three functions defined on $\mathbb{R}^2 \times \mathbb{R}^2$ are metrics:

- $(a, b) \mapsto |a_1 - b_1| + |a_2 - b_2|$;
- (EUCLIDEAN METRIC ON $\mathbb{R}^2$) $(a, b) \mapsto \sqrt{(a_1 - b_1)^2 + (a_2 - b_2)^2}$;
- $(a, b) \mapsto \max\{|a_1 - b_1|, |a_2 - b_2|\}$.

Distance

Definition 6.1.8

Suppose (X, d) is a semimetric space, $A, B \subseteq X$ and $x, y \in X$. We define the DISTANCE from x to y to be $d(x, y)$, the DISTANCE $\operatorname{dist}(x, A)$ from x to A to be ∞ if $A = \varnothing$ and $\inf\{d(x, a) \mid a \in A\}$ otherwise, and the DISTANCE $\operatorname{dist}(A, B)$ from A to B to be ∞ if $A = \varnothing$ or $B = \varnothing$ and $\inf\{d(a, b) \mid a \in A, b \in B\}$ otherwise. We define the DIAMETER of A to be $\operatorname{diam}(A) = \sup\{d(r, s) \mid r, s \in A\}$.

Example 6.1.9

Suppose $S \subseteq \mathbb{R}$, where $\mathbb{R}$ is endowed with its Euclidean metric. We have seen that S is bounded in $\mathbb{R}$ if and only if S is order-bounded in $\mathbb{R}$ (4.2.9). And now, if $s = \sup S$ and $z = \operatorname{dist}(s, S)$, then $s - z$ is an upper bound for S and, by the definition of supremum, $s \leq s - z$, which yields $z = 0$, since $z \in \mathbb{R}^{\oplus}$. So $\operatorname{dist}(\sup S, S) = 0$. A similar calculation yields $\operatorname{dist}(\inf S, S) = 0$.

Balls

Geometric shapes emerge naturally from geometric structure. The most primitive of these are *balls*, the natural generalizations of bounded intervals to arbitrary semimetric spaces. As geometric shapes, balls are somewhat elusive, because the most natural way to picture a ball may or may not correspond to our intuitive notion, depending on the semimetric which spawns it.

Definition 6.1.10

Suppose (X, d) is a semimetric space. A subset B of X is called a BALL of X if and only if $\{b \in X \mid d(a, b) < r\} \subseteq B \subseteq \{b \in X \mid d(a, b) \leq r\}$ for some $a \in X$ and $r \in \mathbb{R}^+$; the point a is called a CENTRE of the ball B and the number r is called a RADIUS. For each $a \in X$ and $r \in \mathbb{R}^+$, the ball $\{b \in X \mid d(a, b) < r\}$, which we shall denote by $\flat[a \,; r)$, is called the OPEN BALL of X CENTRED at a with RADIUS r and the ball $\{b \in X \mid d(a, b) \leq r\}$, denoted by $\flat[a \,; r]$, is called the CLOSED BALL of X CENTRED at a with RADIUS r.

Example 6.1.11

The open balls of $\mathbb{R}$ with the Euclidean metric are the bounded open intervals; the closed balls are the non-degenerate bounded closed intervals; singleton subsets of $\mathbb{R}$ are not balls. The open balls of the Complex Plane are its open discs and the closed balls are its closed discs (4.2.11).

Example 6.1.12

When $\mathbb{R}^2$ has the metrics of 6.1.7 and is represented as a plane with axes arranged in the usual way, the balls of radius 1 centred at $(0, 0)$ are the regions pictured in the following diagrams, where the dot indicates the point $(1, 0)$. In each space, all other balls have the same shape and differ only in size and location; this feature is not shared by all metrics on $\mathbb{R}^2$ (see 6.1.13).

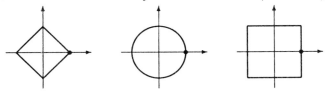

Example 6.1.13

It is tempting to think that the triangle inequality ensures that metrics on linear spaces always produce convex balls; but this is not true. Let e denote the Euclidean metric on $\mathbb{R}^2$ and define a metric d on $\mathbb{R}^2$ by $d(a, b) = e(a, b) + 1$ if exactly one of a and b is $(0, 0)$ and $d(a, b) = e(a, b)$ otherwise. Set $z = (1, 0)$; then $\flat_d[z \,; 2) = \flat_e[z \,; 2) \setminus \{(0, 0)\}$, being PUNCTURED at $(0, 0)$ is not convex. Not all translates of $\flat_d[z \,; 2)$ are punctured and, for $\lambda \in (1, \infty)$, $\lambda \flat_d[z \,; 2) \neq \flat_d[z \,; 2\lambda)$.

Example 6.1.14

We shall discuss open and closed balls of radius r and mean sets of the type $\flat[a\,;r)$ and $\flat[a\,;r]$ respectively. It should be borne in mind, however, that, unlike the diameter, neither the centre nor the radius need be unique. Moreover, the diameter of a ball may be less than a stated radius. If, for example, X is a discrete metric space and $x \in X$, then the singleton set $\{x\}$, whose diameter is 0, can be described as $\flat[x\,;r)$ for any $r \in (0,1)$; and the whole space X can be described as $\flat[x\,;r)$ for any $r \in (1,\infty)$.

Bounded Sets and Functions

We extend the concept of a bounded set (4.2.9, 4.2.11) to arbitrary semimetric spaces.

Definition 6.1.15

Suppose (X,d) is a metric space. A subset S of X is said to be BOUNDED if and only if it is included in some ball of X.

We have seen (4.2.9) that, in $\mathbb{R}$, a set is bounded with respect to the metric if and only if it is order-bounded. But $\mathbb{R}$ is not typical in this respect. An arbitrary set X may be endowed both with a partial ordering $<$ and with a metric d; then some sets are bounded in the metric space (X,d) and some are order-bounded in $(X,<)$, and there is no reason why these two concepts should necessarily be related to one another. The set $\mathbb{N}$, for example, is not order-bounded in $\mathbb{R}$ with the usual ordering, but is bounded when $\mathbb{R}$ is given the discrete metric. A less trivial example is given in 6.1.16.

Example 6.1.16

Endow $\mathbb{R}$ with the partial order relation $\{(a,b) \in \mathbb{R} \times \mathbb{R} \mid a - \lfloor a \rfloor < b - \lfloor b \rfloor\}$. The interval $(0,1)$, though bounded in $\mathbb{R}$ with the Euclidean metric, is not order-bounded; and conversely, the set $\{n + 1/2 \mid n \in \mathbb{Z}\}$ though unbounded in $\mathbb{R}$ with the Euclidean metric, is order-bounded, being bounded below by 0 and above by 3/4.

Definition 6.1.17

Suppose (X,d) is a semimetric space and f is a function into X. Then f is said to be BOUNDED if and only if $\operatorname{ran}(f)$ is a bounded subset of X. If $x \in X$, then f is said to be BOUNDED AWAY FROM x if and only if $\operatorname{dist}(x\,,\operatorname{ran}(f)) > 0$.

Example 6.1.18

Suppose (X, d) is a metric space and S is a non-empty set. The set X^S has a subset $B(S, X)$ whose members are the bounded functions from S to X; the map $(f, g) \mapsto \sup\{d(f(s), g(s)) \mid s \in S\}$ is a metric on $B(S, X)$.

Example 6.1.19

A sequence (x_n) in a metric space X is bounded if and only if its range $\{x_n \mid n \in \mathbb{N}\}$ is bounded in X. The set of bounded sequences in X has as its standard metric that of 6.1.18, namely $(x, y) \mapsto \sup\{d(x_n, y_n) \mid n \in \mathbb{N}\}$.

Open and Closed Subsets

The terms *open* and *closed* have been used to describe intervals, discs, balls and certain subsets of the complex plane. Definition 6.1.20 below unifies and generalizes these concepts.

Definition 6.1.20

Suppose (X, d) is a semimetric space. A subset A of X is said to be OPEN, or d-open, in X if and only if A can be expressed as a union of open balls of X. A subset B of X is said to be CLOSED, or d-closed, in X if and only if $X \backslash B$ is open in X. Note that both $\varnothing = \bigcup \varnothing$ and X, the union of all open balls of X, are open in X; they are therefore both closed in X as well.

Example 6.1.21

Suppose (X, d) is a semimetric space. Every open ball of X is certainly an open subset of X; every closed ball $\flat[a \, ; r]$ is a closed subset of X because its complement can be expressed as the union $\bigcup\{\flat[x \, ; d(x, a) - r) \mid x \in X \backslash \flat[a \, ; r]\}$ of open balls. If d is a metric, then a similar argument shows that every singleton subset of X is closed in X; but this is not true for any non-metric semimetric.

Example 6.1.22

Bounded open intervals of $\mathbb{R}$ are open balls of $\mathbb{R}$ and therefore open subsets of $\mathbb{R}$; unbounded open intervals of $\mathbb{R}$ are also open subsets of $\mathbb{R}$ because they can be expressed as unions of bounded open intervals; for example, for $a \in \mathbb{R}$, $(a, \infty) = \bigcup\{(a, b) \mid a, b \in \mathbb{R}, \, a < b\}$. The complement of a closed interval is empty or is an unbounded open interval or is a union of two unbounded open intervals; in any case the complement is open and so the interval itself is closed.

Theorem 6.1.23

Suppose (X, d) is a semimetric space and $U \subseteq X$. Then U is open in X if and only if, for each $a \in U$, there exists $s \in \mathbb{R}^+$ such that $b[a\,;s) \subseteq U$.

Proof

Suppose firstly that U is open in X and that $a \in U$. Then there exists an open ball $b[b\,;r)$ of X such that $a \in b[b\,;r) \subseteq U$. Let $s = r - d(b, a)$. Then $s > 0$ and $b[a\,;s) \subseteq b[b\,;r) \subseteq U$. Conversely, if the condition is satisfied, U can be expressed as the union of all the open balls it includes, so is open.   □

Corollary 6.1.24

Suppose (X, d) is a semimetric space and F is a non-empty subset of X. Then F is closed in X if and only if, for each $x \in X$, $\mathrm{dist}(x\,, F) = 0 \Rightarrow x \in F$.

Proof

This result follows from 6.1.23 and the observation that, for each $x \in X$, $\mathrm{dist}(x\,, F) = 0$ if and only if there is no open ball of X centred at x and included in $X \backslash F$.   □

Example 6.1.25

We must be careful about extrapolating from 6.1.24. It is true that if A is a closed subset of a metric space X and $x \in X \backslash A$, then $\mathrm{dist}(x\,, A) > 0$. It is not true, however, that the distance between two closed disjoint subsets of X is necessarily strictly positive. A counterexample in $\mathbb{R}^2$ with the Euclidean metric is given by the disjoint subsets $\{z \in \mathbb{R}^2 \mid z_2 = 0\}$ and $\{z \in \mathbb{R}^2 \mid z_1 > 0,\ z_1 z_2 = 1\}$. It is easy to show that each of these has open complement in $\mathbb{R}^2$ and is therefore closed in $\mathbb{R}^2$, and that the distance between them is 0.

Semimetric and Metric Topologies

There are key concepts in the analysis of metric spaces, such as *convergence*, *continuity* and *compactness* which depend not on the specific metrics involved but on the collections of open subsets derived from them. Notice, for example, that the definition we gave in 4.3.9 of a convergent complex sequence does not refer explicitly to distance. With a view to beginning a study of this phenomenon in Chapter 7, we introduce metric *topologies*.

Definition 6.1.26

Suppose (X, d) is a semimetric space. The collection of subsets of X which are open with respect to d is called the TOPOLOGY on X induced by d, and is referred to as a SEMIMETRIC TOPOLOGY or METRIC TOPOLOGY as appropriate.

Theorem 6.1.27

Suppose (X, d) is a semimetric space and $\mathcal{O}$ is the topology on X induced by d. Then $X \in \mathcal{O}$ and $\varnothing \in \mathcal{O}$ and $\mathcal{O}$ is closed under arbitrary unions and under finite intersections.

Proof

X, being the union of all open balls of X, is open; and $\varnothing$, being the trivial union, is open. Every union of members of $\mathcal{O}$ is a union of open balls of X, so is open. It remains to show that $\mathcal{O}$ is closed under finite intersections. An empty intersection is the trivial union, so is open. Suppose $\mathcal{U}$ is a non-empty finite collection of open subsets of X and $x \in \bigcap \mathcal{U}$. For each $U \in \mathcal{U}$, there exists $r_U \in \mathbb{R}^+$ such that $\flat[x\,;r_U) \subseteq U$ (6.1.23). $\mathcal{U}$ is finite; let $m = \min\{r_U \mid U \in \mathcal{U}\}$; then $m > 0$. So $\flat[x\,;m) \subseteq \bigcap \mathcal{U}$. Therefore $\bigcap \mathcal{U}$ is open by 6.1.23. $\square$

Example 6.1.28

A metric topology need not be closed under infinite intersections. Consider for example the collection of open intervals $\{(-r, r) \mid r \in \mathbb{R}\backslash\{0\}\}$ of $\mathbb{R}$, whose intersection is the singleton set $\{0\}$.

Comparison of Semimetrics and of Metrics

Definition 6.1.29

Suppose X is a non-empty set and d and d' are semimetrics on X. We say that d is STRONGER than d' and that d' is WEAKER than d if and only if the topology induced by d includes that induced by d'. If d is neither stronger than nor weaker than d' then d and d' are NOT COMPARABLE. We say that d and d' are EQUIVALENT if and only if they induce the same topology.

Theorem 6.1.30

Suppose X is a non-empty set and d and d' are semimetrics on X. Then d is stronger than d' if and only if every open ball of (X, d') includes an open ball of (X, d) with the same centre.

Proof

Suppose firstly that d is stronger than d' and that B is a d'-open ball centred at a. Then B is d'-open in X and hence also d-open in X, whence B includes a d-open ball of X centred at a, by 6.1.23. Conversely, suppose that the condition is satisfied and that U is d'-open in X. Let $\mathcal{B}$ be the collection of all d-open balls of X which are included in U and let $\mathcal{B}'$ be the collection of all d'-open balls of X which are included in U. Since U is d'-open, 6.1.23 ensures that, for each $a \in U$, there is some member of $\mathcal{B}'$ centred at a; our hypothesis then implies that there is some member of $\mathcal{B}$ centred at a, and hence that $U \subseteq \bigcup \mathcal{B}$. Since certainly $\bigcup \mathcal{B} \subseteq U$, we have $U = \bigcup \mathcal{B}$, so that U is d-open in X. □

Example 6.1.31

Suppose (X, d) is a metric space. Let $d'(x, y) = d(x, y)/(1 + d(x, y))$ for each $x, y \in X$ (see 6.1.4). It is not hard to verify that, for each $a \in X$ and $r \in \mathbb{R}^+$, we have $\flat'[a\,;r/(1 + r)) \subseteq \flat[a\,;r) \subseteq \flat'[a\,;r)$, so that d' is equivalent to d, despite the fact that d' is bounded and d may or may not be.

Isometric Maps

We have discussed the concept of isomorphism in relation to algebraic structures; the corresponding notion for geometric structures is that of isometry. A function between metric spaces which preserves distance is *isometric*.

Definition 6.1.32

Suppose (X, d) and (Y, d') are metric spaces. A function $f: X \to Y$ is said to be ISOMETRIC if and only if $d'(f(a), f(b)) = d(a, b)$ for every $a, b \in X$. That d is a metric forces f to be injective; if f is also surjective, then it is called an ISOMETRY. Evidently, the inverse of an isometry is also an isometry. X and Y are said to be ISOMETRIC spaces if and only if there exists an isometry from one onto the other.

Isometric spaces are, apart from the labelling of the points, indistinguishable from one another as metric spaces, and may often be treated as identical when structure other than that dependent on the metric is not under consideration.

Example 6.1.33

$\mathbb{C}$ and $\mathbb{R}^2$ are isometric spaces. The multiplicative structure of $\mathbb{C}$, which is not usually defined on $\mathbb{R}^2$, is quite independent of the metric structure.

Example 6.1.34

Suppose (X, d) is a metric space and Y is a set which is equinumerous with X. Suppose $f \colon X \to Y$ is a bijective map. Then $(a, b) \mapsto d(f^{-1}(a), f^{-1}(b))$ is a metric on Y. Endowed with it, Y is isometric to X and f is an isometry.

EXERCISES

Q 6.1.1 Suppose (X_1, d_1) and (X_2, d_2) are metric spaces and $p > 1$. Define e on $(X_1 \times X_2) \times (X_1 \times X_2)$ by $e(a, b) = ((d_1(a_1, b_1))^p + (d_2(a_2, b_2))^p)^{1/p}$ for each $a = (a_1, a_2)$ and $b = (b_1, b_2)$ in $X_1 \times X_2$. Show that e is a metric on $X_1 \times X_2$.

Q 6.1.2 Suppose (X, d) is a metric space and (Y, d) is a subspace of X. Suppose V is an open subset of Y. Show that there exists an open subset U of X such that $V = Y \cap U$.

Q 6.1.3 Suppose (X, d) is a metric space and Z is a closed subset of X. Suppose F is closed in (Z, d). Show that F is closed also in X.

Q 6.1.4 Suppose X is a set and d and d' are metrics on X. Suppose there exists a positive real number r which satisfies $r\, d(x, y) \le d'(x, y)$ for all $x, y \in X$. Show that d' is stronger than d.

Q 6.1.5 Prove the following result, complementary to 6.1.27: if X is a metric space and $\mathcal{F}$ is the collection of closed subsets of X, then $\mathcal{F}$ is closed under arbitrary intersections and under finite union.

Q 6.1.6 Show that all the metrics on $\mathbb{R}^2$ given in 6.1.7 are equivalent.

6.2 Seminorms and Norms

In every linear space X there is a distinguished point, namely the origin. If a metric is defined on X, then the concept of *length* of a vector might be captured in its distance from the origin. But arbitrary metrics are not well enough behaved for this purpose (see 6.1.13); instead we use *norms*.

Definition 6.2.1

Suppose X is a linear space. A function $\|\cdot\| \colon X \to \mathbb{R}$ is called a SEMINORM on X if and only if, for every $x, y \in X$ and for every scalar λ, the following two conditions are satisfied:

- $\|\lambda x\| = |\lambda| \, \|x\|$;
- (TRIANGLE INEQUALITY) $\|x + y\| \le \|x\| + \|y\|$.

Then $(X, \|\cdot\|)$, or more simply X, is called a SEMINORMED LINEAR SPACE and, for each $x \in X$, the image of x under $\|\cdot\|$ is denoted by $\|x\|$. Such a seminorm $\|\cdot\|$ is called a NORM on X if and only if, for each $x \in X \backslash \{0\}$, we have $\|x\| \ne 0$; in this case, $(X, \|\cdot\|)$, or more simply X, is called a NORMED LINEAR SPACE. Any linear subspace M of X normed or seminormed by the restriction of $\|\cdot\|$ is called a SEMINORMED or NORMED LINEAR SUBSPACE of X, as appropriate; and X is then a SEMINORMED or NORMED LINEAR SUPERSPACE of M. Most norms will be denoted by $\|\cdot\|$; duplication of notation is unlikely to cause confusion.

Because $0 = \|0\| = \|x - x\| \le \|x\| + \|-x\| = 2\|x\|$, the values of a seminorm are non-negative; moreover, the triangle inequality can be extended by induction, in the sense that, if S is a finite subset of a seminormed linear space, then $\|\sum_{x \in S} x\| \le \sum_{x \in S} \|x\|$. It is also easily established that each seminorm $\|\cdot\|$ induces a semimetric $(a, b) \mapsto \|a - b\|$ and that this is a metric if and only if $\|\cdot\|$ is a norm; thus seminormed linear spaces are semimetric spaces. All definitions made for semimetric spaces and all results proved for them apply to seminormed linear spaces through this induced semimetric. Distances and diameters in a seminormed linear space are defined to be the distances and diameters with respect to the related semimetric. Similarly, balls, open subsets, closed subsets, and so on, of a seminormed linear space X, are understood to be the appropriate subsets of the related semimetric space X, and the topology induced by the seminorm is understood to be the topology induced by the related semimetric. Seminormed linear subspaces are semimetric subspaces; and, although a non-empty subset of a seminormed linear space is not a linear subspace unless it is closed under the algebraic operations, it is always a semimetric space with the semimetric induced by the seminorm.

Example 6.2.2

The prototype for norms is the modulus function on $\mathbb{C}$; it induces the usual metrics of $\mathbb{R}$ and $\mathbb{C}$. But every non-trivial linear space admits many metrics which are not induced by any norm. The discrete metric is an example.

Example 6.2.3

Every seminorm on a one dimensional space is clearly determined entirely by its value at a single non-zero point; it follows that the zero map is the only seminorm which is not a norm and that all norms are scalar multiples of each other. A norm on a space of more than one dimension is not determined by its values at a basis (see 6.2.13).

Example 6.2.4

Every linear space admits a norm. Suppose X is a linear space. Let B be a Hamel basis for X. For each $x \in X$, set $\|x\|$ to be $\max\{|\alpha_b| \mid b \in B\}$ where the finite sum $\sum_{b \in B} \alpha_b b$ expresses x as a linear combination of members of B, each α_b being scalar, and all except a finite number of them zero.

Example 6.2.5

Suppose S is a non-empty set and X is a normed linear space. $B(S, X)$, the set of bounded maps from S to X, is a linear subspace of X^S with addition and scalar multiplication defined pointwise. $f \mapsto \sup\{\|f(s)\| \mid s \in S\}$ is the SUPREMUM or UNIFORM NORM on $B(S, X)$ and may be denoted by $\|\cdot\|_\infty$; it induces the metric of 6.1.18.

Example 6.2.6

Suppose $n \in \mathbb{N}$ and $p \in \mathbb{R}^+ \cup \{\infty\}$. Consider the function $a \mapsto \|a\|_p$ defined on $\mathbb{K}^n \times \mathbb{K}^n$. If $p < 1$, then the triangle inequality does not hold (4.4.8) and this is not a norm; but if $1 \le p \le \infty$, Minkowski's result (4.4.7) establishes the inequality, and the function is a norm on $\mathbb{K}^n$. The resulting structure is usually denoted by $\ell_p^n(\mathbb{K})$ or simply by ℓ_p^n. The space $\ell_2^n(\mathbb{K})$ is called n-DIMENSIONAL EUCLIDEAN SPACE, real or complex depending on whether $\mathbb{K}$ is $\mathbb{R}$ or $\mathbb{C}$; its norm $\|\cdot\|_2$ is the EUCLIDEAN NORM.

Example 6.2.7

Let $p \in [1, \infty]$ and $X = \{x \in \mathbb{K}^\mathbb{N} \mid \|x\|_p < \infty\}$. Minkowski's inequality ensures both that X is a linear space (5.1.4) and that the triangle inequality holds for the map $x \mapsto \|x\|_p$. That the other norm properties hold is easy to show, and the resulting normed linear space is denoted by $\ell_p(\mathbb{K})$ or simply by ℓ_p.

Example 6.2.8

In the particular case of 6.2.5 with $X = \mathbb{K}$ and $S = \mathbb{N}$, $B(S, X)$ is the linear space of bounded sequences of scalars endowed with the supremum norm; this is precisely the normed linear space $\ell_\infty(\mathbb{K})$ of 6.2.7.

Example 6.2.9

Suppose X and Y are linear spaces over $\mathbb{K}$ and $T \in \mathcal{L}(X, Y)$. Then, if Y is normed, linearity of T ensures that the map $x \mapsto \|Tx\|$ is a seminorm on X, and is a norm if and only if T is injective (3.4.13). In that case, the norm is called the NORM INDUCED BY T. More generally, if $\mathcal{C}$ is a collection of linear maps with the same domain X and with co-domains which are normed, and if $\mathcal{C}$ has the property that, for each $x \in X$, the set $\{\|Tx\| \mid T \in \mathcal{C}\}$ is bounded, then $x \mapsto \sup\{\|Tx\| \mid T \in \mathcal{C}\}$ is a seminorm on X which is a norm if and only if $\bigcap\{\ker(T) \mid T \in \mathcal{C}\} = \{0\}$. This norm is called the NORM INDUCED BY $\mathcal{C}$.

Example 6.2.10

Since every convergent sequence of scalars is bounded (Q 4.3.4), the linear space of convergent sequences in $\mathbb{K}$ (5.1.3) is a linear subspace of $\ell_\infty(\mathbb{K})$; it is denoted by $c(\mathbb{K})$ or simply by c. This normed linear space has other important subspaces: $c_0(\mathbb{K})$, or simply c_0, is the normed linear space of scalar sequences which converge to 0; and $c_{00}(\mathbb{K})$, or simply c_{00}, is the normed linear space of those sequences which are EVENTUALLY ZERO in the sense that $\{0\}$ is a tail.

Unit Balls

6.2.12 below shows that each open ball of a normed linear space is a translate of a scalar multiple of the convex ball $\flat[0\,;1)$. Thus all the norm structure of the space is encapsulated in this one ball. Moreover, the potential wildness inherent in arbitrary metrics (6.1.13) is curbed in norms.

Definition 6.2.11

Suppose X is a seminormed linear space. The ball $\flat[0\,;1)$ is called the OPEN UNIT BALL of X and the ball $\flat[0\,;1]$ is called the CLOSED UNIT BALL of X.

Theorem 6.2.12

Suppose X is a seminormed linear space. Then the open unit ball of X is convex and $\flat[z\,;r) = \{z\} + r\flat[0\,;1)$ for all $z \in X$ and $r \in \mathbb{R}^+$.

Proof

For $a, c \in \flat[0\,;1)$ and $t \in [0,1]$, we have $\|ta + (1-t)c\| \le t\|a\| + (1-t)\|c\| < 1$, so that $\flat[0\,;1)$ is convex. The second assertion follows from the equivalences $x \in \flat[z\,;r) \Leftrightarrow \|x - z\| < r \Leftrightarrow r^{-1}(x - z) \in \flat[0\,;1) \Leftrightarrow x \in \{z\} + r\flat[0\,;1)$. $\qquad\square$

Example 6.2.13

The unit balls of $\ell_1^2(\mathbb{R})$, $\ell_2^2(\mathbb{R})$ and $\ell_\infty^2(\mathbb{R})$ are pictured in 6.1.12. The unit balls of $\ell_p^2(\mathbb{R})$ for other values of $p \in [1,\infty]$ are sets which include the first of these

and are included in the last. If we were to extend the process
and draw the 'unit balls' of $\ell_p^2(\mathbb{R})$ for $p \in (0,1)$, we should find
that these are proper subsets of the unit ball of $\ell_1^2(\mathbb{R})$ which are
not convex; that for $p = 1/2$ is pictured at the right. This lack
of convexity is a spatial realization of the failure of the triangle
inequality for norms (but see 6.1.13).

Bounded Linear Maps

Suppose X and Y are linear spaces and Y is seminormed by $\|\cdot\|$. Then the
only maps in $\mathcal{L}(X,Y)$ which are bounded in the usual sense (6.1.17) are those
which the seminorm does not distinguish from the zero map. Specifically, if
$T \in \mathcal{L}(X,Y)$ and there exists $x \in X$ with $\|Tx\| \neq 0$, then $\{\|T(nx)\| \mid n \in \mathbb{N}\}$
is unbounded. This suggests that we might use the term *bounded* more con-
structively in this context; in fact we shall say that T is bounded with respect
to a seminorm if its restriction to the appropriate unit ball is bounded.

Definition 6.2.14

Suppose X and Y are seminormed linear spaces and $T \in \mathcal{L}(X,Y)$. Then T is
called a BOUNDED LINEAR MAP or BOUNDED LINEAR OPERATOR if and only if
$T(\flat_X[0\,;1))$ is bounded in Y.

In like manner to 6.2.5, the collection of bounded linear maps from X to Y is
a linear subspace of $\mathcal{L}(X,Y)$. Restrictions of such maps to the open unit ball of
X are, by definition, members of $B(\flat_X[0\,;1)\,,Y)$; so the linear space of bounded
linear maps from X to Y is seminormed by the map $T \mapsto \left\|T|_{\flat_X[0\,;1)}\right\|_\infty$; this
seminorm is clearly a norm if the seminorm on Y is a norm. The resulting
seminormed linear space will be denoted by $\mathcal{B}(X,Y)$, or by $\mathcal{B}(X)$ if $X = Y$. It
is easy to check that $\|T\| = \sup\{\|Tx\| \mid \|x\| < 1\}$ and that $\|Tx\| \leq \|T\|\,\|x\|$ for
all $T \in \mathcal{B}(X,Y)$ and $x \in X$. There may or may not exist $z \in X$ with $\|z\| \neq 0$
such that $\|Tz\| = \|T\|\,\|z\|$. It is of some interest that, if Y is normed, then
$\{T|_{\flat_X[0\,;1)} \mid T \in \mathcal{B}(X,Y)\}$ is closed in $B(\flat_X[0\,;1)\,,Y)$ (6.2.16).

Example 6.2.15

$\mathbb{K}$ has the modulus function as its norm. So, if X is a seminormed linear space,
then the space $\mathcal{B}(X,\mathbb{K})$ of BOUNDED LINEAR FUNCTIONALS is a normed linear
space when each member f has norm defined by $\|f\| = \sup\{|f(x)| \mid \|x\| < 1\}$.

Theorem 6.2.16

Suppose X is a seminormed linear space and Y is a normed linear space. Let $U = \flat_X[0\,;1)$. Then $Z = \{S|_U \mid S \in \mathcal{B}(X,Y)\}$ is closed in $B(U,Y)$.

Proof

Suppose $f \in B(U,Y)$ and $\text{dist}(f\,,Z) = 0$. Let $\epsilon \in \mathbb{R}^+$. There exists $L \in \mathcal{B}(X,Y)$ such that $\|f - L|_U\|_\infty < \epsilon$. Suppose $u \in U$ and $\lambda \in \mathbb{K}$ with $\lambda u \in U$. If $\|u\| = 0$, then $\|Lu\| = 0$, whence $L(\lambda u) = Lu = 0$ and so $\|f(u)\| < \epsilon$ and $\|f(\lambda u)\| < \epsilon$, yielding $f(\lambda u) = 0 = \lambda f(u)$. On the other hand, if $\|u\| \neq 0$, we have both $\|f(u) - Lu\| < \epsilon$ and $\|f(\lambda u) - \lambda Lu\| < \epsilon$, whence $\|\lambda f(u) - f(\lambda u)\| < (1 + |\lambda|)\epsilon$. Since $|\lambda| < \|u\|^{-1}$ and ϵ is arbitrary, it follows here too that $\lambda f(u) = f(\lambda u)$. Let T be the unique extension of f to X for which $T(\lambda u) = \lambda Tu$ for all $u \in U$ and $\lambda \in \mathbb{K}$. If $a,b \in X$, then $\|Ta - La\| \leq \epsilon\|a\|$ and $\|Tb - Lb\| \leq \epsilon\|b\|$ and, furthermore, $\|T(a+b) - La - Lb\| \leq \epsilon\|a+b\|$, from which it follows that $\|T(a+b) - Ta - Tb\| \leq \epsilon(\|a\| + \|b\| + \|a+b\|)$. Since ϵ is arbitrary, we have $T(a+b) = Ta + Tb$, so that $T \in \mathcal{L}(X,Y)$. But $T|_U = f \in B(U,Y)$; therefore $T \in \mathcal{B}(X,Y)$ and $f \in Z$. So Z is closed in $B(U,Y)$ by 6.1.24. □

Theorem 6.2.17

Suppose X is a seminormed linear space, Y is a normed linear space and $T \in \mathcal{B}(X,Y)$. Then $\ker(T)$ is closed in X.

Proof

Suppose $z \in X$, $\text{dist}(z\,,\ker(T)) = 0$ and $\epsilon \in \mathbb{R}^+$. Then there exists $x \in \ker(T)$ such that $\|T\|\,\|z - x\| < \epsilon$, whence $\|Tz\| = \|T(z-x)\| \leq \|T\|\,\|z - x\| < \epsilon$, and, since ϵ is arbitrary, $\|Tz\| = 0$. Because Y is normed, this implies that $z \in \ker(T)$. So $\ker(T)$ is closed by 6.1.24. □

Characterization of Unit Balls

There are four properties which determine that a subset B of a linear space is an open unit ball related to some seminorm—that B is convex; that B is *balanced*; that $\langle B \rangle = X$ and that $\mathfrak{I}(B) = B$.

Definition 6.2.18

Suppose X is a linear space and S is a non-empty subset of X. Then S is said to be BALANCED if and only if $\alpha S \subseteq S$ for all $\alpha \in \mathbb{K}$ with $|\alpha| \leq 1$.

Example 6.2.19

The origin belongs to every balanced subset of a linear space. The unit circle $\mathbb{T}$ of $\mathbb{C}$ is not balanced, but the closed unit disc $\mathbb{D}$ is. The line segment $[-1, 1]$ in $\mathbb{C}$ is balanced if $\mathbb{C}$ is regarded as a 2-dimensional real linear space, but not if $\mathbb{C}$ is regarded as a 1-dimensional complex linear space; the line segment $(-1, 1]$ is not balanced in either case. All subspaces of a linear space are balanced.

Theorem 6.2.20

Suppose $(X, \|\cdot\|)$ is a seminormed linear space. Then the open unit ball B of X is convex and balanced; and $\mathbb{R}^+ B = X$ and $\mathfrak{I}(B) = B$. Moreover, $\|\cdot\|$ is a norm if and only if B includes no line of X.

Proof

B is convex by 6.2.12. Suppose $x \in X$. Then $\|\alpha x\| = |\alpha| \|x\| \leq \|x\|$ for all $\alpha \in \mathbb{K}$ with $|\alpha| \leq 1$; so B is balanced. If $\|x\| \neq 0$, then $B \cap \mathbb{R}x = (-x/\|x\|, x/\|x\|)$; if $\|x\| = 0$, then $\mathbb{R}x \subseteq B$. So $\mathbb{R}^+ B = X$. Lastly, suppose $a \in B$; if $\|x\| = 0$ then $a + \mathbb{R}x \subseteq B$; otherwise, for $t = (1 - \|a\|)/\|x\|$, we have $(-tx, tx) \subseteq B - \{a\}$, whence $\mathfrak{I}(B) = B$. $\square$

Theorem 6.2.21

Suppose X is a linear space. Let $\mathcal{B}$ denote the set of convex balanced subsets S of X for which $\langle S \rangle = X$ and $\mathfrak{I}(S) = S$; and let $\mathcal{B}'$ denote the subset of $\mathcal{B}$ consisting of those members which include no line of X. Then the map associating each seminorm on X with the unit ball it defines is a bijection onto $\mathcal{B}$; and its restriction to the collection of norms on X is a bijection onto $\mathcal{B}'$.

Proof

We have shown in 6.2.20 that all unit balls are members of $\mathcal{B}$ and that those determined by norms are the ones that do not include any line of X. It is clear that no two seminorms determine the same unit ball, so that the stated functions are injective. We need only show therefore that every member of $\mathcal{B}$ is the open unit ball determined by some seminorm. Towards this, suppose $S \in \mathcal{B}$. For each $z \in X = \langle S \rangle$, there exists a finite subset A of S and a family $(\alpha_s)_{s \in A}$ of non-zero scalars such that $z = \sum_{s \in A} \alpha_s s$. Let $\beta = \sum_{s \in A} |\alpha_s|$. Then either $z = 0$ and $\mathbb{R}z \subseteq S$, or $\beta > 0$ and, because S is balanced, $\beta^{-1}z = \sum_{s \in A}(|\alpha_s|/\beta)(\alpha_s s/|\alpha_s|)$ is a convex combination of members of S and is therefore in S, because S is convex. Define $\|z\| = \inf\{r \in \mathbb{R}^+ \mid r^{-1}z \in S\}$. We show that $\|\cdot\|$ is a seminorm on X whose associated open unit ball is S. Certainly $\|0\| = 0$. Suppose $\lambda \in \mathbb{K}\setminus\{0\}$, $x \in X$ and $t \in \mathbb{R}^+$. Then $t^{-1}x \in S$ if and only if $(t|\lambda|)^{-1}(\lambda x) \in S$, because S is balanced. So $\|\lambda x\| = |\lambda| \|x\|$.

Now suppose $a, b \in X$. If $\|a\| > 0$ and $\|b\| > 0$, then, for each $k \in (1, \infty)$, we have $(k\|a\|)^{-1} a \in S$ and $(k\|b\|)^{-1} b \in S$ because S is balanced. So, by convexity of S, we have $t(k\|a\|)^{-1} a + (1 - t)(k\|b\|)^{-1} b \in S$ for all $t \in [0, 1]$. In particular, this is true for $t = \|a\| / (\|a\| + \|b\|)$; so $(k(\|a\| + \|b\|))^{-1}(a + b) \in S$ for all $k \in (1, \infty)$, whence $\|a + b\| \leq \|a\| + \|b\|$. If either $\|a\| = 0$ or $\|b\| = 0$, the argument is easily modified to achieve the same result. So $\|\cdot\|$ is a seminorm on X. Finally, for $z \in X$, we have $\|z\| < 1 \Rightarrow z \in S \Rightarrow \|z\| \leq 1$, the first implication because S is balanced, the second by definition; then, if $\|z\| = 1$, we have $\mathbb{R}^+ z \cap (S - \{z\}) = \varnothing$, so that, since $\langle S \rangle = X$ by hypothesis, we have $z \notin \mathfrak{I}(S)$ and, since $\mathfrak{I}(S) = S$ by hypothesis, $z \notin S$. Therefore $S = \{z \in X \mid \|z\| < 1\}$. $\quad\square$

Comparison of Seminorms and of Norms

Like semimetrics, seminorms are deemed equivalent if and only if they induce the same topology. But there is a much nicer criterion for equivalence of seminorms than the one which we obtained for semimetrics.

Definition 6.2.22

Suppose X is a linear space and $\|\cdot\|$ and $\|\cdot\|'$ are seminorms on X. We say that $\|\cdot\|$ is STRONGER than $\|\cdot\|'$ and that $\|\cdot\|'$ is WEAKER than $\|\cdot\|$ if and only if the topology induced by $\|\cdot\|$ includes that induced by $\|\cdot\|'$. If $\|\cdot\|$ is neither stronger than nor weaker than $\|\cdot\|'$ we say that they are NOT COMPARABLE. $\|\cdot\|$ and $\|\cdot\|'$ are said to be EQUIVALENT if and only if they induce the same topology.

Two metrics on a given set may be equivalent even if a ball of radius 1 with respect to one of them includes balls of all radii with respect to the other. This phenomenon occurs in the case of the two metrics of 6.1.31, for example. But it can never happen with equivalent seminorms; this is a consequence of 6.2.23 below.

Theorem 6.2.23

Suppose X is a linear space and $\|\cdot\|$ and $\|\cdot\|'$ are seminorms defined on X. Then $\|\cdot\|$ is stronger than $\|\cdot\|'$ if and only if there exists $t \in \mathbb{R}^+$ such that, for every $x \in X$, we have $t\|x\|' \leq \|x\|$.

Proof

Let B denote the open unit ball of $(X, \|\cdot\|)$ and B' that of $(X, \|\cdot\|')$. Notice that the stated condition is equivalent to the proposition that there exists $t \in \mathbb{R}^+$

such that $tB \subseteq B'$. If $\|\cdot\|$ is stronger than $\|\cdot\|'$ such an inclusion obtains by 6.1.30. Conversely, if there exists $t \in \mathbb{R}^+$ such that $tB \subseteq B'$, then, for each $z \in X$ and $r \in \mathbb{R}^+$, we have $\flat[z\,;tr] = trB + \{z\} \subseteq rB' + \{z\} = \flat'[z\,;r]$, from 6.2.12; the proof is completed by invoking 6.1.30 again. $\qquad\square$

Corollary 6.2.24

Suppose X is a linear space and $\|\cdot\|$ and $\|\cdot\|'$ are seminorms defined on X. Then $\|\cdot\|$ is equivalent to $\|\cdot\|'$ if and only if there exist $s, t \in \mathbb{R}^+$ such that, for every $x \in X$, $t\|x\|' \leq \|x\| \leq s\|x\|'$.

Quotient Spaces

Theorem 6.2.25

Suppose $(X, \|\cdot\|)$ is a seminormed linear space and M is a linear subspace of X. Then the function $x/M \mapsto \operatorname{dist}(x\,, M)$ defined on the linear space X/M is a seminorm on X/M. It is a norm if and only if M is closed in X. It is called the QUOTIENT SEMINORM or NORM as appropriate.

Proof

We shall use $\|\cdot\|$ to denote not only the seminorm on X but also the specified function on X/M. For $x, y \in X$, we have $\|\alpha x/M\| = |\alpha|\,\|x/M\|$ and

$$
\begin{aligned}
\|(x/M) + (y/M)\| &= \|(x+y)/M\| \\
&= \inf\{\|x + y + m\| \mid m \in M\} \\
&= \inf\{\|x + a + y + b\| \mid a, b \in M\} \\
&\leq \inf\{\|x + a\| + \|y + b\| \mid a, b \in M\} \\
&= \|x/M\| + \|y/M\|.
\end{aligned}
$$

Therefore $\|\cdot\|$ is a seminorm on X/M. Lastly, $\|x/M\| = 0 \Leftrightarrow \operatorname{dist}(x\,, M) = 0$, and we conclude, using 6.1.24, that $\|\cdot\|$ is a norm on X if and only if M is closed in X. $\qquad\square$

Example 6.2.26

Suppose X is a seminormed linear space. Then $K = \ker\|\cdot\|$ is certainly a subspace of X. If $x \in X$ and $\operatorname{dist}(x\,, K) = 0$, then, for each $\epsilon \in \mathbb{R}^+$, there exists $z \in K$ such that $\|x - z\| < \epsilon$; so $\|x\| \leq \|z\| + \epsilon = \epsilon$. Since ϵ is arbitrary in $\mathbb{R}^+$, it follows that $x \in K$. So K is closed in X by 6.1.24. Therefore $x/K \mapsto \operatorname{dist}(x\,, K)$ defines a norm on X/K. Notice that $\|x/K\|$ is precisely $\|x\|$.

The Riesz Lemma

If M is a linear subspace of a seminormed linear space X, then, for each $x \in X$, $\|x/M\|$ as defined in 6.2.25 is $\mathrm{dist}(x, M)$; in particular, $\|x/M\|$ cannot be larger than $\|x\|$. That this upper bound is approached to any degree of proximity for some $x \in X$ with $\|x\| = 1$ is guaranteed by the important result due to Riesz.

Lemma 6.2.27

Suppose X is a seminormed linear space, $a \in X$, $r \in \mathbb{R}^+$ and M is a subspace of X. Then $\pi(\flat_X[a\,;r]) = \flat_{X/M}[a/M\,;r)$, where $\pi \colon X \to X/M$ is the quotient map; also $\pi^{-1}(\flat_{X/M}[a/M\,;r)) = \flat_X[a\,;r) + M$.

Proof

Suppose $c \in \flat_X[a\,;r)$; then $\|c/M - a/M\| \leq \|c - a\| < r$, and it follows that $\pi(c) \in \flat_{X/M}[a/M\,;r)$. Conversely, suppose $b \in X$ and $\|b/M - a/M\| < r$; then there exists $m \in M$ such that $\|b - a - m\| < r$, whence $b - m \in \flat_X[a\,;r)$ and $b/M = \pi(b - m) \in \pi(\flat_X[a\,;r))$. This proves the first assertion and the second follows immediately. □

Theorem 6.2.28 THE RIESZ LEMMA

Suppose X is a seminormed linear space and M is a linear subspace of X. Then the quotient map $\pi \colon X \to X/M$ is in $\mathcal{B}(X, X/M)$ and $\|\pi\|$ is either 0 or 1.

Proof

It is clear from its construction that π is linear and bounded and that $\|\pi\| \leq 1$. Set $B = \flat_X[0\,;1)$ and let $r \in (0,1)$. If $B \subseteq rB + M$, then, by induction, $B \subseteq r^n B + M$ for all $n \in \mathbb{N}$, so that, for all $x \in B$, $\|\pi(x)\| = \mathrm{dist}(x, M) = 0$, whence $\|\pi\| = 0$. Otherwise, there exists $z \in B \backslash (rB + M)$; then $\|z\| < 1$ and, by 6.2.27, $\|\pi(z)\| \geq r$. So $\|\pi\| > r$. Since r is arbitrary in $(0,1)$ and $\|\pi\| \leq 1$, this yields $\|\pi\| = 1$. □

Product Spaces

There are many ways in which norms can be defined on products of normed spaces, or, for infinite products, on suitable subsets of the product. We have already seen, for example, various norms on sequence spaces (6.2.7) following from Minkowski's inequality.

Example 6.2.29

Suppose $(X_i)_{i \in I}$ is a family of normed linear spaces and $P = \prod_i X_i$. Let $S = \{x \in P \mid \sup\{\|x_i\| \mid i \in I\} < \infty\}$. Define $\|x\|_\infty = \sup\{\|x_i\| \mid i \in I\}$ for each $x \in S$. It is easy to show that this is a norm on S. The resulting normed linear space is called the DIRECT PRODUCT of the family (X_i).

Example 6.2.30

Suppose $(X_n)_{n \in \mathbb{N}}$ is a sequence of normed linear spaces and $P = \prod_n X_n$. Let $S = \{x \in P \mid \sum_{n=1}^\infty \|x_n\|^p < \infty\}$. For each $x \in S$, we define the p-NORM of x by the equation $\|x\|_p = (\sum_{n=1}^\infty \|x_n\|^p)^{1/p}$. It follows from Minkowski's inequality that this is a norm on S.

Normed Algebras

Some normed linear spaces are endowed with a multiplication which makes them into algebras. But the term *normed algebra* is reserved for algebras endowed with *submultiplicative* norms.

Definition 6.2.31

Suppose A is an algebra and $\|\cdot\|$ is a seminorm on A for which $\|xy\| \leq \|x\| \, \|y\|$ for all $x, y \in A$. Then $(A, \|\cdot\|)$, or more simply A, is called a SEMINORMED ALGEBRA, or a NORMED ALGEBRA if $\|\cdot\|$ is a norm. A seminormed algebra which has unity 1 with $\|1\| = 1$ is called a UNITAL SEMINORMED ALGEBRA. An inductive argument shows that, if S is a finite subset of A, then $\|\prod_{a \in S} a\| \leq \prod_{a \in S} \|a\|$ for every total ordering of S.

Example 6.2.32

Suppose S is a non-empty set and X is a normed algebra. Then $B(S, X)$ is a normed algebra with pointwise operations and the supremum norm. Moreover, if X has involution $*$, then $B(S, X)$ has the involution $f \mapsto f^*$ where $f^*(s) = (f(s))^*$ for all $s \in S$. And if the involution on X is isometric, i.e., if $\|x^*\| = \|x\|$ for all $x \in X$, then the involution on $B(S, X)$ is also isometric.

Example 6.2.33

ℓ_∞ is a unital normed algebra with isometric involution $(x_n) \mapsto (\overline{x_n})$.

Example 6.2.34

Suppose X is a seminormed linear space. Then $\mathcal{B}(X)$ is a subalgebra of $\mathcal{L}(X)$, where the multiplication is composition (3.4.10). If $T, S \in \mathcal{B}(X)$, then certainly $\|TS\| \leq \|T\| \|S\|$, so that $\mathcal{B}(X)$ is a seminormed algebra. Unless the seminorm is zero, $\|I\| = 1$ and $\mathcal{B}(X)$ is unital.

Isometric Isomorphisms

Definition 6.2.35

Normed linear spaces X and Y are said to be ISOMETRICALLY ISOMORPHIC if and only if there exists an isometry $\phi: X \to Y$ which is also a linear space isomorphism. Such a map is called an ISOMETRIC ISOMORPHISM.

Example 6.2.36

Let $p \in [1, \infty]$ and $n \in \mathbb{N}$. Then the linear space ℓ_p^n is isomorphic to the linear subspace of ℓ_p which consists of those sequences whose $(n+1)^{th}$ tail is $\{0\}$; the natural isomorphism is clearly an isometric isomorphism.

Example 6.2.37

Suppose $(X, \|\cdot\|)$ is a normed linear space, (Y, d) is a metric space and $\phi: X \to Y$ is an isometry. Then ϕ, being a bijective map, induces algebraic operations on Y which make it into a linear space (see 3.4.8); moreover, it is easy to check that the map $y \mapsto \|\phi^{-1}(y)\|$ is a norm on the linear space Y which, because ϕ is an isometry, induces the metric d. Then the normed linear space Y is isometrically isomorphic to X and ϕ is an isometric isomorphism. There are metrics (6.1.13) which are not induced by norms; a metric space with such a metric cannot therefore be isometric to a normed linear space.

Bounded Linear Functionals

Lemma 6.2.38

Suppose X is a complex linear space, S is a balanced subset of X, and $f: S \to \mathbb{C}$ satisfies $f(\mu s) = \mu f(s)$ for all $s \in S$ and $\mu \in \mathbb{D}$. Then $|\Re f|(S) = |f|(S)$.

Proof

Let $u = \Re f$. Certainly $f(0) = 0 = u(0)$. Suppose $s \in S$ and $f(s) = \mu \neq 0$; set $\beta = \overline{\mu}/|\mu|$; then $\beta s \in S$ and $f(\beta s) = |\mu| \in \mathbb{R}$, so that $u(\beta s) = |\mu|$. Therefore $|f|(S) \subseteq |u|(S)$. On the other hand, if $|u(s)| = r$ and $r \neq 0$, then $r \leq |f(s)|$.

Set $\gamma = r/|f(s)|$, so that $\gamma \in \mathbb{D}$ and $|f(\gamma s)| = r$; we then have the reverse inclusion $|u|(S) \subseteq |f|(S)$. $\square$

Theorem 6.2.39 ANALYTIC HAHN–BANACH THEOREM

Suppose X is a seminormed linear space and M is a seminormed linear subspace of X. Suppose f is a bounded linear functional on M. Then there exists a bounded linear functional $\tilde{f}$ on X with $\|\tilde{f}\| = \|f\|$ such that $\tilde{f}|_M = f$.

Proof

Let $u = \Re f$. For each $x \in X$, define $p(x) = \|u\| \, \|x\|$. Then p is a sublinear map and u is dominated on M by $p|_M$. By 5.3.6, there exists $\tilde{u} \in \mathcal{L}(X, \mathbb{R})$ such that $\tilde{u}|_M = u$ and $\tilde{u} \leq p$. It follows that $\tilde{u} \in \mathcal{B}(X, \mathbb{R})$ and $\|\tilde{u}\| = \|u\|$. Invoking 5.3.2, let $\tilde{f}$ be the unique member of $\mathcal{L}(X, \mathbb{K})$ with $\Re \tilde{f} = \tilde{u}$. Then $\tilde{f}|_M = f$ and, by 6.2.38, $\tilde{f}$ is bounded and $\|\tilde{f}\| = \|\tilde{u}\| = \|u\| = \|f\|$. $\square$

Corollary 6.2.40

Suppose X is a seminormed linear space, M is a subspace of X and $z \in X \backslash M$. Then $\mathrm{dist}(z, M) > 0$ if and only if there exists a bounded linear functional f on X such that $f(M) = \{0\}$ and $f(z) \neq 0$.

Proof

Firstly, let $\mathrm{dist}(z, M) = r$ and suppose that $r > 0$. Define g on $M + \mathbb{K}z$ by $g(m + \lambda z) = \lambda$ for each $m \in M$ and $\lambda \in \mathbb{K}$; then $g(M) = \{0\}$, $g(z) = 1$ and, since $\|m + \lambda z\| \geq r|\lambda|$, g is bounded with $\|g\| \leq r^{-1}$. Then g has a bounded linear extension to X by the Hahn–Banach Theorem. Towards the converse, suppose there is a bounded linear functional f on X such that $f(M) = \{0\}$ and $f(z) \neq 0$. Then $\|f\| \neq 0$ and $|f(z)| = |f(m + z)| \leq \|f\| \, \|m + z\|$ for all $m \in M$, whence $\|m + z\| \geq \|f\|^{-1} |f(z)|$. Therefore $\mathrm{dist}(z, M) \geq \|f\|^{-1} |f(z)| > 0$. $\square$

Corollary 6.2.41

Suppose X is a seminormed linear space and $f \in \mathcal{L}(X, \mathbb{K})$. Then f is bounded if and only if $\ker(f)$ is closed in X.

Proof

If f is bounded, then $\ker(f)$ is closed by 6.2.17. For the converse, suppose $\ker(f)$ is closed. If $f = 0$, then certainly f is bounded; otherwise, there exists $z \in X \backslash \ker(f)$ and, by 6.1.24, $\mathrm{dist}(z, \ker(f)) > 0$. It follows from 6.2.40 that there exists $g \in \mathcal{B}(X, \mathbb{K})$ such that $g(\ker(f)) = \{0\}$ and $g(z) \neq 0$. Since X is $\ker(f) + \mathbb{K}z$, it follows that $f = f(z)g/g(z)$ and hence that f is bounded. $\square$

Corollary 6.2.42

Suppose $(X, \|\cdot\|)$ is a seminormed linear space and $z \in X \backslash \ker\|\cdot\|$. Then there exists a bounded linear functional f on X with $\|f\| = 1$ such that $f(z) = \|z\|$.

Proof

Define f on $\mathbb{K}z$ to be $\alpha z \mapsto \alpha\|z\|$ and extend to X by 6.2.39.   □

Example 6.2.43

Let X be a seminormed linear space and $\mathcal{C} = \{f \in \mathcal{B}(X, \mathbb{K}) \mid \|f\| \leq 1\}$. Certainly, for each $x \in X$, $\{|f(x)| \mid f \in \mathcal{C}\}$ is bounded. It follows from 6.2.42 that the seminorm induced on X by $\mathcal{C}$ (6.2.9) is the original seminorm.

EXERCISES

Q 6.2.1 Suppose X is a real linear space and p is sublinear functional on X. Show that the function q defined on X as $x \mapsto p(x) + p(-x)$ is a seminorm on X.

Q 6.2.2 Suppose $r, s \in \mathbb{R}$ and $1 \leq r < s < \infty$. Show that ℓ_r is a proper linear subspace of ℓ_s which is in turn a proper linear subspace of ℓ_∞. (These spaces are endowed with quite different norms and are not normed linear subspaces of one another).

Q 6.2.3 Suppose $(X, \|\cdot\|)$ is a seminormed linear space. Show that $\|\cdot\|$ is a norm if and only if the singleton subsets of X are closed in X.

Q 6.2.4 Give an example of an unbounded linear map between normed linear spaces whose kernel is closed.

Q 6.2.5 Suppose X is a normed linear space, A and B are subsets of X, and A is open in X. Show that $A + B$ is open also.

Q 6.2.6 Let B be the closed unit ball of ℓ_2. Find a closed bounded subset S of ℓ_2 such that $B + S$ is not closed and $B + S \neq \{z \in \ell_2 \mid \text{dist}(z, S) \leq 1\}$.

Q 6.2.7 Show that a maximal wedge W of a real normed linear space X is closed in X if and only if W includes a ball of X.

Q 6.2.8 (HAHN–BANACH SEPARATION THEOREM II) Suppose X is a normed linear space, A and G are non-empty disjoint convex subsets of X and G is open in X. Show that there exists $f \in \mathcal{B}(X, \mathbb{K})$ and $t \in \mathbb{R}$ such that $\Re f(g) > t \geq \Re f(a)$ for all $g \in G$ and $a \in A$.

6.3 Sesquilinear Forms and Inner Products

Definition 6.3.1

Suppose X is a linear space over $\mathbb{K}$. A function $\langle \cdot , \cdot \rangle \colon X \times X \to \mathbb{K}$ which satisfies

- $\langle x + y , z \rangle = \langle x , z \rangle + \langle y , z \rangle$
- $\langle \alpha x , y \rangle = \alpha \langle x , y \rangle$
- $\langle x , y + z \rangle = \langle x , y \rangle + \langle x , z \rangle$
- $\langle x , \alpha y \rangle = \overline{\alpha} \langle x , y \rangle$

for each $x, y, z \in X$ and $\alpha \in \mathbb{K}$ is called a SESQUILINEAR FORM on X; if $\mathbb{K} = \mathbb{R}$, $\langle \cdot , \cdot \rangle$ is also called a BILINEAR FORM. A sesquilinear form is called a POSITIVE FORM if and only if $\langle x , x \rangle \in \mathbb{R}^{\oplus}$ for each $x \in X$ and a POSITIVE DEFINITE FORM if and only if also $\langle x , x \rangle > 0$ for all $x \neq 0$. A sesquilinear form is called a HERMITIAN FORM if and only if $\langle x , y \rangle = \overline{\langle y , x \rangle}$ for each $x, y \in X$; if $\mathbb{K} = \mathbb{R}$, the term SYMMETRIC FORM may be used for a hermitian form. A positive definite hermitian form is called an INNER PRODUCT, qualified as real or complex if necessary. The Latin word *sesqui* means *one and a half times*; thus the term *sesquilinear* describes succinctly a form which is linear in the first variable (the first two properties) and CONJUGATE-LINEAR in the second (the third and fourth properties).

Example 6.3.2

Suppose X is a linear space with Hamel basis S. For each $a = \sum_{s \in S} \alpha_s s$ and $b = \sum_{s \in S} \beta_s s$ in X, define $\langle a , b \rangle = \sum_{s \in S} \alpha_s \overline{\beta}_s$. Then $\langle \cdot , \cdot \rangle$ is an inner product.

Example 6.3.3

There are positive forms on real linear spaces which are not hermitian (Q 6.3.1). But every positive form on a complex linear space is necessarily hermitian. Suppose X is a complex linear space and $\langle \cdot , \cdot \rangle$ is a positive form on X. Suppose $x, y \in X$. The identity $\langle x + y , x + y \rangle = \langle x , x \rangle + \langle x , y \rangle + \langle y , x \rangle + \langle y , y \rangle$ yields $\Im \langle x , y \rangle = -\Im \langle y , x \rangle$. Replacing x by ix, we get $\Re \langle x , y \rangle = \Re \langle y , x \rangle$.

Definition 6.3.4

Suppose X is a linear space with inner product $\langle \cdot , \cdot \rangle$ and $A \subseteq X$ is non-empty. Then $A^{\perp} = \{ x \in X \mid \langle a , x \rangle = 0 \text{ for all } a \in A \}$ is called the ORTHOGONAL COMPLEMENT of $\langle A \rangle$ in X. If B is also a non-empty subset of X, then A and B are said to be ORTHOGONAL if and only if $B \subseteq A^{\perp}$, in which case we have also $A \subseteq B^{\perp}$. Vectors $a, b \in X$ are said to be ORTHOGONAL if and only if $\langle a , b \rangle = 0$.

Parallelogram Law and Schwarz Inequality

Theorem 6.3.5 PARALLELOGRAM LAW

Suppose X is a linear space and $\langle \cdot , \cdot \rangle$ is a sesquilinear form on X. Suppose $x, y \in X$. Then $\langle x + y , x + y \rangle + \langle x - y , x - y \rangle = 2\langle x , x \rangle + 2\langle y , y \rangle$.

Proof

Addition applied to $\langle x + y , x + y \rangle = \langle x , x \rangle + \langle x , y \rangle + \langle y , x \rangle + \langle y , y \rangle$ and $\langle x - y , x - y \rangle = \langle x , x \rangle - \langle x , y \rangle - \langle y , x \rangle + \langle y , y \rangle$ gives the result. $\qquad\square$

Example 6.3.6

The usual inner product on $\mathbb{R}^2$ is given by $\langle x , y \rangle = x_1 y_1 + x_2 y_2$; the Parallelogram Law yields $\|x + y\|^2 + \|x - y\|^2 = 2\|x\|^2 + 2\|y\|^2$, where $\|\cdot\|$ is the Euclidean norm. Note that $\|x + y\|$ and $\|x - y\|$ are the lengths of the diagonals of the parallelogram in the plane determined by x and y; so the Parallelogram Law expresses the fact that the sum of the squares of the lengths of the diagonals of a parallelogram is equal to the sum of the squares of the lengths of its sides.

Theorem 6.3.7 THE SCHWARZ INEQUALITY

Suppose X is a linear space, $\langle \cdot , \cdot \rangle$ is a positive form on X and $x, y \in X$. Then $\langle x , y \rangle \langle y , x \rangle \in \mathbb{R}$ and $\langle x , y \rangle \langle y , x \rangle \leq \langle x , x \rangle \langle y , y \rangle$ with equality if $x = y$ or $\{x, y\}$ is linearly dependent. If $\langle \cdot , \cdot \rangle$ is positive definite, equality occurs in no other case.

Proof

Certainly, equality holds if $x = 0$, $y = 0$ or $\mathbb{K}x = \mathbb{K}y$. Suppose otherwise. Then $\langle y , y \rangle (\langle x , x \rangle \langle y , y \rangle - \langle x , y \rangle \langle y , x \rangle) = \langle \langle x , y \rangle y - \langle y , y \rangle x , \langle x , y \rangle y - \langle y , y \rangle x \rangle$, which is in $\mathbb{R}^\oplus$, and in $\mathbb{R}^+$ if $\langle \cdot , \cdot \rangle$ is positive definite, because of the condition on x an y. Everything follows if $\langle y , y \rangle \neq 0$, or, similarly, if $\langle x , x \rangle \neq 0$. Lastly, if $\langle x , x \rangle = 0 = \langle y , y \rangle$, 6.3.5 gives $\langle x + \alpha y , x + \alpha y \rangle = 0$ and hence $\alpha \langle y , x \rangle = -\overline{\alpha} \langle x , y \rangle$ for every scalar α. If $\mathbb{K} = \mathbb{R}$, this yields $\langle x , y \rangle \langle y , x \rangle \leq 0$; if $\mathbb{K} = \mathbb{C}$, it yields $\langle x , y \rangle = 0 = \langle y , x \rangle$. In either case, the result holds. $\qquad\square$

Theorem 6.3.8

Suppose X is a linear space and $\langle \cdot , \cdot \rangle$ is a positive hermitian form on X. Then the map $\|\cdot\|$ defined as $x \mapsto \sqrt{\langle x , x \rangle}$ on X is a seminorm which is a norm if and only if $\langle \ , \ \rangle$ is an inner product.

Proof

The triangle inequality follows from the Schwarz Inequality by the calculation
$$\|x+y\|^2 = \langle x+y\,,x+y \rangle = \|x\|^2 + 2\,\Re\langle x\,,y \rangle + \|y\|^2 \le \|x\|^2 + 2\|x\|\,\|y\| + \|y\|^2,$$
valid for all $x,y \in X$, and the other properties are easy to verify.   □

Definition 6.3.9

Suppose H is a linear space and $\langle \cdot\,,\cdot \rangle$ is an inner product on H. Then $(H,\langle \cdot\,,\cdot \rangle)$, or more simply H, will be called an INNER PRODUCT SPACE. In this case, the norm of 6.3.8 is called the NORM INDUCED BY THE INNER PRODUCT.

Example 6.3.10

The map $(a,b) \mapsto \sum_{n=1}^{\infty} a_n \overline{b_n}$ defined on $\ell_2 \times \ell_2$ is an inner product which induces the norm $\|\cdot\|_2$ on ℓ_2.

Theorem 6.3.11

Suppose $(H,\langle \cdot\,,\cdot \rangle)$ is an inner product space. For each $g \in H$, define $\phi_g\colon H \to \mathbb{K}$ to be $x \mapsto \langle x\,,g \rangle$. Then the map $g \mapsto \phi_g$ is a norm preserving conjugate-linear map from H into $\mathcal{B}(H,\mathbb{K})$.

Proof

For each $g \in H$, the map ϕ_g is clearly linear; and, by the Schwarz Inequality, $|\phi_g(x)| = |\langle x\,,g \rangle| \le \|x\|\,\|g\|$, so that $\phi_g \in \mathcal{B}(H,\mathbb{K})$ and $\|\phi_g\| \le \|g\|$; that $\|\phi_g\| = \|g\|$ is got by noting that $\phi_g(g) = \|g\|^2$. For $x,y,z \in H$ and $\alpha \in \mathbb{K}$, we have $\phi_{\alpha x}y = \langle y\,,\alpha x \rangle = \overline{\alpha}\langle y\,,x \rangle = \overline{\alpha}\phi_x(y)$, so that $\phi_{\alpha x} = \overline{\alpha}\phi_x$; we also have $\phi_{x+y}(z) = \langle z\,,x+y \rangle = \langle z\,,x \rangle + \langle z\,,y \rangle = (\phi_x + \phi_y)(z)$, so that $\phi_{x+y} = \phi_x + \phi_y$. Therefore the map $g \to \phi_g$ is conjugate-linear.   □

EXERCISES

Q 6.3.1 In contrast to 6.3.3, give an example of a non-hermitian positive definite bilinear form on a real linear space.

Q 6.3.2 Suppose $(X,\langle \cdot\,,\cdot \rangle)$ is an inner product space and S is a non-empty subset of X. Show that $S^\perp$ is a closed linear subspace of X.

Q 6.3.3 Suppose $(X,\|\cdot\|)$ is a complex normed linear space which satisfies the Parallelogram Law: $\|x+y\|^2 + \|x-y\|^2 = 2\|x\|^2 + 2\|y\|^2$ for every $x,y \in X$. Show that the norm is determined by an inner product. (A similar result holds for real spaces, but appears to be harder to prove.)

7

Topological Structure

We remarked in 6.1 that some important analytical concepts—such as *convergence*, *continuity* and *compactness*—which emerge from a study of distance, flow equally naturally from a study of open sets. This is not merely a curiosity; these concepts depend ultimately not on metrics but on topologies, in the sense that exactly the same theory of convergence, continuity or compactness will emerge in a particular metric space if the metric is replaced by any other equivalent metric. This suggests that we build a general theory of analysis on open sets rather than on metrics. Such a theory, though more general than that of metrics, is not intended to displace it; where a distance function induces the topology, it will in general hold more information than the topology.

7.1 Topologies

The collection of open subsets of a metric space was called a metric topology in 6.1.26. We now generalize this concept, taking the fundamental properties of the metric topology (6.1.27) as the basis for a definition; consistency in our use of the words *topology*, *open* and *closed* is thereby assured.

Definition 7.1.1
Suppose X is a set and $\mathcal{O}$ is a collection of subsets of X. Then $\mathcal{O}$ is called a TOPOLOGY on X if and only if $X \in \mathcal{O}$ and $\mathcal{O}$ is closed under arbitrary unions

(which implies incidentally that $\varnothing = \bigcup \varnothing \in \mathcal{O}$) and under finite intersections; in that case, the pair $(X, \mathcal{O})$ is called a TOPOLOGICAL SPACE. We shall usually say simply that X is a topological space endowed with the topology $\mathcal{O}$. The members of $\mathcal{O}$ are called OPEN SUBSETS of X and their complements in X are called CLOSED SUBSETS of X; we say that such sets are OPEN IN X and CLOSED IN X, respectively; if there is a possibility of confusion regarding the topology, we may say they are $\mathcal{O}$-open or $\mathcal{O}$-closed.

Example 7.1.2

The topology induced by the usual metric on $\mathbb{R}$ is called the EUCLIDEAN TOPOLOGY or USUAL TOPOLOGY ON $\mathbb{R}$; it is the collection of all subsets of $\mathbb{R}$ which can be expressed as unions of open intervals of $\mathbb{R}$. The USUAL TOPOLOGY ON $\mathbb{C}$ is that induced by the usual metric. The USUAL TOPOLOGY ON $\tilde{\mathbb{R}}$ is the collection of subsets of $\tilde{\mathbb{R}}$ which can be expressed as unions of open intervals of $\tilde{\mathbb{R}}$. When we consider any of $\mathbb{R}$, $\mathbb{C}$ or $\tilde{\mathbb{R}}$ as a topological space, we assume it to be endowed with the usual topology unless the contrary is stated.

Example 7.1.3

The empty set can be endowed with just one topology, $\{\varnothing\}$. Let $X = \{1, 2, 3\}$. Then $\{\varnothing, \{1\}, \{1, 2\}, X\}$ is a topology on X. The collection $\{\varnothing, \{2, 3\}, X\}$ is another topology on X. The set $\{\varnothing, \{1\}, \{2\}, X\}$ is not a topology on X.

Example 7.1.4

An arbitrary set X can be endowed with the DISCRETE TOPOLOGY, $\mathcal{P}(X)$; or with the INDISCRETE or TRIVIAL topology $\{\varnothing, X\}$. Every topology on X is included in the former and includes the latter. Endowed with the first, X is called a DISCRETE SPACE; with the second, it is called an INDISCRETE SPACE.

Example 7.1.5

Suppose X is a set. The collection $\mathcal{O} = \{S \subseteq X \mid |X \backslash S| < \infty\} \cup \{\varnothing\}$ is closed under arbitrary unions and under finite intersections. Since $X \in \mathcal{O}$, it follows that $\mathcal{O}$ is a topology on X. This topology is called the FINITE COMPLEMENT TOPOLOGY on X. If X is finite, then it is identical with the discrete topology.

Example 7.1.6

Suppose X is a set. Then the set $\mathcal{O} = \{S \subseteq X \mid |X \backslash S| \leq \infty\} \cup \{\varnothing\}$ is a topology on X. It is called the COUNTABLE COMPLEMENT TOPOLOGY on X; if X is countable, then it is identical with the discrete topology.

Example 7.1.7

Suppose X is a non-empty set and suppose $x \in X$. Then the collection $\{S \subseteq X \mid x \in S\} \cup \{\varnothing\}$ is a topology on X; it is called the PARTICULAR POINT TOPOLOGY associated with the point x. The collection $\{S \subseteq X \mid x \notin S\} \cup \{X\}$ is also a topology on X; it is called the EXCLUDED POINT TOPOLOGY associated with the point x.

Example 7.1.8

Here is a curious example. If ω is an ordinal, then ω_+ is a topology on ω: certainly, $\omega \in \omega_+$ and $\varnothing \in \omega_+$; and if $x \subseteq \omega_+$ and $x \neq \varnothing$, then $\bigcap x = \min x \in \omega_+$, so that ω_+ is closed under intersections, and $\bigcup x \subset \omega_+$, so that, since $\bigcup x$ is an ordinal by 1.4.8, $\bigcup x \in \omega_+$, and ω_+ is closed under unions.

Example 7.1.9

Every set admits a semimetric which in turn induces a topology; indeed, except in trivial cases, such a semimetric is not unique because equivalent semimetrics can be constructed in abundance as in 6.1.31. In order to be sure that the theory of topological spaces is more general than that of semimetric spaces, we must know that there are topologies which are not determined by semimetrics. Consider $\mathbb{N}$ with the excluded point topology (7.1.7) associated with the point 1. This topology is not induced by any semimetric d, for, if it were, then, for each $r \in \mathbb{R}^+$, the open ball $\flat[1\,;r)$ would be $\mathbb{N}$, since $\mathbb{N}$ is the only open set to which 1 belongs, giving $d(1,n) = 0$ for all $n \in \mathbb{N}$ and hence, by the triangle inequality, $d = 0$; but the topology induced by the zero semimetric is the indiscrete topology.

Definition 7.1.10

A topological space is said to be METRIZABLE if and only if there exists a metric which induces its topology.

Open Sets and Closed Sets

Open sets exhibit the properties of intersection and union stipulated in the definition of a topology. Closed sets display corresponding properties.

Theorem 7.1.11

Suppose X is a topological space. Then every intersection of closed subsets of X is closed in X and every finite union of closed subsets of X is closed in X.

Proof

Since closed sets are complements of open sets, it follows from De Morgan's Laws (1.1.13) that an intersection of closed sets is the complement of a union of open sets and that a union of closed sets is the complement of an intersection of open sets. So this result follows from the definition of a topology.   □

Example 7.1.12

Suppose $(X, \mathcal{O})$ is a topological space. Then $\varnothing$ is open because $\bigcup \varnothing = \varnothing$; moreover, $\varnothing$ is closed because $X \in \mathcal{O}$. That the empty set is both open and closed implies that its complement X is also both open and closed. If $X = \mathbb{C}$, then no other subset is both open and closed (Q 4.3.5). In general topological spaces, however, there may or may not be other subsets which have this property; in a discrete space, for example, every subset is both open and closed. If there are no subsets of X other than X and $\varnothing$ which are both open and closed, we say that the space $(X, \mathcal{O})$ is *connected*. The concept of connectedness will be discussed in Chapter 9.

Comparison of Topologies

In general, sets can be endowed with many different topologies. It is therefore natural to try to compare such topologies by inclusion. The trivial topology on a set X is included in every topology on X and every topology is included in the discrete topology; there will usually be many topologies in between these two extremes. Any given pair of topologies on X may exhibit inclusion one way or the other, or may not. The two topologies on X in 7.1.3, for example, do not exhibit inclusion in either direction.

Definition 7.1.13

Suppose X is a set and $\mathcal{O}$ and $\mathcal{P}$ are topologies on X. Then $\mathcal{P}$ is said to be STRONGER than $\mathcal{O}$ and $\mathcal{O}$ is said to be WEAKER than $\mathcal{P}$ if and only if $\mathcal{O} \subseteq \mathcal{P}$. We say that $\mathcal{O}$ and $\mathcal{P}$ are NOT COMPARABLE if and only if $\mathcal{O}$ is neither stronger nor weaker than $\mathcal{P}$. Each topology is both stronger and weaker than itself. The converse is also true: if $\mathcal{O}$ is both stronger and weaker than $\mathcal{P}$, then $\mathcal{O} = \mathcal{P}$.

Example 7.1.14

The finite complement topology on $\mathbb{R}$ is weaker than the usual topology: every singleton subset is closed in $\mathbb{R}$ by 6.1.21; it follows that every finite subset is closed and hence that every subset of $\mathbb{R}$ with finite complement is open.

Theorem 7.1.15

Suppose X is a set and $\mathfrak{T}$ is a non-empty collection of topologies on X. Then $\bigcap \mathfrak{T}$ is the strongest topology on X which is weaker than every $\mathcal{O} \in \mathfrak{T}$.

Proof

Certainly $X \in \bigcap \mathfrak{T}$. Suppose $\mathcal{D} \subseteq \bigcap \mathfrak{T}$. Then $\mathcal{D} \subseteq \mathcal{O}$ for each $\mathcal{O} \in \mathfrak{T}$. So, since each $\mathcal{O}$ is closed under arbitrary unions and under finite intersections, it follows that $\bigcup \mathcal{D} \in \bigcap \mathfrak{T}$ and, if $\mathcal{D}$ is finite, that $\bigcap \mathcal{D} \in \bigcap \mathfrak{T}$. Therefore $\bigcap \mathfrak{T}$ is a topology on X; that it is weaker than each of the members of $\mathfrak{T}$ is clear, and that it is stronger than any other topology with this property is also clear.   □

Bases and Subbases for Topologies

The open subsets of a metric space are those which can be expressed as unions of open balls of the space. We say that the open balls form a *base* for the topology or that they *generate* it. In this subsection, we formalize and generalize these ideas.

Definition 7.1.16

Suppose $(X, \mathcal{O})$ is a topological space. A subset $\mathcal{B}$ of $\mathcal{O}$ is called a BASE for $\mathcal{O}$ if and only if each member of $\mathcal{O}$ is a union of members of $\mathcal{B}$. A subset $\mathcal{S}$ of $\mathcal{O}$ is called a SUBBASE for $\mathcal{O}$ if and only if the set of non-empty finite intersections of members of $\mathcal{S}$ is a base for $\mathcal{O}$; it is called the base DETERMINED by $\mathcal{S}$.

Example 7.1.17

The collection of bounded open intervals of $\mathbb{R}$ forms a base, called the USUAL BASE, for the usual topology. More generally, the usual base for a metric topology is the collection of open balls determined by the metric.

Example 7.1.18

Every base for a topology is a subbase for the same topology, but the converse is not generally true. Consider $\mathbb{N}$ with the finite complement topology $\mathcal{O}$. Let $\mathcal{S} = \{\mathbb{N} \backslash \{n\} \mid n \in \mathbb{N}\}$. Each non-empty member of $\mathcal{O}$ is of the form $\mathbb{N} \backslash F$ where F is some finite subset of $\mathbb{N}$. Provided $F \neq \varnothing$, $\mathbb{N} \backslash F = \bigcap \{\mathbb{N} \backslash \{n\} \mid n \in F\}$; it follows that $\mathcal{S}$ is a subbase for $\mathcal{O}$. $\mathcal{S}$ is not a base for $\mathcal{O}$; in fact, the base determined by $\mathcal{S}$ is very nearly the whole topology.

Definition 7.1.19

Suppose X is a set and C is a collection of subsets of X. The intersection of all topologies on X which include C is called the topology GENERATED BY C.

Example 7.1.20

Suppose $(S, <)$ is a totally ordered set. The topology generated by all upper and lower segments of S is called the ORDER TOPOLOGY on S. The usual topology on $\mathbb{R}$, namely the topology induced by the Euclidean metric on $\mathbb{R}$ (7.1.2), is the same as the order topology on $\mathbb{R}$ (Q 7.1.1); this expresses in the most succinct way the very close relationship which exists between order and metric in $\mathbb{R}$.

Theorem 7.1.21

Suppose X is a set and $\mathcal{S}$ is a collection of subsets of X. Then $\mathcal{S}$ is a subbase for the topology generated by $\mathcal{S}$ if and only if $\bigcup \mathcal{S} = X$.

Proof

Since X is open, $\mathcal{S}$ cannot be a subbase unless $\bigcup \mathcal{S} = X$. Suppose now that that condition is satisfied. The topology generated by $\mathcal{S}$ includes $\mathcal{S}$ and therefore includes every union of finite intersections of members of $\mathcal{S}$. So it is sufficient to show that the collection $\mathcal{O}$ of arbitrary unions of finite intersections of members of $\mathcal{S}$ is a topology on X. But $X = \bigcup \mathcal{S} \in \mathcal{O}$ and $\mathcal{O}$ is closed under arbitrary unions; and the Generalized Distributive Law (3.1.10) ensures that $\mathcal{O}$ is closed under finite intersections, which completes the proof. $\square$

EXERCISES

Q 7.1.1 Show that the usual topology on $\mathbb{R}$ is the same as the order topology.

Q 7.1.2 Show that a union of two topologies need not be a topology.

Q 7.1.3 Show that a topology need not be closed under all intersections.

Q 7.1.4 Give an example to demonstrate that an arbitrary union of closed sets in a topological space need not itself be closed.

Q 7.1.5 Suppose X is a set and $\mathfrak{T}$ is a collection of topologies on X. Show that the intersection of all topologies on X which are stronger than every $\mathcal{O} \in \mathfrak{T}$ is the weakest topology on X with that property.

Q 7.1.6 Suppose X is a set and $\mathfrak{N}$ is a nest of topologies on X. Is $\bigcup \mathfrak{N}$ necessarily a topology on X?

7.2 Neighbourhoods

Each point x of a topological space X belongs to at least one open subset of X, namely X itself. In general there may be many more such open subsets to which x belongs. They are called open neighbourhoods of x.

Definition 7.2.1

Suppose $(X, \mathcal{O})$ is a topological space, $x \in X$ and $S \subseteq X$. Then S is called a NEIGHBOURHOOD of x in X if and only if there exists $U \in \mathcal{O}$ such that $x \in U \subseteq S$. The collection of neighbourhoods of x in X will be denoted by $\mathrm{Nbd}(x)$, and the collection $\mathcal{O} \cap \mathrm{Nbd}(x)$ of OPEN NEIGHBOURHOODS of x in X will be denoted by $\mathrm{nbd}(x)$.

Example 7.2.2

Neighbourhoods may be large or small. If $(X, \mathcal{O})$ is a topological space, then X is a neighbourhood of each point of X; if the space is discrete, then, for each $x \in X$, the singleton set $\{x\}$ is a neighbourhood of x.

Example 7.2.3

For each $x \in \mathbb{R}$, $\mathrm{Nbd}(x) = \{N \subseteq \mathbb{R} \mid \exists I \text{ an open interval of } \mathbb{R} : x \in I \subseteq N\}$. To check this, note that every open subset of $\mathbb{R}$ can be expressed as a union of open intervals; if x is a member of an open set, then it must be a member of such an interval.

Example 7.2.4

Suppose x is a point of a topological space X and $\mathcal{B}$ is a base for the topology on X. Then every neighbourhood of x includes some member of $\mathcal{B}$. Note, however, that a neighbourhood need not include a member of a subbase; the set $\mathbb{N}\backslash\{1, 2\}$ is an open neighbourhood of 3 in the finite complement topology on $\mathbb{N}$, but it includes no member of the subbase $\{\mathbb{N}\backslash\{n\} \mid n \in \mathbb{N}\}$.

Boundary, Interior and Exterior

In a topological space $(X, \mathcal{O})$, each subset A of X gives rise to a partition of X consisting of three subsets of X, namely the *interior* of A, the *exterior* of A and the *boundary* of A.

Definition 7.2.5

Suppose X is a topological space and A is a subset of X.

- A point x of X is called a BOUNDARY POINT of A in X if and only if, for every $U \in \mathrm{nbd}(x)$ we have both $A \cap U \neq \varnothing$ and $A^c \cap U \neq \varnothing$. The set of boundary points of A is called the BOUNDARY of A and is denoted by ∂A; it is evident from the symmetrical nature of the definition that $\partial(A^c) = \partial A$.

- The set $A \backslash \partial A$ is called the INTERIOR of A and is denoted by A° or by $\mathrm{Int}(A)$; its members are called INTERIOR POINTS of A.

- The interior of A^c is called the EXTERIOR of A and will be denoted by $A^{c\circ}$; its members are called EXTERIOR POINTS of A.

Example 7.2.6

Common parlance allows different meanings for the terms *exterior* and *interior* from those given in 7.2.5. Careful consideration of the examples given here should dispel any confusion. The boundary of the closed unit disc $\mathbb{D}$ of $\mathbb{C}$ is the UNIT CIRCLE $\mathbb{T}$; the interior of $\mathbb{D}$ is the open unit disc $\mathbb{D} \backslash \mathbb{T}$; and the exterior of $\mathbb{D}$ is $\mathbb{C} \backslash \mathbb{D}$. But the boundary of $\mathbb{T}$ is $\mathbb{T}$; the interior of $\mathbb{T}$ is $\varnothing$; and the exterior of $\mathbb{T}$ is $\mathbb{C} \backslash \mathbb{T}$. The boundary of $\mathbb{R}$ in $\mathbb{C}$ is $\mathbb{R}$; the boundary of $\mathbb{R}$ in $\mathbb{R}$ is $\varnothing$.

Theorem 7.2.7

Suppose X is a topological space and A is a subset of X. Then $\{A^\circ, \partial A, A^{c\circ}\}$ is a partition of X.

Proof

$A^\circ = A \backslash \partial A$ and $A^{c\circ} = A^c \backslash \partial(A^c) = A^c \backslash \partial A$.   □

Example 7.2.8

It is easy to construct large sets with empty interior and small sets with empty exterior, and it is also possible for a proper non-trivial subset of a topological space to have empty boundary. Indeed strange phenomena abound in the most familiar spaces: every open interval of $\mathbb{R}$ has in it some rational and some irrational number (4.2.3, 4.2.7); therefore the boundary of $\mathbb{Q}$ in $\mathbb{R}$ is $\mathbb{R}$ itself, and it follows that both the interior and the exterior of $\mathbb{Q}$ are empty.

Theorem 7.2.9

Suppose X is a topological space, $A \subseteq X$ and $x \in X$. Then

- x is an interior point of A if and only if $A \in \mathrm{Nbd}(x)$;
- x is an exterior point of A if and only if $A^c \in \mathrm{Nbd}(x)$.

Proof

Note that $x \notin \partial A$ if and only if there exists an open neighbourhood of x which is included either in A or in A^c, and the assertions follow using 7.2.7. □

Corollary 7.2.10

Suppose $(X, \mathcal{O})$ is a topological space and suppose A is a subset of X. Let $\mathcal{U} = \{U \in \mathcal{O} \mid U \subseteq A\}$ and $\mathcal{V} = \{V \in \mathcal{O} \mid V \cap A = \varnothing\}$. Then $A^\circ = \bigcup \mathcal{U}$, which is the largest open subset of X which is included in A; and $A^{c\circ} = \bigcup \mathcal{V}$, which is the largest open subset of X which is disjoint from A.

Proof

If $B \in \mathcal{U}$, then A is a neighbourhood of each point of B and 7.2.9 ensures that $B \subseteq A^\circ$. So $\bigcup \mathcal{U} \subseteq A^\circ$. But $A^\circ \backslash \bigcup \mathcal{U}$ is empty, since each $x \in A^\circ$ has an open neighbourhood which is disjoint from $X \backslash A$. So $A^\circ = \bigcup \mathcal{U}$, and this is certainly the largest open set included in A. The second assertion follows immediately because the exterior of A is $(X \backslash A)^\circ$. □

Corollary 7.2.11

Suppose X is a topological space and $A \subseteq X$. Then ∂A is closed in X.

Proof

Both the interior and the exterior of A are open by 7.2.10. So their union is open; and the complement, ∂A, of that union is closed. □

Corollary 7.2.12

Suppose X is a topological space and $A \subseteq X$. Then A is open in X if and only if, for each $a \in A$, there is an open neighbourhood of a which is included in A.

Of course, *open neighbourhood* in 7.2.12 can be replaced by *basic open neighbourhood*—that is an open neighbourhood which is also a member of some specified base for the topology. This change yields an analogue of 6.1.23, in which the open balls form a base for the metric topology.

Closure

Corresponding to the concept of *interior* is that of *closure*. The interior of a subset A of a topological space X is the largest open subset of X included in A. In contrast, the closure is the smallest closed subset of X which includes A.

Definition 7.2.13

Suppose X is a topological space and A is a subset of X. We define the CLOSURE of A in X to be the intersection of all closed subsets of X which include A. It will be denoted by $\overline{A}$ or by $\mathrm{Cl}(A)$. Note the $\overline{A}$ is closed in X by 7.1.11.

Since X is closed and includes A, the intersection of 7.2.13 is non-trivial. We infer the highly desirable inclusion $A \subseteq \overline{A}$. So $\overline{A}$ is the smallest closed subset of X which includes A. In particular, A is closed if and only if $A = \overline{A}$.

Theorem 7.2.14

Suppose X is a topological space. Then, for each subset A of X, we have the equations $\overline{A} = X \backslash A^{c\circ} = A^{\circ} \cup \partial A = A \cup \partial A$.

Proof

$A^{c\circ} = (X \backslash A)^{\circ}$ is the largest open subset of X included in $X \backslash A$; so its complement in X is the smallest closed subset of X which includes A, namely $\overline{A}$. Therefore $\overline{A} = X \backslash A^{c\circ}$; the other equations follow using 7.2.7. $\square$

Corollary 7.2.15

Suppose X is a topological space and A is a subset of X. Then A is closed in X if and only if $\partial A \subseteq A$.

Example 7.2.16

Metrics and semimetrics provide a much nicer way of determining closures than do topologies. Suppose X is a semimetric space and $A \subseteq X$. Consider the set $C = \{x \in X \mid \mathrm{dist}(x, A) = 0\}$. For each $z \in X \backslash C$, we have $\mathrm{dist}(z, A) > 0$ and hence $r = \mathrm{dist}(z, C) > 0$; then $\flat[z\,;r) \subseteq X \backslash C$. So $X \backslash C$ is open in X and C is closed in X. Since $A \subseteq C$ and $\overline{A}$ is the smallest closed subset of X which includes A, we have $\overline{A} \subseteq C$. But $A \subseteq \overline{A}$ and $\overline{A}$ is closed in X, so that $C \subseteq \{x \in X \mid \mathrm{dist}(x, \overline{A}) = 0\} = \overline{A}$ by 6.1.24. Therefore $\overline{A} = C$.

Example 7.2.17

Suppose M is a linear subspace of a seminormed linear space X. Then $\overline{M}$ is also a linear subspace of X. Suppose $a, b \in \overline{M}$ and $\epsilon \in \mathbb{R}^{+}$; then there exist $x, y \in M$ with $\|a - x\| < \epsilon$ and $\|b - y\| < \epsilon$, so that $\|a + b - (x + y)\| < 2\epsilon$ and, for $\lambda \in \mathbb{K}$, $\|\lambda a - \lambda x\| \leq |\lambda|\,\epsilon$. Since $x + y \in M$ and $\lambda x \in M$ and ϵ is arbitrary, we have $\mathrm{dist}(a + b, M) = 0$ and $\mathrm{dist}(\lambda a, M) = 0$, whence $a + b \in \overline{M}$ and $\lambda a \in \overline{M}$ by 7.2.16.

Example 7.2.18

The closure of each interval of $\mathbb{R}$ is the corresponding closed interval of $\mathbb{R}$. But the closure of an open ball of a metric space need not be a closed ball of the space. Consider the sets $S = \{z \in \mathbb{R}^2 \mid z_1 \geq 0, z_1^2 + z_2^2 < 1\}$, $V_a = \{z \in \mathbb{R}^2 \mid z_2 = 1\}$ and $V_b = \{z \in \mathbb{R}^2 \mid z_2 = -1\}$. Let Y denote the metric subspace $\{(-1, 0)\} \cup S \cup V_a \cup V_b$ of $\mathbb{R}^2$ with the Euclidean metric. Then $\flat_Y[0\,;1) = S$, and the closure of this ball in Y is $S \cup \{(0, -1), (0, 1)\}$; it is an easy exercise to show that this set cannot be expressed as a closed ball of Y with any centre.

Even when the closure of an open ball is a closed ball, it may not be the ball we expect; the best we can be sure of is that $\overline{\flat[a\,;r)} \subseteq \flat[a\,;r]$. Suppose X is a discrete metric space with more than one point and let $x \in X$. Then the open ball $\flat[x\,;1)$ is $\{x\}$ and its closure is also $\{x\}$, which can be represented as the closed ball $\flat[x\,;r]$ for any $r \in (0, 1)$. It is not, however, the same as the closed ball $\flat[x\,;1]$, which is X itself.

Density

By the Density Theorem (4.2.3), we have $r = \sup_{\mathbb{R}}\{q \in \mathbb{Q} \mid q < r\}$ for each $r \in \mathbb{R}$. This, together with the explicit relationship between metric and order established in 6.1.9, gives $\mathrm{dist}(r, \mathbb{Q}) = 0$ and thereby $r \in \overline{\mathbb{Q}}$. So the density of $\mathbb{Q}$ in $\mathbb{R}$ yields the equation $\overline{\mathbb{Q}} = \mathbb{R}$. This informs our generalization of the concept of density.

Definition 7.2.19

Suppose X is a topological space and A is a subset of X. We say that A is DENSE in X if and only if $\overline{A} = X$. We say that A is NOWHERE DENSE in X if and only if $X \backslash \overline{A}$ is dense in X.

Theorem 7.2.20

Suppose X is a topological space and $A \subseteq X$. Then A is dense in X if and only if $A \cap U \neq \varnothing$ for every non-empty open subset U of X; and A is nowhere dense in X if and only if $\overline{A}$ has empty interior in X.

Proof

Suppose U is a non-empty open subset of X and $A \cap U = \varnothing$. Then $A \subseteq X \backslash U$, so that, since $X \backslash U$ is closed, we have also $\overline{A} \subseteq X \backslash U$; hence A is not dense in X. Conversely, suppose $\overline{A} \neq X$; then $X\backslash\overline{A}$ is non-empty and open and $A \cap (X\backslash\overline{A}) = \varnothing$. Towards the second assertion: $X\backslash\overline{A}$ is dense in X if and only if every open subset of X has non-empty intersection with $X\backslash\overline{A}$, which occurs if and only if no open subset of X is included in $\overline{A}$. □

Corollary 7.2.21

Suppose X is a metric space, $\epsilon \in \mathbb{R}^+$, U and V are non-empty open subsets of X, and U is dense in X. Then there exists an open ball B of X of radius less than ϵ such that $\overline{B} \subseteq U \cap V$.

Proof

$V \cap U$ is open and is non-empty by 7.2.20. Let $x \in V \cap U$. Because $X\backslash(V \cap U)$ is closed, $\gamma = \text{dist}(x, X\backslash(V \cap U)) > 0$. Let $r \in (0, \min\{\epsilon, \gamma\})$. Then we have the inclusions $\overline{\flat[x\,;r)} \subseteq \flat[x\,;r] \subseteq \flat[x\,;\gamma) \subseteq V \cap U$. □

Definition 7.2.22

A topological space $(X, \mathcal{O})$ will be called a BAIRE SPACE if and only if every non-trivial countable intersection of dense open subsets of X is dense in X.

Example 7.2.23

We shall show later that $\mathbb{R}$ is a Baire space (12.1.19). But $\mathbb{Q}$ with its usual metric topology induced from $\mathbb{R}$ is not, because the collection $\{\mathbb{Q}\backslash\{q\} \mid q \in \mathbb{Q}\}$ of dense open subsets of $\mathbb{Q}$ is countable and has empty intersection.

Isolated Points and Accumulation Points

A subset A of a topological space may have members which are topologically *isolated* from the rest of the set. Every point of $\overline{A}$ which is not an isolated point of A is called an *accumulation point* of A, whether or not it is a member of A.

Definition 7.2.24

Suppose X is a topological space, $A \subseteq X$ and $x \in X$.

- x is called an ISOLATED POINT of A if and only if there exists $U \in \text{nbd}(x)$ with $U \cap A = \{x\}$. The set of all isolated points of A is denoted by $\text{iso}(A)$.
- x is called an ACCUMULATION POINT of A in X if and only if, for every $U \in \text{nbd}(x)$, we have $U \cap A\backslash\{x\} \neq \varnothing$. The set of all accumulation points of A in X is denoted by $\text{acc}(A)$.

Example 7.2.25

Suppose X is a topological space and $x \in X$. Then x is an isolated point of X if and only if $\{x\}$ is open in X.

Example 7.2.26

The set of accumulation points of the interval $(0, 1)$ in $\mathbb{R}$ is the closed interval $[0, 1]$. The set of isolated points of $(0, 1)$ is empty. Every point of $\mathbb{N}$ is isolated in $\mathbb{R}$. On the other hand, $\mathrm{acc}_{\mathbb{R}}(\mathbb{N})$ is empty.

Theorem 7.2.27

Suppose X is a topological space and $A \subseteq X$. Then

- $\{\mathrm{acc}(A), \mathrm{iso}(A)\}$ is a partition of $\overline{A}$;
- $\overline{A} = A \cup \mathrm{acc}(A)$;
- A is closed in X if and only if $\mathrm{acc}(A) \subseteq A$.

Proof

$\mathrm{acc}(A) \cap \mathrm{iso}(A) = \varnothing$ by definition. Also, $x \in \mathrm{iso}(A) \cup \mathrm{acc}(A)$ if and only if every neighbourhood of x contains some point of A, which occurs if and only if $x \notin A^{c\circ}$, which occurs if and only if $x \in \overline{A}$ by 7.2.14. The rest is easy.   □

EXERCISES

Q 7.2.1 Let X be a topological space and A and B be subsets of X. Show that
- $\partial B \subseteq A \subseteq B \Rightarrow \partial B \subseteq \partial A$;
- ∂A need not equal $\partial(\overline{A})$;
- $\mathrm{acc}(A \cup B) = \mathrm{acc}(A) \cup \mathrm{acc}(B)$;
- $\overline{A \cup B} = \overline{A} \cup \overline{B}$;
- $\overline{A \cap B} \subseteq \overline{A} \cap \overline{B}$, and that this inclusion may be proper;
- $\partial A = \overline{A} \cap \overline{X \backslash A}$;
- $\partial(A \cup B) \subseteq \partial A \cup \partial B$, and that this inclusion may be proper;
- $\partial(A \cap B) \subseteq \partial A \cup \partial B$, and that this inclusion may be proper.

Q 7.2.2 Let $X = \{1, 2, 3\}$ and $\mathcal{O} = \{\varnothing, \{1\}, \{1, 2\}, X\}$. Then $(X, \mathcal{O})$ is a topological space. For each $A \subseteq X$, write down $\overline{A}$, A° and $\mathrm{acc}(A)$.

Q 7.2.3 A set A is called a PERFECT set if and only if $A = \mathrm{acc}(A)$. Show that perfect sets are those which are closed and have no isolated points.

Q 7.2.4 Suppose X is a topological space and F is closed in X. Show that F is nowhere dense in X if and only if $X \backslash F$ is dense in X.

Q 7.2.5 Give an example of a topological space X and a subset S of X which is not nowhere dense in X but whose complement is dense in X.

Q 7.2.6 Suppose X is a topological space. Show that X is a Baire space if the union of every countable collection of nowhere dense closed subsets of X is nowhere dense in X, but that the converse is false.

Q 7.2.7 Suppose X is a Baire space and $\mathcal{F}$ is a countable collection of nowhere dense subsets of X. Show that $\bigcup \mathcal{F} \neq X$.

Q 7.2.8 Give an example of an uncountable collection of dense open subsets of $\mathbb{R}$ which has empty intersection.

Q 7.2.9 Suppose X is a normed linear space and S is a subset of X. Show that if S is convex, then $\overline{S}$ is convex; that if S is balanced, then $\overline{S}$ is balanced; that if S is a wedge, then $\overline{S}$ is wedge; that if X is a normed algebra and S is an ideal of X, then $\overline{S}$ is an ideal of X; and that if S is a cone, $\overline{S}$ need not be a cone.

7.3 Cardinality and Topology

Most interesting topological spaces are not countable. It is often possible, however, to associate with a topology some countable set and to use the desirable properties of countability to learn about the space.

Countability Criteria

We consider three important countability criteria associated with topological spaces, namely *first countability*, *second countability* and *separability*. We shall see shortly that first countability plays an important rôle in deciding when sequence arguments are valid for testing various topological properties. Second countability plays a rôle in deciding whether or not a space is metrizable. Separability has more wide-ranging applications.

Definition 7.3.1
Suppose X is a topological space and $x \in X$. A set $\mathcal{P}$ of open neighbourhoods of x is called a LOCAL BASE at x or a NEIGHBOURHOOD BASE at x if and

only if every neighbourhood of x includes a member of $\mathcal{P}$. Note that, given any countable local base $\mathcal{C}$ at x, we can recursively define an infinite sequence whose terms are nested and also form a local base at x: let $(V_n)_{n\in\mathbb{N}}$ be a surjective sequence in $\mathcal{C}$; then $(\bigcap\{V_k \mid 1 \le k \le n\})_{n\in\mathbb{N}}$ is such a sequence.

Definition 7.3.2

A topological space $(X, \mathcal{O})$ is said to be

- FIRST COUNTABLE if and only if there is a countable local base at each point of X;

- SECOND COUNTABLE if and only if there exists a countable base for $\mathcal{O}$;

- SEPARABLE if and only if X has a countable dense subset.

Example 7.3.3

Every semimetric space (X, d) is first countable: for each $x \in X$, the set of balls $\mathcal{B}_x = \{b[x\,;1/n) \mid n \in \mathbb{N}\}$ is a countable local base at x. And if (X, d) is separable, then it is second countable: if D is dense in X and U is open, then, for each $u \in U$, there are $m \in \mathbb{N}$ and $v \in D$ with $d(v, u) < 1/m < \mathrm{dist}(u\,, X \backslash U)\,/2$; then $u \in b[v\,;1/m) \subseteq U$. So U is a union of members of $\bigcup(\mathcal{B}_x)_{x\in D}$. But not all metric spaces are second countable or separable; indeed $\mathbb{R}$ with the discrete metric is neither, because $\mathbb{R}$ is uncountable and every singleton subset is open.

Example 7.3.4

The open intervals form a base for the usual topology on $\mathbb{R}$, but they do not form a countable set. Nonetheless, $\mathbb{R}$ is second countable. The set of intervals $\{(a, b) \mid a, b \in \mathbb{Q}\}$ forms a base for the topology, because $\mathbb{Q}$ is dense in $\mathbb{R}$.

Example 7.3.5

The finite complement topology on $\mathbb{N}$ is countable by 2.4.7. So $\mathbb{N}$ with this topology is both first and second countable.

Example 7.3.6

$\mathbb{Q}$ is dense in $\mathbb{R}$ and $\{a + ib \mid a, b \in \mathbb{Q}\}$ is dense in $\mathbb{C}$; so $\mathbb{R}$ and $\mathbb{C}$ are separable. Similarly, $\mathbb{R}^n$ and $\mathbb{C}^n$ are separable for all $n \in \mathbb{N}$; and, for each $p \in [1, \infty)$, ℓ_p is separable, a countable dense subset consisting of eventually constant sequences whose terms have rational real and imaginary parts. Similarly, c_0 is separable. But ℓ_∞ is not: suppose $(s_n)_{n\in\mathbb{N}}$ is a sequence of bounded sequences; for each $n \in \mathbb{N}$, set $z_n = 0$ if the nth term of s_n exceeds 1 and $z_n = 2$ otherwise; then $z \in \ell_\infty$ and $\mathrm{dist}(z\,, \{s_n \mid n \in \mathbb{N}\}) \ge 1$.

Countability Criteria in Relation to One Another

We show below that every second countable space is both first countable and separable. But neither separability nor first countability implies second countability. There are, indeed, spaces which are both separable and first countable but not second countable.

Theorem 7.3.7

Every second countable space is both first countable and separable.

Proof

Suppose X is a second countable space and let $\mathcal{B}$ be a countable base for the topology on X. Then, for each $x \in X$, every open neighbourhood of x is a union of members of $\mathcal{B}$, and so includes at least one member of $\mathcal{B}$ which contains x; therefore the set $\mathcal{B} \cap \mathrm{nbd}(x)$ is a countable local base at x. So X is first countable. To show that X is also separable, we proceed as follows. By the Axiom of Choice, there exists a choice function f for $\mathcal{B}$; then $f(\mathcal{B}\backslash\{\varnothing\})$ is a countable subset of X which has non-empty intersection with every member of $\mathcal{B}\backslash\{\varnothing\}$. So $X\backslash\overline{f(\mathcal{B}\backslash\{\varnothing\})}$ is an open subset of X which does not include any member of $\mathcal{B}$ and is therefore empty. So $X = \overline{f(\mathcal{B}\backslash\{\varnothing\})}$ is separable. □

Example 7.3.8

A particular point topology makes $\mathbb{R}$ into a separable first countable space which is not second countable. Let X be the space consisting of the set $\mathbb{R}$ with the particular point topology associated with 0. The singleton set $\{0\}$ is dense in X; so X is separable. Also, for each $x \in \mathbb{R}$, the set $\{\{x,0\}\}$ is a local base at x; so X is first countable. There is, however, no countable base for the topology because $\mathbb{R}$ is uncountable and $\{x,0\}$ is open for each $x \in \mathbb{R}$.

EXERCISES

Q 7.3.1 Let X be a set endowed with the finite complement topology. Show that this space is second countable if and only if X is countable.

Q 7.3.2 Show that a topological space X is second countable if its topology has a countable subbase.

Q 7.3.3 Suppose X is a second countable space. Show that iso(X) is countable.

7.4 Separation

The relationships between points and subsets of topological spaces are characterized by a number of different *separation* properties (not to be confused with separability), four of which we describe briefly below. There are many other such properties, but the importance of these ones can be stated quite simply: T_1 spaces are those in which singleton sets are closed (7.4.4); T_2 spaces are those in which all *limits* are uniquely determined (10.2.6); T_4 spaces are those in which we can guarantee a plentiful supply of *continuous* scalar functions (8.1.13); and *normal* spaces are those which exhibit all of these properties (7.4.5).

Definition 7.4.1

Suppose X is a topological space. X is called

- a T_1 SPACE if and only if, for each pair of distinct points $a, b \in X$, there exist $U \in \mathrm{nbd}(a)$ and $V \in \mathrm{nbd}(b)$ such that $a \notin V$ and $b \notin U$;

- a HAUSDORFF SPACE or a T_2 SPACE if and only if, for every $a, b \in X$ with $a \neq b$, there exist $U \in \mathrm{nbd}(a)$ and $V \in \mathrm{nbd}(b)$ such that $U \cap V = \varnothing$;

- a T_4 SPACE if and only if, for all closed disjoint subsets A and B of X, there are disjoint open subsets U and V of X with $A \subseteq U$ and $B \subseteq V$;

- a NORMAL SPACE if and only if X is both T_1 and T_4.

Example 7.4.2

Every discrete space is normal, since every subset is both open and closed.

Example 7.4.3

Clearly, a semimetric is a metric if and only if the semimetric topology is T_1. In fact, every metric space X is normal. Suppose A and B are disjoint closed subsets of X; for each $x \in A \cup B$, let $\delta_x = \mathrm{dist}(x, B)$ if $x \in A$ and $\delta_x = \mathrm{dist}(x, A)$ if $x \in B$. For each $a \in A$ and $b \in B$, we have $\delta_a + \delta_b \leq 2d(a, b)$. Therefore $U = \bigcup \{\flat[x; \delta_x/2) \mid x \in A\}$ and $V = \bigcup \{\flat[x; \delta_x/2) \mid x \in B\}$ are disjoint open subsets of X which include A and B respectively.

The four separation properties given in 7.4.1 are distinct, but not quite hierarchical. Every normal space is both T_4 and T_1 by definition, and certainly every T_2 space is T_1. Any non-singleton set with the trivial topology is T_4 but not T_1; and any infinite set with the finite complement topology is T_1 but neither T_2 nor T_4. There are also T_2 spaces which are not T_4, though they are harder to find (8.3.20). To complete our catalogue, every normal space is T_2, as we show in 7.4.5.

Theorem 7.4.4

A topological space is T_1 if and only if every singleton subset is closed.

Proof

Suppose X is a T_1 space and $x \in X$. Let U denote the union of all open subsets of X which do not contain x. Since X is T_1, $U = X \backslash \{x\}$. But U is open; so $\{x\}$ is closed. Conversely, suppose each singleton subset of X is closed. Then, for each pair of distinct elements a, b of X, the sets $X \backslash \{b\}$ and $X \backslash \{a\}$ are open neighbourhoods of a and b respectively.   □

Corollary 7.4.5

Suppose X is a normal topological space. Then X is Hausdorff.

Proof

Suppose $a, b \in X$ with $a \neq b$. Since X is T_1, $\{a\}$ and $\{b\}$ are closed by 7.4.4; then, since X is T_4, there exist disjoint open subsets U and V of X such that $a \in \{a\} \subseteq U$ and $b \in \{b\} \subseteq V$.   □

Theorem 7.4.6

Suppose X is a topological space. Then X is T_4 if and only if, for every closed subset F of X and open subset W of X with $F \subseteq W$, there is an open subset U of X such that $F \subseteq U \subseteq \overline{U} \subseteq W$.

Proof

If the conclusion holds, then it is clear that disjoint closed sets can be separated. Conversely, if X is T_4, there exist disjoint open sets U and V such that $F \subseteq U$ and $X \backslash W \subseteq V$. So $X \backslash V \subseteq W$. Since $X \backslash V$ is closed and includes U, it includes also the closure of U and the result follows.   □

EXERCISES

Q 7.4.1 Suppose X is a non-empty set. Show that the finite complement topology is the weakest topology on X which makes X into a T_1 space.

Q 7.4.2 Show that $\tilde{\mathbb{R}}$ with its usual topology is normal.

Continuity and Openness

Thy crown does sear mine eye-balls. And thy hair,
Thou other gold-bound brow, is like the first.
A third is like the former. Filthy hags!
Why do you show me this? A fourth? Start, eyes.
What, will the line stretch out to th' crack of doom?
Another yet? A seventh? I'll see no more.
And yet the eighth appears, who bears a glass
Which shows me many more; and some I see
That two-fold balls and treble sceptres carry.   *Macbeth, IV,i.*

Perhaps the most important attribute a function might display is *continuity*. The mathematical concept of continuity is likely to be less restrictive than our intuitive one if only because the mathematical definition of a function is so broad. Even so, it is hardly a natural expectation that mathematical continuity should be defined as a rather subtle topology-preserving property (8.1.1).

8.1 Preservation of Topological Structure

An isomorphism, as we have seen, is a bijection which preserves algebraic structure; and an isometry is a bijection which preserves metrics. The corresponding notion for topological spaces—a bijective map which preserves topological structure—is called a *homeomorphism*. Homeomorphisms, as well as being bijective, exhibit two quite distinct properties, *continuity* and *openness*, which we shall consider in turn.

Continuous Functions

Definition 8.1.1
Suppose $(X, \mathcal{O})$ and $(Y, \mathcal{P})$ are topological spaces and $f: X \to Y$. Then f is said to be CONTINUOUS if and only if $f^{-1}(\mathcal{P}) \subseteq \mathcal{O}$.

Example 8.1.2

Before we can talk about continuity of a function, we must have in mind topologies on both the domain and a specific co-domain. So, for example, the identity function from a set X to X is continuous if and only if the intended topology on X as co-domain is weaker than the intended topology on X as domain. But every constant function is continuous irrespective of which topologies are given to the domain and any given co-domain. To verify this, suppose X and Y are topological spaces and $c \in Y$, and let $f: X \to Y$ be the function given by $x \mapsto c$. Suppose V is open in Y. If $c \in V$, then $f^{-1}(V) = X$; otherwise $f^{-1}(V) = \varnothing$. In either case, $f^{-1}(V)$ is open in X. So f is continuous.

Example 8.1.3

Suppose X and Y are sets, $\mathcal{O}_s$ and $\mathcal{O}_w$ are topologies on X with $\mathcal{O}_s$ stronger than $\mathcal{O}_w$, and $\mathcal{P}_s$ and $\mathcal{P}_w$ are topologies on Y with $\mathcal{P}_s$ stronger than $\mathcal{P}_w$. If $f: X \to Y$ is continuous when X is endowed with the topology $\mathcal{O}_w$ and Y with $\mathcal{P}_s$, then f is continuous when X is endowed with $\mathcal{O}_s$ and Y with $\mathcal{P}_w$. In particular, if either X has the discrete topology or Y has the trivial topology, then every function from X to Y is continuous.

Example 8.1.4

The most straightforward compositions of continuous functions are continuous. Suppose $(X, \mathcal{O})$, $(Y, \mathcal{P})$ and $(Z, \mathcal{Q})$ are topological spaces and $f: X \to Y$ and $g: Y \to Z$ are continuous functions. Then, for each $U \in \mathcal{Q}$, we have $g^{-1}(U) \in \mathcal{P}$ and then $(g \circ f)^{-1}(U) = f^{-1}(g^{-1}(U)) \in \mathcal{O}$. So $g \circ f$ is continuous. There is a further detail which must be considered (see 8.3.13) in cases in which $\operatorname{dom}(g) \neq \operatorname{codom}(f)$.

Using Subbases to Test for Continuity

In testing a function for continuity, it is sufficient to examine inverse images of basic, or simply of subbasic, open subsets of the co-domain. A function whose co-domain is a metric space, for example, is usually tested for continuity by examining the inverse images of the open balls of the space.

Theorem 8.1.5

Suppose $(X, \mathcal{O})$ and $(Y, \mathcal{P})$ are topological spaces and $f: X \to Y$. Suppose $\mathcal{S}$ is a subbase and $\mathcal{B}$ is a base for $\mathcal{P}$. Then we have equivalent statements: f is continuous $\Leftrightarrow f^{-1}(\mathcal{B}) \subseteq \mathcal{O} \Leftrightarrow f^{-1}(\mathcal{S}) \subseteq \mathcal{O}$.

Proof

We show that the third criterion implies the first; the rest follows immediately. Suppose that $V \in \mathcal{P}$. Then there exists a collection $\mathfrak{F}$ of finite subsets of S such that $V = \bigcup \{\bigcap \mathcal{F} \mid \mathcal{F} \in \mathfrak{F}\}$. By Q 1.2.1, $f^{-1}(V) = \bigcup \{\bigcap f^{-1}(\mathcal{F}) \mid \mathcal{F} \in \mathfrak{F}\}$; this is a member of $\mathcal{O}$ if the third condition holds.   □

Example 8.1.6

All seminorms are continuous. Suppose $(X, \|\cdot\|)$ is a seminormed linear space. Then $\|\cdot\|$ is a function from X into $\mathbb{R}$; X has the topology induced by $\|\cdot\|$ and $\mathbb{R}$ has its usual topology. The inverse image of a bounded open interval (a, b) of $\mathbb{R}$ under $\|\cdot\|$ is $\{x \in X \mid a < \|x\| < b\}$, which is empty if $b \leq 0$, is an open ball if $a < 0 < b$, is $\flat[0 ; b) \setminus \{0\}$ if $a = 0$, and is $\flat[0 ; b) \setminus \flat[0 ; a]$ otherwise; in any case, it is open in X. So $\|\cdot\|$ is continuous by 8.1.5.

Example 8.1.7

Suppose X is a seminormed linear space. It follows from 6.2.12 and 8.1.5 that, for each $z \in X$, the translation $x \mapsto x + z$ is continuous (and has continuous inverse $x \mapsto x - z$). This phenomenon is described by saying that the topology induced by the seminorm on X is TRANSLATION INVARIANT. If a linear space M is endowed with a translation invariant topology and U and S are non-empty subsets of M with U open in M, then $U + S$ is also open in M, because $U + S = \bigcup \{U + \{z\} \mid z \in S\}$.

Example 8.1.8

It was proved in 4.3.26 that every complex polynomial function *pulls back* open discs to open sets, in the sense that the inverse image of each open disc is an open set. So, by 8.1.5, those functions are all continuous.

Continuity at a Point

A function may fail to be continuous and yet satisfy a similar property at many points of its domain. In defining this concept of *continuity at a point*, we use open neighbourhoods of a point $f(z)$, but it is easy to check that they may be replaced, using 8.1.5 and Q 1.2.1, by basic or subbasic open neighbourhoods.

Definition 8.1.9

Suppose $(X, \mathcal{O})$ and $(Y, \mathcal{P})$ are topological spaces, $z \in X$ and $f \colon X \to Y$. Then f is said to be CONTINUOUS AT THE POINT z if and only if, for each

$V \in \mathrm{nbd}(f(z))$ there exists $U \in \mathrm{nbd}(z)$ such that $f(U) \subseteq V$. If f is not continuous at z, we say that f is DISCONTINUOUS at z.

Theorem 8.1.10

Suppose X and Y are topological spaces and $f\colon X \to Y$. Then f is continuous if and only if f is continuous at each point of X.

Proof

Suppose firstly that f is continuous and let $x \in X$. Suppose $V \in \mathrm{nbd}(f(x))$. Then $f^{-1}(V) \in \mathrm{nbd}(x)$; moreover, $f(f^{-1}(V)) \subseteq V$. So f is continuous at x. Towards the converse, suppose that f is continuous at every point of X and that V is an open subset of Y. Let $\mathcal{C}$ be the collection of all open subsets U of X for which $f(U) \subseteq V$. Then, certainly, $\bigcup \mathcal{C} \subseteq f^{-1}(V)$; moreover, if $x \in f^{-1}(V)$, then $f(x) \in V$ and, by hypothesis, there exists $U \in \mathrm{nbd}(x)$ with $f(U) \subseteq V$, so that $x \in U \in \mathcal{C}$. So $f^{-1}(V) \subseteq \bigcup \mathcal{C}$. Therefore $f^{-1}(V) = \bigcup \mathcal{C}$, which is open. So f is continuous.   □

Example 8.1.11

When a linear space is endowed with a translation invariant topology, the neighbourhood structure around any one point is identical to that around any other. This fact has a very useful corollary for linear maps: suppose that X and Y are linear spaces with translation invariant topologies and that $T \in \mathcal{L}(X,Y)$; then T is continuous if and only if T is continuous at any one point of X. The proof is easy. Suppose T is continuous at $a \in X$ and let $b \in X$ be arbitrary. Let $B \in \mathrm{nbd}(Tb)$; then, by translation invariance in Y, $B + \{Ta - Tb\} \in \mathrm{nbd}(Ta)$. Since T is continuous at a, there exists $U \in \mathrm{nbd}(a)$ such that $T(U) \subseteq B + \{Ta - Tb\}$. Then $T(U) + \{Tb - Ta\} \subseteq B$. By linearity of T, $T(U + \{b - a\}) = T(U) + \{Tb - Ta\} \subseteq B$ and, by translation invariance in X, $U + \{b - a\} \in \mathrm{nbd}(b)$. So T is continuous at b. Since b is arbitrary in X, 8.1.10 completes the proof.

Example 8.1.12

Suppose that X and Y are seminormed linear spaces and that $T \in \mathcal{L}(X,Y)$. Then T is continuous if and only if T is bounded. Suppose firstly that T is continuous and let $B = \flat_Y[0\,;1)$. Then $T^{-1}(B)$ is open in X. Since also $0 \in T^{-1}(B)$, it follows from 6.1.23 that there exists $r \in \mathbb{R}^+$ such that $\flat_X[0\,;2r) \subseteq T^{-1}(B)$. Then $T(\flat_X[0\,;r]) \subseteq B$, and, using 6.2.12, $T(\flat_X[0\,;1]) \subseteq r^{-1}B$. Towards the converse, suppose $T \in \mathcal{B}(X,Y)$. If $\|T\| = 0$, then T is certainly continuous; we suppose $\|T\| \neq 0$. Let $U \in \mathrm{nbd}_Y(0)$. Then there exists $r \in \mathbb{R}^+$ such that

$\flat_Y[0\,;r) \subseteq U$. Then $T(\flat_X[0\,;r/\|T\|)) \subseteq \flat_Y[0\,;r) \subseteq U$. So T is continuous at 0, and hence continuous as in 8.1.11. This gives us an important example of a phenomenon which we shall meet again: that concepts and structures which are defined differently may, sometimes quite surprisingly, turn out to be identical.

Real Continuous Functions

The condition required to ensure a good supply of non-constant real continuous functions on a given topological space is that the space be T_4. This result of Urysohn formed a key step in the characterization of separable metrizable spaces as those topological spaces which are both normal and second countable.

Theorem 8.1.13 URYSOHN'S LEMMA
Suppose X is a topological space. Then X is T_4 if and only if, for each pair A and B of non-empty disjoint closed subsets of X, there exists a continuous function from X onto the interval $[0, 1]$ such that $f(A) = \{0\}$ and $f(B) = \{1\}$.

Proof
Suppose A and B are non-empty disjoint closed subsets of X. Let $\mathcal{U}$ be the collection of all open subsets U of X for which $A \subseteq U \subseteq \overline{U} \subseteq X \backslash B$. Impose an enumerative well ordering on $\mathbb{Q} \cap (0, 1]$ in which 1 is the minimum element. Then 7.4.6 and the Recursive Choice Theorem ensure that there exists a function $h: \mathbb{Q} \cap (0, 1] \to \mathcal{U}$ such that, for each $q \in \mathbb{Q} \cap (0, 1)$, we have

$$\bigcup\left\{\overline{h(r)} \mid r \in \dot{q} \cap (0, q)\right\} \subseteq h(q) \subseteq \overline{h(q)} \subseteq \bigcap\{h(r) \mid r \in \dot{q} \cap (q, 1]\},$$

where $\dot{q}$ refers to the segment in the specified ordering. In verifying that this is a valid recursive use of 7.4.6 it must be noted that, for $q \in \mathbb{Q} \cap (0, 1)$, the segment $\dot{q}$ is non-empty and finite; then the left hand side of the displayed inclusion is a finite union of closed sets, and therefore closed, and the right hand side is a non-trivial finite intersection of open supersets of A in $X \backslash B$ and is therefore itself an open superset of A in $X \backslash B$.

Define $f: X \to [0, 1]$ to be $x \mapsto \sup\{q \in \mathbb{Q} \cap [0, 1) \mid q = 0 \text{ or } x \notin h(q)\}$, the supremum being with respect to the usual ordering. The definitions of $\mathcal{U}$ and h ensure that $f(A) = \{0\}$ and $f(B) = \{1\}$. It remains to show that f is continuous. Towards this, we have, for each $a \in (0, 1)$, the two identities

$$f^{-1}([0, a)) = \{x \mid f(x) < a\} = \{x \in X \mid \exists q < a, x \in \overline{h(q)}\} = \bigcup\{h(q) \mid 0 < q < a\}$$
$$f^{-1}((a, 1]) = \{x \mid f(x) > a\} = \left\{x \in X \mid \exists q > a, x \notin \overline{h(q)}\right\} = \bigcup\left\{X \backslash \overline{h(q)} \mid q > a\right\}$$

which show that $f^{-1}([0, a))$ and $f^{-1}((a, 1])$ are both open. Since the collection of all such intervals is a subbase for the metric topology on $[0, 1]$, it follows from

8.1.5 that f is continuous. This proves one half of the theorem. The converse is easy: if such a function f exists, then $f^{-1}([0, 1/2))$ and $f^{-1}((1/2, 1])$ are disjoint open sets which include A and B respectively.   □

Continuity in Metric Spaces

There is a more familiar, though less neat, formulation for continuity of a function between metric spaces.

Theorem 8.1.14

Suppose (X, d) and (Y, d') are metric spaces, $z \in X$ and $f: X \to Y$. Then f is continuous at z if and only if, for every $\epsilon \in \mathbb{R}^+$, there exists $\delta \in \mathbb{R}^+$ such that, for all $x \in X$, we have $d(z, x) < \delta \Rightarrow d'(f(z), f(x)) < \epsilon$.

Proof

Suppose firstly that f is continuous at z and let $\epsilon \in \mathbb{R}^+$. Then there exists an open subset U of X such that $z \in U \subseteq f^{-1}(\flat[f(z)\,;\epsilon))$, and consequently there exists an open ball B of X such that $z \in B \subseteq U \subseteq f^{-1}(\flat[f(z)\,;\epsilon))$; let $\delta = \mathrm{dist}(z\,, X\backslash B)$; then $\delta > 0$ because $X\backslash B$ is closed, and, for each $x \in X$, $d(x, z) < \delta \Rightarrow f(x) \in \flat[f(z)\,;\epsilon)$. Towards the converse, suppose that the condition is satisfied and that V is an open neighbourhood of $f(z)$ in Y; then $\epsilon = \mathrm{dist}(f(z), Y\backslash V) > 0$ because $Y\backslash V$ is closed; and there exists $\delta \in \mathbb{R}^+$ such that $x \in \flat[z\,;\delta) \Rightarrow f(x) \in \flat[f(z)\,;\epsilon)$; set $U = \flat[z\,;\delta)$ and we have $f(U) \subseteq V$.   □

It follows from 8.1.14 and 8.1.10 that a function $f: X \to Y$, where (X, d) and (Y, d') are metric spaces, is continuous if and only if, for every $z \in X$ and every $\epsilon \in \mathbb{R}^+$, there exists $\delta \in \mathbb{R}^+$ such that $d(z, x) < \delta \Rightarrow d'(f(x), f(z)) < \epsilon$ for all $x \in X$. In general, δ will depend on z as well as on ϵ. But some functions have better continuity properties than others; in some cases, the same δ satisfies the criterion for all $z \in X$. If f is an isometry, for example, then for each $z \in X$ and $\epsilon \in \mathbb{R}^+$, we can choose $\delta = \epsilon$ to establish the continuity of f at z.

Definition 8.1.15

Suppose (X, d) and (Y, d') are metric spaces and $f: X \to Y$. Then f is called

- a UNIFORMLY CONTINUOUS FUNCTION if and only if, for every $\epsilon \in \mathbb{R}^+$, there exists $\delta \in \mathbb{R}^+$ such that $d(z, x) < \delta \Rightarrow d'(f(z), f(x)) < \epsilon$ for all $z, x \in X$;
- a CONTRACTION if and only if $d'(f(x), f(z)) \leq d(x, z)$ for all $z, x \in X$;

- a STRONG CONTRACTION if and only if there exists $k \in (0,1)$ such that $d'(f(z), f(x)) \leq kd(z, x)$ for all $z, x \in X$. In this case k is called a LIPSCHITZ CONSTANT for the contraction f.

Example 8.1.16

It is evident that there is a hierarchy: strong contractions are contractions; contractions are uniformly continuous; and uniformly continuous functions are continuous. Moreover, the concepts are distinct: the function $x \mapsto x^2$ defined on $\mathbb{R}$ is continuous but not uniformly continuous; the function $x \mapsto x^2$ defined on $[0, 1]$ is uniformly continuous but is not a contraction; and the identity function on $\mathbb{R}$ is a contraction which is not strong.

Example 8.1.17

Careful scrutiny of the proof of 4.3.26 reveals that restrictions of complex polynomial functions to bounded subsets of $\mathbb{C}$ are uniformly continuous functions. This is typical of complex functions which are continuous at every point of $\mathbb{C}$.

Example 8.1.18

Suppose X and Y are seminormed linear spaces and $T \in \mathcal{L}(X, Y)$. Since seminorms induce translation invariant topologies (8.1.7), 8.1.11 implies that continuity of T at a single point implies continuity at all points and therefore continuity of T. In fact it implies also its uniform continuity. Suppose T is continuous at 0 and let $\epsilon \in \mathbb{R}^+$. Then there exists $\delta \in \mathbb{R}^+$ such that, for all $x \in X$, $\|x\| < \delta \Rightarrow \|Tx\| < \epsilon$. Let $z, w \in X$; then, if $\|z - w\| < \delta$ we have $\|Tz - Tw\| = \|T(z - w)\| < \epsilon$.

Open Maps

Open mappings are those which carry open sets onto open sets. The definitions of continuous maps and open maps are thus similar and it is to be expected that some of the properties of continuous functions have natural analogues which hold for open mappings. But there are dissimilarities which are perhaps unexpected. For example, although we can test for continuity using only subbasic open sets, we cannot in general do the same for openness.

Definition 8.1.19

Suppose $(X, \mathcal{O})$ and $(Y, \mathcal{P})$ are topological spaces and $f \colon X \to Y$. Then f is said to be OPEN if and only if $f(\mathcal{O}) \subseteq \mathcal{P}$.

Example 8.1.20

Suppose X and Y are sets, $\mathcal{O}_s$ and $\mathcal{O}_w$ are topologies on X with $\mathcal{O}_s$ stronger than $\mathcal{O}_w$, and $\mathcal{P}_s$ and $\mathcal{P}_w$ are topologies on Y with $\mathcal{P}_s$ stronger than $\mathcal{P}_w$. Suppose $f: X \to Y$ is open when X is endowed with $\mathcal{O}_s$ and Y with $\mathcal{P}_w$; then f is open when X is endowed with $\mathcal{O}_w$ and Y with $\mathcal{P}_s$.

Example 8.1.21

Only one half of the analogue for openness of the example at the end of 8.1.3 is true. Suppose X and Y are sets endowed with topologies. If Y has the discrete topology, then every function from X to Y is open; but if Y has any topology other than the discrete topology and X is non-empty and has any topology, then there is a map from X to Y which is not open: just consider a constant function whose range is a singleton set which is not open in Y.

Example 8.1.22

The most straightforward compositions of open maps are open. Suppose $(X, \mathcal{O})$, $(Y, \mathcal{P})$ and $(Z, \mathcal{Q})$ are topological spaces, and $f: X \to Y$ and $g: Y \to Z$ are open mappings; then, for each $U \in \mathcal{O}$, we have $f(U) \in \mathcal{P}$ and then $(g \circ f)(U) = g(f(U)) \in \mathcal{Q}$, so $g \circ f$ is open. Another detail must be considered (see 8.3.13) in cases in which $\operatorname{dom}(g) \neq \operatorname{codom}(f)$.

Theorem 8.1.23

Suppose $(X, \mathcal{O})$ and $(Y, \mathcal{P})$ are topological spaces and $f: X \to Y$. Suppose $\mathcal{B}$ is a base for $\mathcal{O}$. Then f is an open mapping if and only if $f(\mathcal{B}) \subseteq \mathcal{P}$.

Proof

Certainly, if f is an open mapping, then $f(\mathcal{B}) \subseteq f(\mathcal{O}) \subseteq \mathcal{P}$. Conversely, suppose $f(\mathcal{B}) \subseteq \mathcal{P}$ and U is open in X. Then there exists $\mathcal{C} \subseteq \mathcal{B}$ such that $U = \bigcup \mathcal{C}$, and $f(U) = f(\bigcup \mathcal{C}) = \bigcup f(\mathcal{C})$, which, being a union of members of $\mathcal{P}$, is open.   □

Example 8.1.24

8.1.23 is the counterpart of 8.1.5 for open maps rather than continuous maps. In 8.1.5, however, we proved that a mapping which pulls back subbasic open sets to open sets is continuous. The analogous statement for open maps is true if the function is injective (Q 8.1.2), but not in general. Endow $\mathbb{N}$ with the finite complement topology and let $Y = \{0, 1\}$ have the trivial topology. Consider the function $f: \mathbb{N} \to Y$ for which $f(1) = f(2) = 0$ and $f(n) = 1$ if $n > 2$. Then f is not an open map because $S = \mathbb{N} \setminus \{1, 2\}$ is open in $\mathbb{N}$ but $f(S) = \{1\}$ is not open in Y. However, $f(\mathbb{N} \setminus \{n\}) = Y$ is open in Y for all $n \in \mathbb{N}$.

Homeomorphisms

Since continuous maps pull back open sets to open sets, and open maps carry open sets onto open sets, it follows that a bijective map which is both open and continuous provides a one-to-one identification of domain and range which incorporates also a one-to-one identification of their respective open subsets. The domain and range of such a map are said to be *homeomorphic*; from a purely topological point of view, there is no difference between two such spaces and they may be considered to be identical. There may, of course, be non-topological (even metric) structure which differentiates them from one another.

Definition 8.1.25

Suppose $(X, \mathcal{O})$ and $(Y, \mathcal{P})$ are topological spaces. A bijective map $f: X \to Y$ is called a HOMEOMORPHISM if and only if $f(\mathcal{O}) = \mathcal{P}$. If such a mapping exists, then its inverse is clearly also a homeomorphism and the spaces X and Y are said to be HOMEOMORPHIC.

Example 8.1.26

Every isometry between metric spaces is a homeomorphism, but there are many homeomorphisms which do not preserve distance. For example, the unit disc $\mathbb{D}$ of $\mathbb{C}$ and the unit square $\{z \in \mathbb{R}^2 \mid -1 \le z_1 \le 1, -1 \le z_2 \le 1\}$ are homeomorphic: set $\phi(0) = 0$ and $\phi(w) = |w| / \max\{|x|, |y|\}$ for each $w \in \mathbb{D}\backslash\{0\}$, where $w = x + iy$ in standard form; let $h(w) = (\phi(w)x, \phi(w)y)$; then h is a homeomorphism which does not preserve distance for the usual metrics.

Example 8.1.27

Suppose $(X, \mathcal{O})$ is a topological space and Y is a set which is in one-to-one correspondence with X. Then each bijective function $f: X \to Y$ imposes on Y a topology, namely $f(\mathcal{O})$. When Y is endowed with this topology, f becomes a homeomorphism and the spaces X and Y are homeomorphic.

Example 8.1.28

Every translation on a normed linear space is bijective and its inverse is a translation. Such translations are continuous (8.1.11); so each is a homeomorphism.

Example 8.1.29

Suppose A is a unital normed algebra; then the inverse map $a \mapsto a^{-1}$ on $\mathrm{inv}(A)$ is a homeomorphism. Suppose that $v \in \mathrm{inv}(A)$ and that $\epsilon \in \mathbb{R}^+$. For every $w \in \mathrm{inv}(A)$ with $\|w - v\| < \min\{\epsilon/2\|v^{-1}\|^2, 1/2\|v^{-1}\|\}$, we then have

$2(1 - \|v^{-1}\| \, \|v - w\|)\|v^{-1} - w^{-1}\| \leq 2\|v^{-1}\| \, \|v - w\| \, (\|w^{-1}\| - \|v^{-1} - w^{-1}\|)$
and, since $\|v^{-1}\| \, \|v - w\| < 1/2$ and $\|w^{-1}\| - \|v^{-1} - w^{-1}\| \leq \|v^{-1}\|$, we get $\|v^{-1} - w^{-1}\| \leq 2\|v^{-1}\|^2 \, \|w - v\| < \epsilon$. So the map is continuous; being its own inverse, it is therefore a homeomorphism.

Normed Linear Space Isomorphisms

In our earlier use of the term *isomorphism* (3.4.3) we have meant preservation of whatever structure is under consideration; likewise, in the present context, we reserve the term for linear space isomorphisms which are also homeomorphisms.

Definition 8.1.30

Suppose X and Y are seminormed linear spaces and $T \in \mathcal{L}(X, Y)$. Then T is called a SEMINORMED LINEAR SPACE ISOMORPHISM, or simply an ISOMORPHISM, if and only if T is a bijective homeomorphism. If such a map exists, then X and Y are called ISOMORPHIC seminormed linear spaces. If X and Y are seminormed algebras, then T is called a SEMINORMED ALGEBRA ISOMORPHISM if and only if T is a seminormed linear space isomorphism which also preserves multiplication; if such a map exists, then the seminormed algebras X and Y are said to be ISOMORPHIC.

Example 8.1.31

Suppose X and Y are seminormed linear spaces and $T \in \mathcal{L}(X, Y)$. If T is open then $T(\flat_X[0\,;1))$ is an open neighbourhood of 0 in Y, so that, for each $y \in Y$, there exists $t \in \mathbb{R}^+$ such that $ty \in T(\flat_X[0\,;1))$; so T is necessarily surjective. This implies that we can safely replace *bijective* by *injective* in Definition 8.1.30. But a much more important result (12.2.10) states that every bijective continuous linear map is an isomorphism when certain restrictions are put on X and Y.

EXERCISES

Q 8.1.1 Suppose X and Y are topological spaces. and $f\colon X \to Y$. Show that the following are equivalent:

- f is continuous;
- for each closed subset F of Y, $f^{-1}(F)$ is closed in X;
- for each subset A of X, $f(\overline{A}) \subseteq \overline{f(A)}$;
- for each subset B of Y, $\overline{f^{-1}(B)} \subseteq f^{-1}(\overline{B})$.

Q 8.1.2 Suppose X and Y are topological spaces and $f: X \to Y$ is injective. Let $\mathcal{S}$ be a subbase for the topology on X. Show that f is open if and only if $f(U)$ is open for every $U \in \mathcal{S}$.

Q 8.1.3 Suppose X and Y are topological spaces and $f: X \to Y$ is bijective. Show that f is continuous if and only if f^{-1} is open. Hence show that f is a homeomorphism if and only if both f and f^{-1} are continuous.

8.2 Topologies Defined by Functions

The analyst has a more fundamental interest in functions than in topologies; and a more especial interest in continuous functions than in arbitrary ones. Yet we have presented continuity as a mere part in a process for preserving topologies. Lest our true intent be mistaken, we shall now place the emphasis where it belongs—on functions and on their continuity. For the reality is this: most topologies used in analysis are designed to make certain functions continuous.

Initial Topologies

The question to be addressed here is as follows: given a collection of functions with the same domain, and whose co-domains are endowed with topologies, which topologies on this domain ensure the continuity of all the given functions?

Definition 8.2.1
Suppose X is a set and $\mathcal{F}$ is a collection of functions defined on X whose co-domains are endowed with topologies. Then the topology on X generated by $\mathcal{S} = \{f^{-1}(U) \mid f \in \mathcal{F},\ U \text{ open in } \operatorname{codom}(f)\}$ is called the INITIAL TOPOLOGY or the WEAK TOPOLOGY on X determined by $\mathcal{F}$. If $\mathcal{F} \neq \varnothing$, then, by 7.1.21, $\mathcal{S}$ is a subbase, called the DEFINING SUBBASE, for the initial topology; the base generated by it is called the DEFINING BASE for the initial topology.

Theorem 8.2.2
Suppose X is a set and $\mathcal{F}$ is a collection of functions defined on X whose co-domains are topological spaces. Then a topology on X ensures the continuity of every member of $\mathcal{F}$ if and only if it is stronger than the initial topology on X determined by $\mathcal{F}$. The initial topology is thus the weakest topology on X with this property.

Proof

If $\mathcal{F} = \varnothing$, then $\mathcal{F}$ determines the trivial topology. Otherwise, the initial topology on X has subbase $\{f^{-1}(U) \mid f \in \mathcal{F},\ U \text{ open in } \text{codom}(f)\}$. A topology ensures the continuity of each $f \in \mathcal{F}$ if and only if it includes this set and therefore if and only if it is stronger than the initial topology. $\square$

If $(f_i)_{i \in I}$ is a family of functions, we refer, for simplicity, to a topology determined by the set $\{f_i \mid i \in I\}$ as that determined by the family (f_i). Some topologies are determined by a singleton set; again for simplicity, we talk of the topology determined by the function f rather than by the set $\{f\}$. It should be noted that the initial topology determined by a single function is precisely the same as the defining subbase; specifically, if X is a set, Y is a topological space and $f \colon X \to Y$ is a function, then the defining subbase $\{f^{-1}(V) \mid V \text{ open in } Y\}$ is in fact a topology on X (Q 1.2.1).

Example 8.2.3

Suppose (X, d) is a semimetric space. Then the semimetric topology on X is the initial topology $\mathcal{Q}$ determined by the family $(\delta_z)_{z \in X}$ of functions $x \mapsto d(z, x)$ defined on X. The proof of continuity of these functions with respect to the metric topology follows closely that of 8.1.6: the inverse image under each δ_z of each bounded open interval is either empty or is the difference between an open ball and a closed subset of X; in any case it is open in X. Since each δ_z is continuous, it follows from 8.2.2 that the metric topology is stronger than $\mathcal{Q}$. But the metric topology is also weaker than $\mathcal{Q}$ because each open ball $\flat[w\,;r)$ of X can be expressed as $\delta_w^{-1}(-\infty, r)$.

Example 8.2.4

The seminorm topology on a non-trivial seminormed linear space is not the initial topology determined by the seminorm (see Q 8.2.5). But every open ball $\flat[0\,;r)$ is the inverse image of $(-\infty, r)$ under the seminorm; so the initial topology determined by the seminorm includes the set of all such balls. The seminorm topology is thus the weakest translation invariant topology on X which ensures the continuity of the seminorm.

Theorem 8.2.5

Suppose that Y is a set endowed with the initial topology determined by a collection $\mathcal{F}$ of functions defined on Y. Suppose X is a topological space and $g \colon X \to Y$. Then g is continuous if and only if, for every $f \in \mathcal{F}$, the composition $f \circ g \colon X \to \text{codom}(f)$ is continuous.

Proof

If $\mathcal{F} = \varnothing$, then the topology on Y is trivial and the result is immediate, so we suppose otherwise. If g is continuous, then so are all the compositions $f \circ g$ by 8.1.4. Conversely, suppose all the compositions are continuous. Suppose V is a member of the defining subbase for the topology on Y. Then $V = f^{-1}(S)$ for some $f \in \mathcal{F}$ and some open subset S of codom(f). Therefore we have $g^{-1}(V) = (f \circ g)^{-1}(S)$, which is open in X; so g is continuous by 8.1.5.   □

Final Topologies

Now we address the question: given a collection of functions whose domains are endowed with topologies and whose ranges are all included in some given set, what is the strongest topology which can be placed on that set for which all the given functions are continuous?

Theorem 8.2.6

Suppose Y is a set and $\mathcal{F}$ is a set of functions into Y whose domains are endowed with topologies. Then $\mathcal{P} = \{U \subseteq Y \mid f^{-1}(U) \text{ open in } \mathrm{dom}(f) \text{ for all } f \in \mathcal{F}\}$ is a topology on Y. A topology on Y makes every member of $\mathcal{F}$ continuous if and only if it is weaker than this topology.

Proof

Certainly $Y \in \mathcal{P}$, because $\mathrm{dom}(f)$ is open in $\mathrm{dom}(f)$ for all $f \in \mathcal{F}$ (vacuously if $\mathcal{F}$ is empty). Suppose $\mathcal{S} \subseteq \mathcal{P}$. For each $f \in \mathcal{F}$, we have, by Q 1.2.1, $f^{-1}(\bigcup \mathcal{S}) = \bigcup f^{-1}(\mathcal{S})$, which is open in $\mathrm{dom}(f)$, so that $\bigcup \mathcal{S} \in \mathcal{P}$. Moreover, if we suppose that $\mathcal{S}$ is finite, then, for each $f \in \mathcal{F}$, we have, also by Q 1.2.1, $f^{-1}(\bigcap \mathcal{S}) = \bigcap f^{-1}(\mathcal{S})$, which is open in $\mathrm{dom}(f)$, so that $\bigcap \mathcal{S} \in \mathcal{P}$. So $\mathcal{P}$ is a topology, as claimed, and it is clear from its construction that $\mathcal{P}$ ensures the continuity of every member of $\mathcal{F}$. It is certain that every weaker topology on Y also ensures this continuity. Moreover, if $U \subseteq Y$ and $U \notin \mathcal{P}$, then there exists $f \in \mathcal{F}$ such that $f^{-1}(U)$ is not open in $\mathrm{dom}(f)$, whence f is not continuous when Y is endowed with any topology to which U belongs.   □

Definition 8.2.7

Suppose Y is a set and $\mathcal{F}$ is a set of functions into Y whose domains are topological spaces. Then $\{U \subseteq Y \mid f^{-1}(U) \text{ open in } \mathrm{dom}(f) \text{ for all } f \in \mathcal{F}\}$ is called the FINAL TOPOLOGY or the STRONG TOPOLOGY on Y determined by $\mathcal{F}$. It is, by 8.2.6, the strongest topology on Y which makes every $f \in \mathcal{F}$ continuous.

Theorem 8.2.8

Suppose that Y is a set endowed with the final topology determined by a collection $\mathcal{F}$ of functions into Y. Suppose that Z is a topological space and $g\colon Y \to Z$. Then g is continuous if and only if, for every $f \in \mathcal{F}$, the composition $g \circ f\colon \mathrm{dom}(f) \to Z$ is continuous.

Proof

If g is continuous, then so are all the compositions $g \circ f$ by 8.1.4. Towards the converse, suppose the compositions are all continuous and V is open in Z. Then $f^{-1}(g^{-1}(V)) = (g \circ f)^{-1}(V)$ is open in $\mathrm{dom}(f)$ for each $f \in \mathcal{F}$. This is precisely the condition that ensures that $g^{-1}(V)$ is in the strong topology on Y determined by $\mathcal{F}$, and it follows that g is a continuous function.   □

Openness of Maps which Determine Topologies

Functions which determine initial or final topologies are always, by definition, continuous with respect to those topologies; but they are not usually open maps. There are, however, some exceptions: in particular, a single surjective function which determines an initial topology is necessarily open, and a single injective function which determines a final topology is also open.

Theorem 8.2.9

Suppose $g\colon X \to Y$, where Y is a topological space and X has the initial topology determined by g. Then g is open if and only if $g(X)$ is open in Y.

Proof

If g is an open map, then, since X is open in X, it follows that $g(X)$ is open in Y. Towards the converse, suppose $g(X)$ is open in Y and U is an open subset of X. Recall that the topology on X in this case coincides with the defining subbase, so that $U = g^{-1}(V)$ for some open subset V of Y. Then $g(U) = g(g^{-1}(V)) = g(X) \cap V$, which is open in Y.   □

Corollary 8.2.10

A surjective map which determines the topology on its domain is open.

The case for final topologies is more apparent; 8.2.11 below is little more than a restatement of what it means for a set to belong to a final topology. Its corollary 8.2.12 follows immediately by putting $\mathcal{F} = \{g\}$.

Theorem 8.2.11

Suppose Y is a set endowed with the final topology determined by a collection $\mathcal{F}$ of functions from topological spaces into Y. Suppose $(X, \mathcal{O})$ is a topological space and $g\colon X \to Y$. Then g is open if and only if $f^{-1}(g(U))$ is open in $\mathrm{dom}(f)$ for every $f \in \mathcal{F}$ and $U \in \mathcal{O}$.

Corollary 8.2.12

An injective map which determines the topology on its co-domain is open.

EXERCISES

Q 8.2.1 Suppose $(X, \mathcal{O})$ and $(Y, \mathcal{P})$ are topological spaces and $f\colon X \to Y$ is continuous, open and surjective. Show that $\mathcal{P}$ is the final topology determined (with respect to $\mathcal{O}$) by f.

Q 8.2.2 Suppose $(X, \mathcal{O})$ and $(Y, \mathcal{P})$ are topological spaces and $f\colon X \to Y$ is continuous, open and injective. Show that $\mathcal{O}$ is the initial topology determined (with respect to $\mathcal{P}$) by f.

Q 8.2.3 Suppose X and Y are topological spaces and $f\colon X \to Y$ is bijective. Suppose that either the topology on X is the initial topology determined by f or the topology on Y is the final topology determined by f. Show that f is a homeomorphism.

Q 8.2.4 A topological space Y is said to be REGULAR if and only if, for every $x \in Y$ and $U \in \mathrm{nbd}(x)$, there exists $V \in \mathrm{nbd}(x)$ such that $\overline{V} \subseteq U$. Suppose X is a set, M is a topological space, $S \subseteq M^X$ and X is endowed with the weak topology determined by S. Show that if M is regular, then so is X. Show also that X is Hausdorff if and only if M is Hausdorff and S separates the points of X in the sense that, for all $x, y \in X$ with $x \neq y$, there exists $f \in S$ such that $f(x) \neq f(y)$.

Q 8.2.5 Describe the weak topology determined by an arbitrary norm.

8.3 Derived Topological Spaces

There are standard ways of endowing subsets, quotients and products of topological spaces with topologies. A subset is given the initial topology determined by the inclusion map; a quotient the final topology determined by the quotient map; and a product the initial topology determined by the natural projections.

Subspaces

Definition 8.3.1

Suppose $(X, \mathcal{O})$ is a topological space and S is a subset of X. Then the initial topology on S determined by the inclusion function inc: $S \to X$, namely $\mathcal{P} = \{S \cap U \mid U \in \mathcal{O}\}$, is called the RELATIVE TOPOLOGY or the SUBSPACE TOPOLOGY on S, and the topological space $(S, \mathcal{P})$ is called a TOPOLOGICAL SUBSPACE of $(X, \mathcal{O})$. A subset A of X which is a member of $\mathcal{P}$ is said to be OPEN IN S or RELATIVELY OPEN; its set complement in S, namely $S \backslash A$, is said to be CLOSED IN S.

Example 8.3.2

Every semimetric subspace of a semimetric space is a topological subspace when each is endowed with the topology induced by the semimetric. A similar statement applies to seminormed linear subspaces of seminormed linear spaces.

Example 8.3.3

The relative topology on $[0, 1]$ as a subspace of $\mathbb{R}$ is the metric topology, namely that generated by all intervals of the types $[0, a)$ and $(a, 1]$ for $a \in (0, 1)$. The relative topology on $\mathbb{R}$ considered as a subspace of $\mathbb{\bar{R}}$ is the usual topology on $\mathbb{R}$. The relative topology on $\mathbb{N}$ as a subspace of $\mathbb{R}$ is the discrete topology; this is the usual topology on $\mathbb{N}$. So is the relative topology on $\mathbb{N}$ considered as a subspace of $\tilde{\mathbb{N}}$ where $\tilde{\mathbb{N}}$ has its usual topology, namely the relative topology given to it by $\mathbb{\bar{R}}$. It might be observed that this topology on $\tilde{\mathbb{N}}$ is $\mathcal{P}(\mathbb{N}) \cup \{S \subseteq \tilde{\mathbb{N}} \mid \infty \in S \text{ and } |\mathbb{N} \backslash S| < \infty\}$.

Theorem 8.3.4

Suppose $(X, \mathcal{O})$ is a topological space and $\mathcal{S}$ is a base or subbase for its topology. Suppose $A \subseteq X$. Then $\mathcal{C} = \{A \cap V \mid V \in \mathcal{S}\}$ is a base or subbase, respectively, for the relative topology on A.

Proof

Let $\mathcal{B}$ denote the base generated by $\mathcal{S}$ and suppose $W \in \mathcal{B}$. Then there exists a finite subset $\mathcal{F}$ of $\mathcal{S}$ such that $W = \bigcap \mathcal{F}$. Also $A \cap W = \bigcap \{A \cap F \mid F \in \mathcal{F}\}$, so that $\{A \cap B \mid B \in \mathcal{B}\}$ is included in the topology on A generated by $\mathcal{C}$. Now suppose $U \in \mathcal{O}$, so that there exists a subset $\mathcal{P}$ of $\mathcal{B}$ such that $U = \bigcup \mathcal{P}$. Then we have $A \cap U = \bigcup \{A \cap P \mid P \in \mathcal{P}\}$, so that the relative topology on A is included in that generated by $\mathcal{C}$. Since each member of $\mathcal{C}$ is certainly a member of the relative topology, this completes the proof.   □

Corollary 8.3.5

Suppose X is a set endowed with the initial topology determined by a collection $\mathcal{F}$ of functions defined on X. Suppose $A \subseteq X$. Then the subspace topology on A is the initial topology determined by $\{f|_A \mid f \in \mathcal{F}\}$.

Proof

$\{A \cap f^{-1}(U) \mid f \in \mathcal{F}, U \text{ open in } \operatorname{codom}(f)\}$ is a subbase for the relative topology (8.3.4). But $A \cap f^{-1}(U) = f|_A^{-1}(U)$ for each $f \in \mathcal{F}$ and $U \subseteq \operatorname{codom}(f)$. □

A subspace of a subspace of a topological space X is itself a subspace of X. By this we mean that, if $Z \subseteq Y \subseteq X$, then the relative topology on Z as a subspace of Y, where Y has its relative topology as a subspace of X, is the same as the relative topology on Z as a subspace of X. This is easy to prove directly or as a consequence of 8.3.5, and we shall use it freely without comment. Care should be taken, however, when claiming that a certain set is open or closed in a particular subspace of a topological space. The interval $(0, 1]$, for instance, is closed in $(0, \infty)$ but is not closed in $\mathbb{R}$. The next result is helpful in this regard.

Theorem 8.3.6

Suppose $(X, \mathcal{O})$ is a topological space and $A \subseteq B \subseteq X$.

- If A is open in X, then A is open in B;
- If A is closed in X, then A is closed in B;
- If A is open in B and B is open in X, then A is open in X;
- If A is closed in B and B is closed in X, then A is closed in X.

Proof

For the first claim, if A is open in X, then $A = A \cap B$ is open in B. Towards the second claim, suppose A is closed in X. Then $X \backslash A$ is open in X. So $B \backslash A = B \cap (X \backslash A)$ is open in B, whence A is closed in B. Towards the third claim, suppose A is open in B and B is open in X. Then there exists an open subset U of X such that $B \cap U = A$. But B and U are both open in X; so their intersection, namely A, is also open in X. For the last claim, suppose A is closed in B and B is closed in X. Then $B \backslash A$ is open in B, so that $B \backslash A = B \cap U$ for some open subset U of X. Then $X \backslash A = (X \backslash B) \cup (B \backslash A) = (X \backslash B) \cup U$ which is open in X, since both $X \backslash B$ and U are open in X. So A is closed in X. □

Example 8.3.7

With X and B as in 8.3.6, the third part of 8.3.6 says that, if B is open in X, then inc: $B \to X$ is an open map. The converse is trivially true.

Example 8.3.8

Subspaces of first countable spaces are first countable and subspaces of second countable spaces are second countable. Suppose X is a topological space and Y is a subspace of X. Using 8.3.4, if $x \in Y$ and $\mathcal{V}$ is a countable local base at x in X, then $\{S \cap Y \mid S \in \mathcal{V}\}$ is a countable local base at x in Y; and, if $\mathcal{B}$ is a countable base for the topology on X, then $\{B \cap Y \mid B \in \mathcal{B}\}$ is a countable base for the subspace topology on Y.

Example 8.3.9

Separable spaces can have inseparable subspaces. Endow $X = \mathbb{R} \times \mathbb{R}$ with the SORGENFREY TOPOLOGY generated by the half-open rectangles $[a, b) \times [c, d)$ where $a, b, c, d \in \mathbb{R}$ and $a < b$ and $c < d$. Then $\mathbb{Q} \times \mathbb{Q}$ is easily seen to be a countable dense subspace of X, so that X is separable. And $L = \{(x, -x) \mid x \in \mathbb{R}\}$ is a discrete subspace of X: for each $z \in \mathbb{R}$, $[z, z + 1) \times [-z, -z + 1)$ is open in X and its intersection with L is the singleton set $\{(z, -z)\}$, so that $\{(z, -z)\}$ is an open subset of L in the subspace topology. Being uncountable and discrete, L is certainly not separable.

Example 8.3.10

There are normal spaces which have non-normal subspaces (Q 11.3.4); but every subspace of a Hausdorff space is Hausdorff. Suppose X is Hausdorff, Y is a subspace of X and $a, b \in Y$ with $a \neq b$. Then there exist disjoint open subsets A and B of X such that $a \in A$ and $b \in B$. So $A \cap Y$ and $B \cap Y$ are disjoint and open in Y, and $a \in A \cap Y$ and $b \in B \cap Y$.

Continuity and Openness of Restrictions and Compositions

The facts concerning the continuity and openness of restrictions and of the most general compositions (see 8.1.4 and 8.1.22) can now be stated. They are given in this subsection.

Lemma 8.3.11

Suppose X and Y are topological spaces and $f: X \to Y$. Let $h: X \to f(X)$ be the surjective restriction (1.2.18) of f. Endow $f(X)$ with the relative topology as a subspace of Y. Then f is continuous if and only if h is continuous. And f is open if and only if h is open and $f(X)$ is an open subset of Y.

Proof

inc: $f(X) \to Y$ determines the relative topology on $f(X)$; since $\mathrm{inc}\circ h = f$, the first assertion is a special case of 8.2.5. If f is an open map, then h is open by the first part of 8.3.6; conversely, if h is open and $f(X)$ is open in Y, then f is open by the third part of 8.3.6.   □

Theorem 8.3.12

Suppose X and Y are topological spaces and $f: X \to Y$. Suppose V is a subset of X and W is a superset of $f(V)$, each endowed with some topology. Let $\tilde{f}$ denote the restriction of f to V with co-domain W. If f and the inclusion function j from V (with its own topology) into X and the inclusion function i from $f(V)$ (as a subspace of Y) into W are all continuous, then so is $\tilde{f}$; if they are all open then so is $\tilde{f}$.

Proof

Let $h: V \to f(V)$ be the surjective restriction of $f \circ j$; then $\tilde{f} = i \circ h$. So, using 8.3.11, the first assertion follows from 8.1.4 and the second from 8.1.22.   □

Theorem 8.3.13

Suppose that X, Y, W and Z are topological spaces, that $f: X \to Y$ and $g: W \to Z$, and that $f(X) \subseteq W$. If f and g and the inclusion function from $f(X)$ (as a subspace of Y) into W are all continuous, then so is $g \circ f: X \to Z$; if they are all open, then so is $g \circ f$.

Proof

Let $h: X \to f(X)$ be the surjective restriction of f to X and i the inclusion map from $f(X)$ to W. Then $g \circ f = g \circ i \circ h$ and the results follow as in 8.3.12.   □

Example 8.3.14

There is some potential for confusion between continuity of a restriction to a subset of a domain and continuity at all points of that subset. It is easy to show that if S is a subset of the domain of a function f and if f is continuous at every point of S, then $f|_S$ is a continuous function (where, of course, S is endowed with the subspace topology). The converse, however, is not generally true. Indeed, it is possible for a function to be continuous on a subset of its domain but discontinuous at every point of that subset. An easy example of this phenomenon is the function $f: \mathbb{R} \to \mathbb{R}$ whose value is 1 at every rational number and 0 at every irrational number. f is discontinuous at every point of its domain; but it is the union of the two constant functions $f|_{\mathbb{Q}}$ and $f|_{\mathbb{R}\setminus\mathbb{Q}}$.

Product Spaces

There is in general more than one natural way of defining a topology on a non-trivial product of topological spaces, only one of which is called the *product topology*. The product topology is the initial topology determined by the natural projections from the product onto its co-ordinate spaces; these natural projections are therefore all continuous maps.

Definition 8.3.15

Suppose $(X_i)_{i \in I}$ is a non-trivial family of topological spaces. Let $P = \prod_i X_i$. The initial topology Q on P determined by the family (π_i) of natural projections associated with this product is called the PRODUCT TOPOLOGY on P. The defining subbase for the product topology is $\{\pi_j^{-1}(S) \mid j \in I,\ S \text{ open in } X_j\}$. The space (P, Q) is called the PRODUCT SPACE of the family (X_i).

The product topology on the product of a finite number of sets is exactly what one might expect. The usual topology on $\mathbb{R}^2$, for example, is that generated by the OPEN RECTANGLES $I_1 \times I_2$ where I_1 and I_2 are bounded open intervals of $\mathbb{R}$. For each $n \in \mathbb{N}$, the USUAL TOPOLOGY ON $\mathbb{R}^n$ is that generated by the OPEN BOXES $\prod_{i=1}^{n} I_i$, where each I_j is a bounded open interval of $\mathbb{R}$. Where the indexing set is infinite, the situation needs to be examined more closely. It is evident from the definition that the defining base for the product topology, got by taking finite intersections of subbasic sets, is the collection of products $\prod_i U_i$ which satisfy two conditions—firstly that each U_j is open in X_j, and secondly that, for all except a finite number of members j of I, we have $U_j = X_j$. If I is infinite, sets $\prod_i U_i$ which do not satisfy the second condition are not in the product topology. Clearly, there is a stronger topology on $\prod_i X_i$ than the product topology, namely that which has as a base all products $\prod_i U_i$ where each U_j is open in X_j. This stronger topology is called the BOX TOPOLOGY. Where a contrary intention is not evident from the context, a product of topological spaces will be assumed to be endowed with the product topology. Only where the indexing set is finite is this the same as the box topology.

Example 8.3.16

Suppose $(X_i)_{i \in I}$ is a non-trivial family of topological spaces. By 8.3.5, the relative topology on a subset S of $\prod_i X_i$ is the initial topology determined by $\{\pi_j|_S \mid j \in I\}$. Therefore, if $(Y_i)_{i \in I}$ is a family of sets with $Y_j \subseteq X_j$ for each $j \in I$, the subspace topology on $\prod_i Y_i$ coincides with the topology got by endowing each Y_j firstly with the relative topology as a subspace of the corresponding X_j and then endowing the product $\prod_i Y_i$ with its product topology.

Example 8.3.17

Suppose X is a set and Y is a topological space. Then the set Y^X of functions from X to Y is the product $\prod_{x \in X} Y$ and (1.2.26) the natural projections are the point evaluation functions $\hat{x} \colon Y^X \to Y$ given by $\hat{x}(f) = f(x)$ for each $x \in X$ and $f \in Y^X$. So the product topology on Y^X is the initial topology determined by these point evaluation functions.

Example 8.3.18

Products do not necessarily preserve countability criteria. The discrete space $\{0, 1\}$ is trivially second countable, therefore also first countable and separable, by 7.3.7. But the uncountable product $\{0, 1\}^{\mathbb{R}}$ clearly does not satisfy any of the countability criteria.

Example 8.3.19

Products of Hausdorff spaces are Hausdorff. Suppose $(X_i)_{i \in I}$ is a non-empty family of Hausdorff spaces. Let $x, y \in \prod_i X_i$ with $x \neq y$. Then $x_j \neq y_j$ for some $j \in I$. Since X_j is Hausdorff, there exist $U_j \in \mathrm{nbd}(x_j)$ and $V_j \in \mathrm{nbd}(y_j)$ with $U_j \cap V_j = \varnothing$. Then $\pi_j^{-1}(U_j)$ and $\pi_j^{-1}(V_j)$ are disjoint open neighbourhoods in $\prod_i X_i$ of x and y respectively.

Example 8.3.20

Products of normal spaces need not be normal. It is fairly easy to confirm that $\mathbb{R}$ with the HALF-OPEN INTERVAL TOPOLOGY, which has as its defining base the collection of intervals $[a, b)$ of $\mathbb{R}$, is a normal space. But the Sorgenfrey space (8.3.9)—the product of this space with itself—is not normal; indeed, it is separable and has the uncountable closed discrete subspace $\{(x, -x) \mid x \in \mathbb{R}\}$, so is not T_4 (Q 8.3.5). Since normal spaces are Hausdorff, 8.3.19 ensures that the Sorgenfrey space is a Hausdorff space which is not normal.

Example 8.3.21

Suppose X is a topological space and $(Y_i)_{i \in I}$ is a family of topological spaces. By 8.2.5, a function $g \colon X \to \prod_i Y_i$ is continuous if and only if, for each $j \in I$, the function $\pi_j \circ g \colon X \to Y_j$ is continuous, where π_j denotes the natural projection of $\prod_i Y_i$ onto its jth co-ordinate space Y_j.

Example 8.3.22

Metrics and semimetrics are continuous maps. Specifically, suppose (X, d) is a semimetric space. Then d is a function from the product space $X \times X$ into $\mathbb{R}$.

Suppose $a, b \in X$ and $\epsilon \in \mathbb{R}^+$. Then, for $(x, y) \in \flat[a \,; \epsilon/2) \times \flat[b \,; \epsilon/2)$, we have $|d(x, y) - d(a, b)| \leq d(x, a) + d(b, y) < \epsilon$, so that $d \colon X \times X \to \mathbb{R}$ is continuous at (a, b). Since (a, b) is arbitrary in $X \times X$, it follows from 8.1.10 that d is a continuous function.

Example 8.3.23

Suppose X is a seminormed linear space. Then addition is continuous; indeed, it is easy to check directly that the inverse image of an open ball $\flat[x \,; r)$ under addition is $\bigcup\{\flat[x - y \,; t) \times \flat[y \,; r - t) \mid y \in X, \, t \in (0, r)\}$, which is open in the product topology. Scalar multiplication is also continuous: suppose that $\beta \in \mathbb{K}$, that $a \in X$ and that $\epsilon \in \mathbb{R}^+$; then, for each $(\lambda, x) \in \flat_{\mathbb{K}}[\beta \,; \epsilon) \times \flat_X[a \,; \epsilon)$, we have $\|\beta a - \lambda x\| \leq |\beta| \, \|a - x\| + |\beta - \lambda| \, \|x\| < \epsilon(|\beta| + \|a\| + \epsilon)$, and the arbitrary nature of ϵ implies that scalar multiplication is continuous at (β, a). In particular, the algebraic operations on $\mathbb{K}$ itself are continuous.

Example 8.3.24

Suppose X is a non-empty topological space and $f, g \in \mathbb{K}^X$ are continuous; then the map $x \mapsto (f(x), g(x))$ from X to $\mathbb{K} \times \mathbb{K}$ is continuous because the inverse image of $I \times J$, where I and J are open balls of $\mathbb{K}$, is $f^{-1}(I) \cap g^{-1}(J)$, which is open in X. But $f + g$ is the composition of this map with addition of scalars, and the latter is continuous by 8.3.23; so $f + g$ is also continuous. Similar calculations show that the pointwise product fg is continuous and that λf is continuous for each $\lambda \in \mathbb{K}$; it follows that the subset of $\mathbb{K}^X$ which consists of the continuous functions is a subalgebra of $\mathbb{K}^X$. As a special case of 6.2.32, $B(X, \mathbb{K})$ is also a subalgebra of $\mathbb{K}^X$; endowed with the supremum norm, it is a normed algebra. The intersection of these two algebras with the inherited norm—the algebra of continuous bounded functions from X to $\mathbb{K}$—is denoted by $\mathcal{C}(X)$, or by $\mathcal{C}(X, \mathbb{R})$ or $\mathcal{C}(X, \mathbb{C})$. The constant function $x \mapsto 1$ is in $\mathcal{C}(X)$; moreover, conjugation $f \mapsto \bar{f}$ is an isometric map and therefore certainly continuous; so $\mathcal{C}(X)$ is closed under conjugation. $\mathcal{C}(X)$ is therefore a unital normed algebra with isometric involution. And $\mathcal{C}(X)$ is closed in $B(X, \mathbb{K})$: suppose $f \in B(X, \mathbb{K})$ and $\operatorname{dist}(f, \mathcal{C}(X)) = 0$ and let $\epsilon \in \mathbb{R}^+$ and $z \in X$; then there exist $g \in \mathcal{C}(X)$ with $\|f - g\| < \epsilon$ and $U \in \operatorname{nbd}(z)$ with $|g(u) - g(z)| < \epsilon$ for all $u \in U$; it follows that $|f(u) - f(z)| \leq |f(u) - g(u)| + |g(u) - g(z)| + |g(z) - f(z)| < 3\epsilon$, whence f is continuous at z.

Theorem 8.3.25

Suppose $(X_i)_{i \in I}$ is a family of topological spaces. Let $P = \prod_i X_i$. Then the natural projections $\pi_j \colon P \to X_j$ are all open maps.

Proof

Let B be a basic open subset of P; then, for each $j \in I$, $\pi_j(B)$ is either X_j or some other open subset of X_j. The result follows from 8.1.23. $\qquad\square$

Example 8.3.26

Suppose $(X_i)_{i \in I}$ is a family of topological spaces. For each $z \in \prod_i X_i$ and $j \in I$, there is a bijective restriction $\pi_{j,z}$ of the natural projection π_j which identifies a subset $X_{j,z}$ of the product with X_j (1.2.25). When $X_{j,z}$ is endowed with the subspace topology, the map $\pi_{j,z}$ is clearly a homeomorphism. In fact, we can take this a little further: suppose $J \subseteq I$ and define $X_{J,z}$ to be the product $\prod_{i \in I} V_i$ where $V_i = X_i$ if $i \in J$ and $V_i = \{z_i\}$ otherwise; when $X_{J,z}$ is endowed with the relative topology as a subspace of $\prod_{i \in I} X_i$, it is homeomorphic to the product $\prod_{i \in J} X_i$.

Quotient Spaces

Definition 8.3.27

Suppose X is a topological space and an equivalence relation $\sim$ is defined on X. Let $\pi \colon X \to X/\!\sim$ be the quotient map. Then the final topology determined by π, namely $\mathcal{P} = \{S \subseteq X/\!\sim \mid \pi^{-1}(S) \text{ open in } X\}$, is called the QUOTIENT TOPOLOGY on $X/\!\sim$, and $(X/\!\sim, \mathcal{P})$ is called the QUOTIENT SPACE of X by $\sim$.

Example 8.3.28

Suppose X is a seminormed linear space and M is a linear subspace of X; then the quotient seminorm on X/M induces the quotient topology. The quotient map π is in $\mathcal{B}(X, X/M)$ (6.2.28), so is continuous with respect to the seminorms by 8.1.12; therefore the seminorm topology on X/M is weaker than the quotient topology. Towards the converse, suppose that V is in the quotient topology. Then, by definition, $\pi^{-1}(V)$ is open in X; so, for each $u \in X$ with $u/M \in V$, there exists $r \in \mathbb{R}^+$ such that $\flat_X[u\,;r] \subseteq \pi^{-1}(V)$; then, by 6.2.27, $\flat_{X/M}[u/M\,;r] = \pi(\flat_X[u\,;r]) \subseteq V$. So V is in the seminorm topology on X/M.

Example 8.3.29

Define an equivalence relation on $\mathbb{R}$ by saying that $a \sim b$ if and only if $a - b \in \mathbb{Z}$; then each member of $\mathbb{R}$ is equivalent to its non-integer part. The sets $\{x/\!\sim \mid x \in (a,b)\}$ where $0 \le a < b \le 1$ and $\{x/\!\sim \mid x \in [0,1)\backslash[a,b]\}$ where $0 < a \le b < 1$ together form a base for the quotient topology on $\mathbb{R}/\sim$. An appropriate way to picture this space is by bending the interval $[0, 1]$ into a circle on which 0 and 1 coincide. The base described here for the quotient topology is the collection of so-called OPEN INTERVALS on this circle.

Example 8.3.30

Every continuous image of a separable space is separable (Q 8.3.9); it follows, in particular, that every quotient of a separable space is separable, because the quotient map is continuous and surjective. But quotients of second countable spaces need not even be first countable; this in turn implies that quotients of first countable spaces need not be first countable; it also warns us that continuity does not preserve either of these properties. We shall consider, for an example, the NOVOSAD SPACE N; this space is the quotient of the second countable space $\mathbb{R}^2$ with the product topology, with respect to the equivalence relation given by saying that $(a, b) \sim (c, d)$ if and only if either $(a, b) = (c, d)$ or $b = 0 = d$. To show that this space is not first countable, we proceed as follows. Suppose $\mathcal{U}$ is a countable collection of open neighbourhoods of $\pi(0, 0)$ in N. Index $\mathcal{U}$ so that $\mathcal{U} = \{U_n \mid n \in \mathbb{Z}\}$. Then, for each $n \in \mathbb{Z}$, $\pi^{-1}(U_n)$ is an open subset of $\mathbb{R}^2$ which includes the line $H = \{(x, 0) \in \mathbb{R}^2 \mid x \in \mathbb{R}\}$; it follows that the set $\pi^{-1}(U_n) \cap \{(n, y) \mid y \in \mathbb{R}^+\}$ is non-empty. By the Product Theorem, there exists a family $(\alpha_n)_{n \in \mathbb{Z}}$ in $\mathbb{R}^+$ such that $\pi(n, \alpha_n) \in U_n$ for each $n \in \mathbb{Z}$. Let $V = \{(x, y) \in \mathbb{R}^2 \mid 2y < (1 - t)\alpha_n + t\alpha_{n+1} : n = \lfloor x \rfloor, t = x - n\}$. V is clearly open in $\mathbb{R}^2$ and, because $H \subseteq V$, we have $\pi^{-1}\pi(V) = V$. Since the topology on N is the strong topology determined by π, $\pi(V)$ is an open neighbourhood of $\pi(0, 0)$ in N. But $\pi(V) \cap \{(n, \alpha_n) \mid n \in \mathbb{Z}\} = \varnothing$; so $\pi(V)$ does not include U_n for any $n \in \mathbb{Z}$; therefore $\mathcal{U}$ is not a local base at $\pi(0, 0)$ in N.

Example 8.3.31

Quotients of normal or Hausdorff spaces need not even be T_1. Consider $\mathbb{R}$ with its usual topology; this is a normal space. Define an equivalence relation on $\mathbb{R}$ by setting $a \sim b$ if and only if $\lfloor a \rfloor = \lfloor b \rfloor$. Let π denote the quotient map and let $x \in \mathbb{R}$; then $\pi^{-1}((\mathbb{R}/\sim)\backslash\{x/\sim\}) = \mathbb{R}\backslash[\lfloor x \rfloor, \lfloor x \rfloor + 1)$ which is not open in $\mathbb{R}$; it follows that the singleton set $\{x/\sim\}$ is not closed in the quotient space.

Example 8.3.32

Suppose X is a topological space and $\sim$ is an equivalence relation on X. Suppose S is a non-empty subset of X. Then $S/\sim$ is a subset of $X/\sim$ if and only if S can be expressed as a union of equivalence classes with respect to $\sim$. In this case, there are two natural ways of endowing $S/\sim$ with a topology. $S/\sim$ can be endowed with the relative topology as a subspace of $X/\sim$ or it can be endowed with the quotient topology after S has been endowed with the relative topology as a subspace of X. These two topologies are identical because each is the strong topology determined by $\pi|_S$ (see 8.3.5).

Theorem 8.3.33

Suppose X is a topological space and $\sim$ is an equivalence relation on X. Then the quotient map $\pi\colon X \to X/\sim$ is open if and only if, for each open subset U of X, the set $\{x \in X \mid x \sim u \text{ for some } u \in U\}$ is also open.

Proof

The set $\{x \in X \mid x \sim u \text{ for some } u \in U\}$ is $\pi^{-1}(\pi(U))$, which, by definition of the quotient topology, is open in X if and only if $\pi(U)$ is open in $X/\sim$.   □

Example 8.3.34

Suppose X is a linear space with a translation invariant topology. Suppose M is a linear subspace of X. Then the quotient map $\pi\colon X \to X/M$ is open: for each open subset U of X, $\pi^{-1}(\pi(U)) = U + M = \bigcup\{U + \{m\} \mid m \in M\}$, which is open, whence π is open by 8.3.33.

EXERCISES

Q 8.3.1 Show that the topology on a co-ordinate space of a product space is the final topology determined by the appropriate natural projection.

Q 8.3.2 Suppose X is a normed linear space, Y is a normed linear subspace of X and M is a closed linear subspace of Y. Show that Y is a topological subspace of X and that Y/M is a topological subspace of X/M.

Q 8.3.3 Suppose X is a linear space and $\langle \cdot, \cdot \rangle \colon X \times X \to \mathbb{K}$ is an inner product. Show that $\langle \cdot, \cdot \rangle$ is a continuous function with respect to the product topology on $X \times X$.

Q 8.3.4 Show that ℓ_∞ is the same space as $\mathcal{C}(\mathbb{N})$ when $\mathbb{N}$ is given an appropriate topology. What topology is appropriate? How can this space be expressed as a direct product?

Q 8.3.5 Let X be a T_4 space. Suppose D is dense in X and F is a closed discrete subspace of X. Show that $|F| \leq |D|$.

Q 8.3.6 Show that every subspace of a T_1 space is also T_1. Show that a quotient space is T_1 if and only if all equivalence classes are closed sets in the original space.

Q 8.3.7 Suppose X is a topological space and $f: X \to \mathbb{R}$ is continuous. Let $N = \{x \in X \mid f(x) \neq 0\}$ and suppose $N \neq \varnothing$. Show that the function $1/f$, defined on N by $x \mapsto 1/f(x)$, is continuous on N.

Q 8.3.8 In contrast to Q 8.1.1, give an example to show that the image of a closed set under an open map is not necessarily closed.

Q 8.3.9 Show that every continuous image of a separable space is separable. Give an example of a continuous image of a second countable space which is not even first countable.

8.4 Topologies on Linear Spaces

A linear space X may be regarded as the union of its lines through the origin. Each such line L of X has a natural topology—that generated by its open line segments—which makes it into a homeomorphic copy of $\mathbb{R}$ by a homeomorphism which is also an order isomorphism of the type $\lambda \mapsto \lambda v$ defined on $\mathbb{R}$ with $v \in L\backslash\{0\}$. In endowing X with a topology, it does not seem unreasonable to ask at the very least that these maps all be continuous; we shall call any topology which ensures their continuity a LINE TOPOLOGY; the strongest such topology, namely the final topology determined by the stated maps, is called here the STRONG LINE TOPOLOGY and is denoted by $\mathcal{SL}$. We mention $\mathcal{SL}$ here because we have already covertly considered aspects of it: if $S \subseteq X$, then a point z of S is an inside point of S if and only if 0 is an interior point of $S - \{z\}$ in the subspace $\langle S - \{z\}\rangle$ of $(X, \mathcal{SL})$. So, for example, we have a special Hahn–Banach Separation Theorem (5.2.31) in this context: *suppose X is a real linear space with a translation invariant line topology and A and U are non-empty disjoint convex subsets of X with U is open in X; then there is a hyperplane of X which separates U from A.*

Line topologies need not be translation invariant; the weak topology determined by a norm on a non-trivial space is not (Q 8.2.5). For analytic purposes, we generally require considerably more—in particular, the continuity of the algebraic operations. Any topology which ensures such continuity is easily seen to be a translation invariant line topology.

Definition 8.4.1
Suppose X is a linear space. A topology $\mathcal{O}$ on X is called a VECTOR TOPOLOGY on X if and only if addition, $(a, b) \mapsto a + b$ defined on $X \times X$, and scalar multiplication, $(\lambda, a) \mapsto \lambda a$, defined on $\mathbb{K} \times X$, are both continuous functions, where the topologies on $X \times X$ and on $\mathbb{K} \times X$ are the usual product topologies.

Endowed with a vector topology, X is called a TOPOLOGICAL LINEAR SPACE. If, moreover, every neighbourhood of 0 includes a convex neighbourhood of 0, then $\mathcal{O}$ is called a LOCALLY CONVEX topology and $(X, \mathcal{O})$ is called a LOCALLY CONVEX SPACE.

Example 8.4.2

According to our definition, all seminorm topologies are vector topologies (8.3.23) and are, indeed, locally convex because balls determined by seminorms are convex (6.2.12). But metric topologies on linear spaces need not be vector topologies; the metric topology considered in 6.1.13, for example, is not translation invariant.

Dual Spaces

Although there is a much more general theory of duality, we restrict our attention here to normed linear spaces. Even in this context, there are interesting non-norm topologies.

Definition 8.4.3

Suppose X is a normed linear space. The set of continuous linear functionals on X is called the DUAL of X and is denoted by X^*.

The continuous linear functionals on X are precisely the bounded ones (8.1.12); so X^* is normed by $f \mapsto \sup\{|f(x)| \mid \|x\| \le 1\}$ and is then identical to the space $\mathcal{B}(X, \mathbb{K})$.

Since the members of X^* are continuous functions, the topology on X includes the weak topology on X determined by X^*; but these topologies usually differ and the latter is called simply the WEAK TOPOLOGY on X; it is identical to the weak topology on X determined by any basis of X^* because linear combinations of continuous scalar functions are continuous (8.3.24).

X^*, being a normed linear space, has its own dual X^{**} which can also be normed after the same fashion as X^*. This SECOND DUAL of X is not quite as abstract as it may seem; indeed, we are already familiar with some of its members, namely the point evaluation functions $\hat{x} \colon X^* \to \mathbb{K}$ given, for $x \in X$, by $f \mapsto f(x)$. These maps are clearly linear and satisfy $\|\hat{x}\| \le \|x\|$; it is then a consequence of 6.2.42 that $\|\hat{x}\| = \|x\|$. It follows easily that the collection of them, which we denote here by $\{\hat{x} \mid x \in X, \operatorname{dom}(\hat{x}) = X^*\}$, is a normed linear subspace of X^{**} and that the map $x \mapsto \hat{x}$ is an isometric isomorphism of X onto its range. So these point evaluation functionals form an isometric isomorphic copy of X included in X^{**}. It may even happen that the point evaluation functionals are the only bounded linear functionals on X^*; the space X is said to be REFLEXIVE if and only if $X^{**} = \{\hat{x} \mid x \in X, \mathrm{dom}(\hat{x}) = X^*\}$.

X^* has its standard norm topology. But, being a normed linear space, it can also be endowed with the weak topology determined by its dual X^{**}. Moreover, a third natural topology now reveals itself, namely the topology determined by the distinguished subspace $\{\hat{x} \mid x \in X, \mathrm{dom}(\hat{x}) = X^*\}$ of X^{**}—or, equivalently, by a basis of this subspace (8.3.24). It is called the WEAK* TOPOLOGY on X^*. And this topology is also already well known to us, for it is the relative topology on X^* as a subspace of the product space $\mathbb{K}^X$. This follows from the fact (8.3.17) that the product topology on $\mathbb{K}^X$ is the weak topology determined by $\{\hat{x} \mid x \in X, \mathrm{dom}(\hat{x}) = \mathbb{K}^X\}$, and the fact (8.3.5) that the subspace topology is the weak topology determined by the appropriate restrictions of these maps, in this case their restrictions to X^*. It is easy to check that the weak and weak* topologies are locally convex. By Q 8.2.4, these topologies are regular; moreover, also with reference to Q 8.2.4, the Hahn–Banach Theorem ensures that $\bigcap\{\ker(f) \mid f \in X^*\} = \{0\}$ and hence that X^* separates the points of X, and it is trivial that $\{\hat{x} \mid x \in X, \mathrm{dom}(\hat{x}) = X^*\}$ separates the points of X^*, so the weak and weak* topologies are Hausdorff. But they need not be normal.

Finally, X^{**} has its own dual X^{***} which includes an isometric isomorphic copy of X^* consisting of the point evaluation functions $\hat{f}$ corresponding to each $f \in X^*$. The weak* topology on X^{**} is the weak topology determined by these functions; it is often loosely referred to as the weak topology on X^{**} determined by X^*. Note that because the weak* topology on X^{**} is determined by a copy of X^*, the isometric isomorphism $x \mapsto \hat{x}$ of X onto its copy in X^{**} is also a homeomorphism from X with its weak topology onto the copy of X as a subspace of X^{**} with its weak* topology.

Theorem 8.4.4

Suppose X is a normed linear space. Then $\{\hat{x} \mid x \in X, \mathrm{dom}(\hat{x}) = X^*\}$ is dense in X^{**} with the weak* topology.

Proof

Let $\mathcal{F}$ be a Hamel basis for X^*. By 8.3.24, $\{\hat{f} \in X^{***} \mid f \in \mathcal{F}\}$ is a subbase for the weak* topology on X^{**}. Suppose W is a non-empty member of this topology. Then there exist a finite sequence $(f_i)_{i \in \mathbb{I}}$ in $\mathcal{F}$ and a corresponding sequence $(U_i)_{i \in \mathbb{I}}$ of non-empty open subsets of $\mathbb{K}$ such that $\bigcap\{\hat{f}_i^{-1}(U_i) \mid i \in \mathbb{I}\} \subseteq W$. By 3.4.17, there exists $v \in X$ such that $\hat{f}_i(\hat{v}) = f_i(v) \in U_i$ for all $i \in \mathbb{I}$, yielding $\hat{v} \in W$.   □

Corollary 8.4.5

Suppose X is a normed linear space and $\phi \in \mathcal{L}(X^{**}, \mathbb{K})$. Then ϕ is continuous with respect to the weak* topology on X^{**} if and only if $\phi \in \{\hat{f} \mid f \in X^*\}$.

Proof

The functionals $\hat{f}$ for $f \in X^*$ determine the weak* topology on X^{**}, so are certainly continuous. Suppose ϕ is weak*-continuous; then $\phi \in X^{***}$ because the weak* topology is weaker than the norm topology; therefore the restriction of ϕ to the copy of X in X^{**} is norm-continuous; this restriction is therefore identical to the restriction of $\hat{f}$ for some $f \in X^*$. Now $\hat{f}$ is continuous with respect to the weak* topology on X^{**}, by definition of the topology; so $\phi - \hat{f}$ is weak*-continuous. But $\phi - \hat{f}$ is 0 on the copy of X in X^{**}; since this copy is weak*-dense in X^{**} (8.4.4), it follows that $\phi = \hat{f}$.   □

Example 8.4.6

Suppose that $p \in [1, \infty)$ and that $q = \infty$ if $p = 1$ and $1/p + 1/q = 1$ otherwise. Then ℓ_p^* is isometrically isomorphic to ℓ_q. This can be proved as follows. For each $y \in \ell_q$ and $x \in \ell_p$, the series $\sum_{n \in \mathbb{N}} x_n y_n$ converges absolutely, satisfying Hölder's inequality (4.4.5); therefore it converges with limit not exceeding $\|x\|_p \|y\|_q$. For each $y \in \ell_q$, let ψ_y be the map $x \mapsto \sum_{n=1}^{\infty} x_n y_n$ defined on ℓ_p; then $|\psi_y(x)| \leq \|x\|_p \|y\|_q$ and it is easily checked that ψ_y is linear, so that $\psi_y \in \ell_p^*$ with $\|\psi_y\| \leq \|y\|_q$. The map ψ is clearly injective. We now show that ψ is surjective onto ℓ_p^*. Let $g \in \ell_p^*$ and set $w = (g(e_i))$, where, for each $i \in \mathbb{N}$, e_i is the sequence whose i^{th} term is 1 and whose other terms are all 0. We claim that $w \in \ell_q$ and that $\|w\|_q \leq \|g\|$. This is certainly true if $p = 1$. Otherwise, we let u be the sequence for which, for each $i \in \mathbb{N}$, $u_i = 0$ if $g(e_i) = 0$ and $u_i = |g(e_i)|^q / g(e_i)$ otherwise; and, for each $n \in \mathbb{N}$, let v_n be the sequence whose first n terms agree with those of u and whose other terms are all zero. Then $\|v_n\|_p^p = \sum_{i=1}^n |u_i|^p = \sum_{i=1}^n |g(e_i)|^q = \sum_{i=1}^n u_i g(e_i) = g(v_n) \leq \|g\| \|v_n\|_p$, whence $\sum_{i=1}^n |g(e_i)|^q = \|v_n\|_p^p \leq \|g\|^q$; since this is true for all $n \in \mathbb{N}$, we have $w \in \ell_q$ and $\|w\|_q \leq \|g\|$ in this case also. Since $(g - \psi_w)e_i = 0$ for all $i \in \mathbb{N}$, it is now an easy matter to check that $g = \psi_w$. Lastly, we have shown both $\|\psi_w\| \leq \|w\|_q$ and $\|w\|_q \leq \|g\|$; since $g = \psi_w$, these yield $\|\psi_w\| = \|w\|_q$.

Example 8.4.7

For each $p \in (1, \infty)$, the space ℓ_p is reflexive. The map $x \mapsto \hat{x}$ is is certainly a homomorphism, and we showed above, using 6.2.42, that it is isometric; so we need only show that it is surjective. Let ψ be the isometric isomorphism from ℓ_q onto ℓ_p^* displayed in 8.4.6, and let ϕ be the corresponding one from ℓ_p onto

ℓ_q^*. Suppose $g \in \ell_p^{**}$; then $g \circ \psi \in \ell_q^*$; so there exists $x \in \ell_p$ such that $g \circ \psi = \phi_x$. Then, for each $y \in \ell_q$, we have $g(\psi_y) = \phi_x(y) = \sum_{n=1}^{\infty} x_n y_n = \psi_y(x) = \hat{x}(\psi_y)$, and $g = \hat{x}$, as required.

Theorem 8.4.8

Every normed linear space with separable dual is separable.

Proof

Suppose X is a normed linear space and X^* is separable; let (f_n) be a dense sequence in X^*. By the Product Theorem, there exists a sequence (x_n) of unit norm vectors in X with $|f_n(x_n)| \geq \|f_n\|/2$ for each $n \in \mathbb{N}$. Let Y be the subset of X consisting of all linear combinations of the x_n with coefficients which have rational parts. Then Y, being a countable union of countable sets, is countable by 2.4.5. Suppose $f \in X^*$ satisfies $f(\overline{Y}) = \{0\}$. Let $\epsilon \in \mathbb{R}^+$. There exists $m \in \mathbb{N}$ such that $\|f - f_m\| < \epsilon$; thus $|f_m(x_m)| = |(f - f_m)(x_m)| \leq \|f - f_m\| < \epsilon$. Since $|f_m(x_m)| \geq \|f_m\|/2$, this yields $\|f_m\| < 2\epsilon$ and then $\|f\| < 3\epsilon$. Since ϵ is arbitrary, we have $f = 0$. Then $\overline{Y} = X$ by 6.2.40.   □

Example 8.4.9

By 8.4.8, ℓ_∞^* is not isometrically isomorphic to ℓ_1, because ℓ_1 is separable and ℓ_∞ is not (7.3.6). In fact, ℓ_1 is isometrically isomorphic to c_0^* (Q 8.4.5).

EXERCISES

Q 8.4.1 Suppose X is a linear space. Show that every proper linear subspace Y of X is $\mathcal{SL}$-closed.

Q 8.4.2 Show that the $\mathcal{SL}$-closure of a cone is not necessarily a cone.

Q 8.4.3 Suppose X is a linear space and $\mathcal{F}$ is a collection of linear maps from X into topological linear spaces. Show that the weak topology determined on X by $\mathcal{F}$ makes X into a topological linear space.

Q 8.4.4 Suppose X is a real linear space with a translation invariant line topology and $f \in \mathcal{L}(X, \mathbb{R})$. Show that f is continuous if and only if wedge (f) includes a non-empty open subset of X.

Q 8.4.5 Try to modify the proof of 8.4.6 for $p = \infty$ and $q = 1$ and discover where it ultimately fails. Hence verify that c_0^* is isometrically isomorphic to ℓ_1 and deduce that c_0 is not reflexive.

Q 8.4.6 Find a separable normed linear space with inseparable dual.

9
Connectedness

We still have slept together,
Rose at an instant, learn'd, play'd, eat together;
And wheresoe'er we went, like Juno's swans,
Still we went coupled and inseparable.　　　*As You Like It, I,iii.*

Let us now return to the phenomenon of the empty boundary, which we have alluded to earlier as a criterion for *connectedness*. We shall examine briefly the relationship between this and two other concepts of connectedness.

9.1 Connected Spaces

Definition 9.1.1
Suppose X is a topological space. X is said to be DISCONNECTED if and only if there exist non-empty open subsets U and V of X such that $\{U, V\}$ is a partition of X; then $\{U, V\}$ will be called a DISCONNECTION of X. X is said to be CONNECTED if and only if X is not disconnected.

Theorem 9.1.2
Suppose X is a topological space. Then the following are equivalent:

- X is connected;
- X has no proper non-empty subset whose boundary is empty;
- the only subsets of X which are both open and closed in X are $\varnothing$ and X.

Proof
Firstly, suppose U is a subset of X with $\partial U = \varnothing$; then $U = U^\circ$ and $U^c = U^{c\circ}$ are both open; so, if X is connected, either $U = \varnothing$ or $U = X$. Secondly, suppose

V is a subset of X which is both open and closed in X; then $\partial V \subseteq V \cap V^c$, so that $\partial V = \varnothing$. Thirdly, suppose that no proper non-empty subset of X is both open and closed in X and that A and B are open subsets of X such that $\{A, B\}$ is a partition of X; then $A = B^c$ is closed in X; so either $A = \varnothing$ or $A = X$; therefore X is connected. $\square$

Example 9.1.3

$\mathbb{R}$ is a connected space. Let V be any proper non-empty subset of $\mathbb{R}$ and suppose $z \in \mathbb{R}\backslash V$. Set $A = \{r \in V \mid r < z\}$ and $B = \{r \in V \mid z < r\}$. Then either A or B is non-empty. If $A \neq \varnothing$ then $\sup A$ is a boundary point of V in $\mathbb{R}$; similarly, if $B \neq \varnothing$, then $\inf B$ is a boundary point of V in $\mathbb{R}$.

Connected Subsets

Definition 9.1.4

A subset Y of a topological space X is called a CONNECTED SUBSET of X if and only if the subspace Y of X is itself a connected space.

Example 9.1.5

Every non-empty connected subset of $\mathbb{R}$ is an interval, for, if $K \subseteq \mathbb{R}$ and K is not an interval, then there exists $x \in \mathbb{R}\backslash K$ such that $\inf K < x < \sup K$, and it follows that $\{K \cap (-\infty, x), K \cap (x, \infty)\}$ is a disconnection of K. The converse is true, but a little harder to establish. Suppose that I is an interval of $\mathbb{R}$, that U is a non-empty open subset of I and that $I\backslash U$ is also open in I. Let $u \in U$. Let $a = \inf\{y \in I \mid [y, u] \subseteq U\}$ and $b = \sup\{y \in I \mid [u, y] \subseteq U\}$. Because $I\backslash U$ is open in I, we have $a \notin I\backslash U$ and $b \notin I\backslash U$. Since U is open in I, it follows that $a = \inf I$ and $b = \sup I$. So $I\backslash U = \varnothing$.

Theorem 9.1.6

Suppose X is a topological space and $S \subseteq X$. Then S is a connected subset of X if and only if, for every pair U and V of open subsets of X for which $S \subseteq U \cup V$ and $S \cap U \cap V = \varnothing$, we have either $S \cap U = \varnothing$ or $S \cap V = \varnothing$.

Proof

If S is not a connected space, then there exists a disconnection $\{A, B\}$ of S; since A and B are open in S, there exist U and V open in X with $A = S \cap U$ and $B = S \cap V$; then $S = A \cup B \subseteq U \cup V$ and $S \cap U \cap V = A \cap B = \varnothing$ and neither $S \cap U$ nor $S \cap V$ is empty. For the converse, suppose there exist open subsets U and V of X for which $S \subseteq U \cup V$ and $S \cap U \cap V = \varnothing$, with $S \cap U \neq \varnothing$ and $S \cap V \neq \varnothing$; then, clearly, $\{S \cap U, S \cap V\}$ is a disconnection of S. $\square$

Corollary 9.1.7

Suppose that X is a topological space, that S is a connected subset of X and that $A \subseteq X$ with $S \subseteq A \subseteq \overline{S}$. Then A is connected.

Proof

Suppose U and V are open subsets of X with $A \subseteq U \cup V$ and $A \cap U \cap V = \varnothing$. Then $S \subseteq U \cup V$ and $S \cap U \cap V = \varnothing$, and, since S is connected, either $S \cap U = \varnothing$ or $S \cap V = \varnothing$ by 9.1.6; it follows, since U and V are open, that either $U \subseteq X \backslash \overline{S}$ or $V \subseteq X \backslash \overline{S}$, whence either $A \cap U = \varnothing$ or $A \cap V = \varnothing$ because $A \subseteq \overline{S}$. So A is connected by 9.1.6. $\square$

Theorem 9.1.8

Suppose X is a topological space and $\mathcal{C}$ is a collection of connected subsets of X with non-empty intersection. Then $\bigcup \mathcal{C}$ is connected.

Proof

Let $S = \bigcap \mathcal{C}$. Then $S \neq \varnothing$ by hypothesis. Suppose $\bigcup \mathcal{C} = U \cup V$ where U and V are disjoint and open in $\bigcup \mathcal{C}$. We claim that $S \cap V \neq \varnothing \Rightarrow U = \varnothing$. Suppose $S \cap V \neq \varnothing$. For each $A \in \mathcal{C}$, the set $\{A \cap U, A \cap V\}$ is a partition of A into open subsets of A. Since A is connected, one or other of these subsets is empty. But $A \cap V \neq \varnothing$, because $S \subseteq A$ and $S \cap V \neq \varnothing$; therefore $A \cap U = \varnothing$. Since $A \in \mathcal{C}$ is arbitrary, we have $\bigcup \mathcal{C} \cap U = \varnothing$ and hence $U = \varnothing$, and our claim is proved. A similar calculation shows that $S \cap U \neq \varnothing \Rightarrow V = \varnothing$. Then, since at least one of $S \cap U$ and $S \cap V$ is not empty, either $U = \varnothing$ or $V = \varnothing$. Therefore $\bigcup \mathcal{C}$ is connected. $\square$

Corollary 9.1.9

Suppose X is a topological space and $x \in X$. Suppose that, for every $y \in X$, there is a connected subset S of X with $\{x, y\} \subseteq S$. Then X is connected.

Connected Components

Every topological space can be partitioned into its *connected components*, which are the maximal connected subsets of the space; for a connected space, there is, of course, just one component.

Definition 9.1.10

Suppose X is a topological space. A connected subset of X is called a CON-NECTED COMPONENT of X if and only if it is not properly included in any connected subset of X.

Theorem 9.1.11

Suppose X is a topological space. The connected components of X are closed and form a partition of X.

Proof

The components are closed by 9.1.7; also, they are mutually disjoint: suppose U and V are connected components of X and $U \cap V \neq \varnothing$; then $U \cup V$ is connected by 9.1.8, so maximality of U and V ensures that $U = U \cup V = V$. Lastly, we suppose $x \in X$ and show that there is a connected component of X to which x belongs. Let $\mathcal{C}$ be the collection of all connected subsets of X to which x belongs. Since $\{x\}$ is connected, we have $x \in \bigcap \mathcal{C}$. Then $\bigcup \mathcal{C}$ is connected by 9.1.8. The maximality of $\bigcup \mathcal{C}$ as a connected subset of X follows from its definition. So $\bigcup \mathcal{C}$ is a connected component of X which contains x.   □

Example 9.1.12

Are the connected components of a topological space X characterized as the minimal non-empty subsets of X which are both open and closed in X? Certainly, such subsets are connected components of X; and, if the number of components is finite, then the fact that they are all closed implies that each component is also open. The proposition is, however, not true in general. Endow $\mathbb{Q}$ with its usual topology induced from $\mathbb{R}$. Suppose $C \subseteq \mathbb{Q}$ and $|C| > 1$. Let $a, b \in C$ with $a < b$. By 4.2.7, there exists $r \in \mathbb{R} \backslash \mathbb{Q}$ with $a < r < b$, whence $\{C \cap (-\infty, r), C \cap (r, \infty)\}$ is a disconnection of C. So $\mathbb{Q}$ is TOTALLY DISCONNECTED, in the sense that its connected components are its singleton subsets. In particular, $\mathbb{Q}$ is not connected and no component is open.

Continuity and Connectedness

Theorem 9.1.13

Every continuous image of a connected space is connected.

Proof

Suppose X and Y are topological spaces, X is connected, and $f: X \to Y$ is continuous. Suppose $f(X) \subseteq A \cup B$ where A and B are both open in Y and

$f(X) \cap A \cap B = \varnothing$. Then $f^{-1}(A) \cap f^{-1}(B) = \varnothing$. Also $X = f^{-1}(A) \cup f^{-1}(B)$, and these sets are open because f is continuous. Since X is connected, we have either $f^{-1}(A) = \varnothing$ or $f^{-1}(B) = \varnothing$. So either $f(X) \cap A = \varnothing$ or $f(X) \cap B = \varnothing$. Therefore $f(X)$ is connected by 9.1.6. □

Example 9.1.14
Suppose I is an interval of $\mathbb{R}$ and f is a continuous real function defined on I. Then, since I is connected, so is $f(I)$. Therefore $f(I)$ is an interval of $\mathbb{R}$, by 9.1.5; this is the INTERMEDIATE VALUE THEOREM.

Connectedness of Derived Spaces

It is clear that subspaces of connected spaces need not be connected. But both quotients and products of connected spaces are always connected.

Theorem 9.1.15
Every quotient of a connected space is connected.

Proof
This follows from 9.1.13 because the quotient map is continuous. □

Lemma 9.1.16
Every finite product of connected spaces is connected.

Proof
We prove the result for two spaces; the general result follows by induction. Suppose X and Y are connected topological spaces; we assume them non-empty, for otherwise the result is trivial. For each $y \in Y$, the subspace $X \times \{y\}$ of $X \times Y$ is homeomorphic to X (8.3.26) and therefore connected by 9.1.13; similarly, for each $x \in X$, $\{x\} \times Y$ is connected. Let $(a, b), (c, d) \in X \times Y$. Let C be the connected component of $X \times Y$ which contains (a, b); then the connected set $\{a\} \times Y$ is included in C; so $(a, d) \in C$, and therefore C is the connected component of $X \times Y$ which contains (a, d); then C in turn includes $X \times \{d\}$, whence $(c, d) \in C$. Since (c, d) is arbitrary in $X \times Y$, we have $C = X \times Y$. □

Theorem 9.1.17
Every product of connected spaces is connected.

Proof

Suppose $(X_i)_{i \in I}$ is a family of connected topological spaces and set $P = \prod_i X_i$. If $P = \varnothing$, the result is trivial; so we suppose otherwise and let $z \in P$. Let $\mathcal{C}$ denote the collection of all finite subsets of I. For each $K \in \mathcal{C}$, let $X_{K,z} = \prod_{i \in I} W_i$, where $W_i = X_i$ for each $i \in K$ and $W_i = \{z_i\}$ for each $i \in I \backslash K$. Then $X_{K,z}$ is homeomorphic to $\prod_{i \in K} X_i$ (8.3.26) and therefore connected by 9.1.13 and 9.1.16. Moreover $z \in \bigcap \{X_{K,z} \mid K \in \mathcal{C}\}$. Therefore, by 9.1.8, $\bigcup \{X_{K,z} \mid K \in \mathcal{C}\}$ is connected. By construction, this union has non-empty intersection with every basic open subset of P; it is therefore dense in P by 7.2.20. So the component to which z belongs, being closed, is P itself.   □

Example 9.1.18

$\mathbb{C}$, being homeomorphic to $\mathbb{R}^2$, is connected; and, for any set X, both $\mathbb{C}^X$ and $\mathbb{R}^X$ are connected. It follows that, if $X \neq \varnothing$, then the range of any continuous real function $f \colon \mathbb{K}^X \to \mathbb{R}$ is an interval.

EXERCISES

Q 9.1.1 Give $\mathbb{R}$ the topology generated by $\mathcal{C} = \{(a, b] \mid a, b \in \mathbb{R}, \, a < b\}$. Show that the resulting space is totally disconnected (9.1.12).

Q 9.1.2 Can a space with a particular point topology be disconnected?

Q 9.1.3 The CANTOR SET is described as follows. From the closed interval $I_0 = [0, 1]$, delete the open middle third $(\frac{1}{3}, \frac{2}{3})$. Call the resulting set I_1; it is the union of two disjoint closed intervals. Delete the open middle third from each of these two closed intervals to get a new set, called I_2, which is the union of four disjoint closed intervals. Continue this process, inductively defining I_n for each $n \in \mathbb{N}$ to be the union of the 2^n disjoint closed intervals remaining after the nth bout of deletion. The Cantor set is defined to be $C = \bigcap (I_n)_{n \in \mathbb{N}}$. Show that the Cantor set is totally disconnected (9.1.12).

Q 9.1.4 What is the error in this 'proof' that a product $P = \prod_{i \in I} X_i$ of connected topological spaces is connected. Suppose $\{U, V\}$ is a partition of P with U and V open in P. For each $j \in I$, the projection π_j is an open map. So, since $\{\pi_j(U), \pi_j(V)\}$ cannot be a disconnection of X_j, we have $\pi_j(U) = \varnothing$ or $\pi_j(U) = X_j$. Since j is arbitrary, $U = \varnothing$ or $U = P$.

Q 9.1.5 Show that a topological space X is connected if and only if there is no continuous function from X onto the discrete space $\{0, 1\}$.

9.2 Pathwise Connectedness

Definition 9.2.1

Suppose X is a topological space. A continuous function $f:[0,1] \to X$ is called an ARC in X; the points $f(0)$ and $f(1)$ are called the ENDPOINTS of the arc and the image $f[0,1]$ of f is called a PATH in X. X is said to be PATHWISE CONNECTED if and only if, for every $x, y \in X$, there exists an arc in X with endpoints x and y.

Theorem 9.2.2

Every path is connected.

Proof

The interval $[0,1]$ is connected by 9.1.5. So every continuous image of $[0,1]$ is also connected by 9.1.13. □

Theorem 9.2.3

Every pathwise connected topological space is connected.

Proof

Suppose X is a pathwise connected topological space and $x \in X$. Then, for each $y \in X$, there is an arc with endpoints x and y. Its path is connected by 9.2.2. Let $\mathcal{C}$ be the collection of all such paths for all $y \in X$. Since $x \in \bigcap \mathcal{C}$, it follows from 9.1.9 that X is connected. □

Polygonal Connectedness

Polygonal connectedness (5.2.8) does not depend on a topology. But when a linear space is endowed with any translation invariant line topology (8.4), then every polygonally connected subset is pathwise connected (9.2.5). The converse of this is not true (Q 9.2.2). But if we restrict ourselves to considering open subsets of locally convex spaces, we get a complete identification of the three concepts of connectedness we have discussed (9.2.6).

Theorem 9.2.4

Suppose X is a linear space endowed with a translation invariant line topology. Then every polygon in X is a path.

Proof

Suppose P is a polygon in X. If P is a singleton, then a constant function determines that P is a path. Suppose that $(a_i)_{i \in \mathbb{I}}$ is a finite sequence in P of length $n \in \mathbb{N} \backslash \{1\}$ which satisfies $P = \bigcup \{[a_{i-1}, a_i] \mid i \in \mathbb{I} \backslash \{1\}\}$. Then define $f \colon [0,1] \to X$ by $f((k - 2 + t)/(n-1)) = (1-t)a_{k-1} + ta_k$ for each $t \in [0,1)$ and $k \in \mathbb{I} \backslash \{1\}$, and $f(1) = a_n$. If U is any open subset of X, then $f^{-1}(U) = \bigcup \{V_i \mid 1 \le i < n\}$ where V_i is open in $[(i-1)/(n-1), i/(n-1)]$ for each $i \in \mathbb{I} \backslash \{n\}$; and $(i-1)/(n-1) \in V_{i-1}$ if and only if $(i-1)/(n-1) \in V_i$ for each $i \in \mathbb{I} \backslash \{1, n\}$. It follows that $f^{-1}(U)$ is open in $[0,1]$. So f is continuous and its image $f[0,1] = P$ is a path. $\square$

Corollary 9.2.5

Suppose X is a linear space endowed with a translation invariant line topology. Every polygonally connected subset of X is pathwise connected and therefore connected by 9.2.3.

Theorem 9.2.6

Suppose X is a locally convex linear space and A is an open subset of X. Then A is connected if and only if A is polygonally connected.

Proof

One implication follows from 9.2.5. Towards the other, suppose A is connected. If $A = \varnothing$, the result is trivial; we suppose otherwise. Let $z \in A$. Let U be the set of all members u of A for which there is a polygon in A which includes $\{z, u\}$. Suppose $a \in A$; since A is open, there exists a convex open neighbourhood B of a included in A. We show that either $B \subseteq U$ or $B \subseteq A \backslash U$ and hence that both A and $A \backslash U$ are open in A. For each $x \in B$, we have $[x, a] \subseteq B \subseteq A$. If $a \in U$ and P is a polygon in A which includes $\{z, a\}$, then, by 5.2.7, $[x, a] \cup P$ is a polygon in A which includes $\{z, x\}$; hence $x \in U$, yielding $B \subseteq U$, as required. On the other hand, if $a \in A \backslash U$, then $x \in A \backslash U$, for otherwise there would be a polygon Q in A with $\{z, x\} \subseteq Q$, and then the polygon $[a, x] \cup Q$ would include $\{z, a\}$; so, in this case, $B \subseteq A \backslash U$, as required. Therefore U and $A \backslash U$ are both open in A. Since A is connected and $U \ne \varnothing$, we must have $U = A$. But z is arbitrary in A. So A is polygonally connected. $\square$

EXERCISES

Q 9.2.1 Find a connected topological space which is not pathwise connected.

Q 9.2.2 Find a subset of $\mathbb{C}$ which is pathwise, but not polygonally, connected.

10

Convergence

There were drawn
Upon a heap a hundred ghastly women,
Transformed with their fear, who swore they saw
Men, all in fire, walk up and down the streets.
And yesterday the bird of night did sit
Even at noon-day, upon the market-place,
Hooting and shrieking. When these prodigies
Do so conjointly meet, let not men say
'These are their reasons—they are natural',
For I believe they are portentous things
Unto the climate that they point upon.   *Julius Caesar, i,iii.*

Elementary applications of the theory of convergence involve sequences in metric spaces. But there are applications of convergence techniques in non-metric spaces and there are those for which convergence of sequences does not suffice.

10.1 Filters

There are two standard equivalent methods for obtaining a satisfactory theory of convergence in arbitrary topological spaces. The first involves generalizing the concept of a sequence to that of a *net*; the second involves generalizing the notion of a set of tails of a sequence to that of a *filter*. It is the second approach which we adopt here. We introduce the following notation: suppose X is a set and $\mathcal{C} \subseteq \mathcal{P}(X)$; then $[\mathcal{C}]$ will denote the set of all supersets in X of members of $\mathcal{C}$, namely $\{S \subseteq X \mid \exists A \in \mathcal{C} : A \subseteq S\}$; if it is not clear from the context which superset is intended, we shall use a suffix as in $[\mathcal{C}]_X$. Of course $[[\mathcal{C}]] = [\mathcal{C}]$.

Definition 10.1.1

Suppose X is a set and $\mathcal{F}$ is a non-empty subset of $\mathcal{P}(X)$.

- $\mathcal{F}$ is called a FILTER in X if and only if $\varnothing \notin \mathcal{F}$, $\mathcal{F}$ is closed under non-trivial finite intersections and $[\mathcal{F}] = \mathcal{F}$.

- $\mathcal{F}$ is said to have the FINITE INTERSECTION PROPERTY if and only if, for every non-empty finite subset $\mathcal{C}$ of $\mathcal{F}$, we have $\bigcap \mathcal{C} \neq \varnothing$.

215

Example 10.1.2

Every filter has the Finite Intersection Property.

Theorem 10.1.3

Suppose X is a set and $\mathcal{F}$ is a non-empty collection of subsets of X. Then $[\mathcal{F}]$ is a filter in X if and only if $\varnothing \notin \mathcal{F}$ and $\mathcal{F}$ is closed under intersection.

Proof

If $[\mathcal{F}]$ is a filter, then certainly $\mathcal{F}$ has the two stated properties, by definition. Towards the converse, suppose that $\mathcal{F}$ has those properties and that $A, B \in [\mathcal{F}]$; then there exist $U, V \in \mathcal{F}$ such that $U \subseteq A$ and $V \subseteq B$, whence $U \cap V \in \mathcal{F}$ and $U \cap V \subseteq A \cap B$, yielding $A \cap B \in [\mathcal{F}]$. The Principle of Finite Induction is invoked to complete the proof. $\square$

Example 10.1.4

Suppose X is a non-empty set and (x_n) is a sequence in X. For each $n \in \mathbb{N}$, let t_n denote the nth tail $\{x_m \mid n \leq m\}$ of (x_n). Then, by 10.1.3, $[\{t_n \mid n \in \mathbb{N}\}]$ is a filter in X. This filter will be called the FILTER DETERMINED BY THE SEQUENCE (x_n).

Example 10.1.5

Suppose X is a non-empty set and A is a non-empty subset of X. Then the collection $[\{A\}] = \{S \subseteq X \mid A \subseteq S\}$ is a filter in X. In particular, for each $x \in X$, the filter $[\{\{x\}\}] = \{S \subseteq X \mid x \in S\}$, consisting of all subsets of X which contain x, is called the POINT FILTER at x.

Example 10.1.6

Suppose X is a topological space and $x \in X$. Then, since $\mathrm{nbd}(x)$ is closed under intersection and $\varnothing \notin \mathrm{nbd}(x)$, it satisfies the conditions of 10.1.3. So $\mathrm{Nbd}(x) = [\mathrm{nbd}(x)]$ is a filter in X. $\mathrm{Nbd}(x)$ is called the NEIGHBOURHOOD FILTER at x.

Images of Filters

Theorem 10.1.7

Suppose X and Y are non-empty sets, $\mathcal{F}$ is a filter in X and $f: X \to Y$. Then $[f(\mathcal{F})]$ is a filter in Y and $f(\mathcal{F})$ is a filter in $f(X)$.

Proof

Suppose $A, B \in [f(\mathcal{F})]$; then there exist $C, D \in \mathcal{F}$ such that $f(C) \subseteq A$ and $f(D) \subseteq B$, whence $f(C \cap D) \subseteq f(C) \cap f(D) \subseteq A \cap B$. So $A \cap B \in [f(\mathcal{F})]$ because $C \cap D \in \mathcal{F}$. Also $\varnothing \notin f(\mathcal{F})$, because $\varnothing \notin \mathcal{F}$. So, by 10.1.3, $[f(\mathcal{F})]$ is a filter in Y. This is true in particular if $Y = f(X)$; so, to complete the proof we must show that every subset of $f(X)$ which has a subset in $f(\mathcal{F})$ is itself a member of $f(\mathcal{F})$. Towards this, suppose that $S \subseteq f(X)$ and that there exists $G \in \mathcal{F}$ such that $f(G) \subseteq S$. Then $G \subseteq f^{-1}(S)$, so that $f^{-1}(S) \in \mathcal{F}$; and, since $S \subseteq \operatorname{ran}(f)$, we have $S = f(f^{-1}(S)) \in f(\mathcal{F})$. $\qquad\square$

Example 10.1.8

Suppose $(X_i)_{i \in I}$ is a family. If $\mathcal{F}$ is a filter in $\prod_i X_i$, then, for each $j \in I$, the surjectivity of π_j (Q 1.5.2) ensures that $\pi_j(\mathcal{F})$ is a filter in X_j.

Example 10.1.9

Suppose X is a set and $\sim$ is an equivalence relation on X. If $\mathcal{F}$ is a filter in X, then, since the quotient map π is surjective, $\pi(\mathcal{F})$ is a filter in $X/\sim$.

Inverse Images of Filters

The analogue of 10.1.7 for inverse images requires a little more care. Suppose X and Y are non-empty sets, $f \colon X \to Y$ and $\mathcal{F}$ is a filter in Y. It may be that $f^{-1}(S) = \varnothing$ for some $S \in \mathcal{F}$, in which case $f^{-1}(\mathcal{F})$ is not a filter. The following example shows that lack of injectivity of f is a further obstacle to $f^{-1}(\mathcal{F})$ being a filter: suppose $X = \{a, b, c\}$ and $Y = \{a, b\}$ and define $f \colon X \to Y$ by $f(a) = a$ and $f(b) = f(c) = b$; then $\mathcal{F} = \{\{a\}, \{a, b\}\}$ is a filter in Y, and, though $[f^{-1}(\mathcal{F})]$ is a filter in X, $f^{-1}(\mathcal{F}) = \{\{a\}, \{a, b, c\}\}$ is not.

Theorem 10.1.10

Suppose X and Y are non-empty sets, $\mathcal{F}$ is a filter in Y, $f \colon X \to Y$ and $\varnothing \notin f^{-1}(\mathcal{F})$. Then $[f^{-1}(\mathcal{F})]$ is a filter in X. If f is injective, $f^{-1}(\mathcal{F})$ is a filter in X.

Proof

Suppose $A, B \in [f^{-1}(\mathcal{F})]$. Then there exist $C, D \in \mathcal{F}$ such that $f^{-1}(C) \subseteq A$ and $f^{-1}(D) \subseteq B$. Therefore $f^{-1}(C \cap D) = f^{-1}(C) \cap f^{-1}(D) \subseteq A \cap B$, and so $A \cap B \in [f^{-1}(\mathcal{F})]$, because $C \cap D \in \mathcal{F}$. We deduce from 10.1.3 that $[f^{-1}(\mathcal{F})]$

is a filter in X because $\varnothing \notin f^{-1}(\mathcal{F})$. Suppose now that $S \in [f^{-1}(\mathcal{F})]$; then there exists $C \in \mathcal{F}$ such that $f^{-1}(C) \subseteq S$; also $C \cup f(S) \in \mathcal{F}$, so that, if f is injective, $S = f^{-1}(C \cup f(S)) \in f^{-1}(\mathcal{F})$ and the second assertion holds.   □

Example 10.1.11

Suppose X is a set, $\mathcal{F}$ is a filter in X and $Z \subseteq X$. If $S \in \mathcal{F} \Rightarrow Z \cap S \neq \varnothing$, then, because the inclusion function is injective, $\{Z \cap S \mid S \in \mathcal{F}\}$ is a filter in Z.

Comparison of Filters

Definition 10.1.12

Suppose $\mathcal{P}$ and $\mathcal{Q}$ are filters in a set X. $\mathcal{P}$ is said to be FINER or STRONGER than $\mathcal{Q}$, and $\mathcal{Q}$ to be COARSER or WEAKER than $\mathcal{P}$, if and only if $\mathcal{Q} \subseteq \mathcal{P}$.

Example 10.1.13

Suppose X is a non-empty set. Then $\{X\}$ is a filter in X. It is weaker than every filter in X and hence the weakest filter in X.

Example 10.1.14

Suppose X is a topological space and $x \in X$. The point filter at x (10.1.5) is the only filter in X to which $\{x\}$ belongs. It is stronger than the neighbourhood filter $\mathrm{Nbd}(x)$ and equal to it if and only if x is an isolated point of X.

Example 10.1.15

Suppose X and Y are sets, $\mathcal{F}$ is a filter in X and $f \colon X \to Y$. Then $[f^{-1}(f(\mathcal{F}))]$ is a filter in X. It is weaker than $\mathcal{F}$, since, for $S \in \mathcal{F}$, we have $S \subseteq f^{-1}(f(S))$ and hence $f^{-1}(f(S)) \in \mathcal{F}$. If f is injective, the two filters are the same.

Theorem 10.1.16

Suppose X is a set, $S \subseteq X$ and $\mathcal{F}$ is a filter in X for which $S \notin \mathcal{F}$. Then $\mathcal{G} = \{A \backslash R \mid A \in \mathcal{F}, R \subseteq S\}$ is a stronger filter than $\mathcal{F}$ and contains $X \backslash S$.

Proof

$\varnothing \notin \mathcal{G}$ because $\mathcal{P}(S) \cap \mathcal{F} = \varnothing$. Suppose $B, C \in \mathcal{F}$ and $P, Q \in \mathcal{P}(S)$. Then $(B \backslash P) \cap (C \backslash Q) = (B \cap C) \backslash (P \cup Q) \in \mathcal{G}$; and, if $B \backslash P \subseteq H \subseteq X$, we have $H = (H \cup P) \backslash (P \backslash H) \in \mathcal{G}$. So, by 10.1.3, $\mathcal{G}$ is a filter in X. That $\mathcal{F} \subseteq \mathcal{G}$ and $X \backslash S \in \mathcal{G}$ are obvious.   □

Intersections and Unions of Filters

Every non-trivial intersection of topologies on a set is a topology on the same set 7.1.15; an analogous result holds for filters (10.1.17). Moreover, in contrast to Q 7.1.6, the union of a non-empty nest of filters is also a filter (10.1.18).

Theorem 10.1.17

Suppose X is a set and $\mathfrak{F}$ is a non-empty collection of filters in X. Then $\bigcap \mathfrak{F}$ is a filter in X; it is the strongest filter in X which is weaker than every $\mathcal{F} \in \mathfrak{F}$.

Proof

Firstly, note that $X \in \bigcap \mathfrak{F}$. Also, since $\varnothing$ is not a member of any $\mathcal{F} \in \mathfrak{F}$, it is certainly not a member of $\bigcap \mathfrak{F}$. If $A, B \in \bigcap \mathfrak{F}$, then $A \cap B \in \bigcap \mathfrak{F}$. If $S \in [\bigcap \mathfrak{F}]$, then there exists $U \in \bigcap \mathfrak{F}$ such that $U \subseteq S$, so that $S \in \mathcal{G}$ for all $\mathcal{G} \in \mathfrak{F}$, whence $S \in \bigcap \mathfrak{F}$. It follows that $\bigcap \mathfrak{F}$ is a filter in X. That it is the strongest filter which is weaker than each $\mathcal{F} \in \mathfrak{F}$ is clear from its construction. □

Theorem 10.1.18

Suppose X is a set and $\mathfrak{N}$ is a non-empty nest of filters in X. Then $\bigcup \mathfrak{N}$ is a filter in X; it is the weakest filter in X which is stronger than every $\mathcal{F} \in \mathfrak{N}$.

Proof

Certainly $\varnothing \notin \bigcup \mathfrak{N}$. If $C \in [\bigcup \mathfrak{N}]$, then there exist $\mathcal{G} \in \mathfrak{N}$ and $D \in \mathcal{G}$ such that $D \subseteq C$, whence $C \in \mathcal{G}$ and $C \in \bigcup \mathfrak{N}$. If $A, B \in \bigcup \mathfrak{N}$, there exist $\mathcal{F}_1, \mathcal{F}_2 \in \mathfrak{N}$ such that $A \in \mathcal{F}_1$ and $B \in \mathcal{F}_2$, so that $A \cap B$ is a member of the stronger and is therefore a member of $\bigcup \mathfrak{N}$. So $\bigcup \mathfrak{N}$ is a filter in X. That it is the weakest filter in X which is stronger than every $\mathcal{F} \in \mathfrak{N}$ then follows from its definition. □

Ultrafilters

Every topology on a set X is included in the maximum topology $\mathcal{P}(X)$. For filters, the situation is not so clear. The bijective mapping on $\mathcal{P}(X)$ given by $S \mapsto X \backslash S$ takes each member of a filter $\mathcal{F}$ to a member of $\mathcal{P}(X) \backslash \mathcal{F}$. So the concept of a maximal filter is more elusive than that of a maximum topology. But the Axiom of Choice ensures that every filter is included in a maximal one.

Definition 10.1.19

Suppose X is a set and $\mathcal{U}$ is a filter in X. Then $\mathcal{U}$ is called an ULTRAFILTER in X if and only if no filter in X properly includes $\mathcal{U}$.

Example 10.1.20

Suppose X is a set and $x \in X$. The point filter at x is an ultrafilter in X.

Theorem 10.1.21

Suppose X is a non-empty set and $\mathcal{F}$ is a filter in X. Then $\mathcal{F}$ is an ultrafilter in X if and only if, for each subset A of X, either $A \in \mathcal{F}$ or $X \backslash A \in \mathcal{F}$.

Proof

If the condition is satisfied and $S \in \mathcal{P}(X) \backslash \mathcal{F}$, then $X \backslash S \in \mathcal{F}$, so that no filter both includes $\mathcal{F}$ and contains S; so, since S is arbitrary in $\mathcal{P}(X) \backslash \mathcal{F}$, it follows that $\mathcal{F}$ is an ultrafilter in X. For the converse: if $\mathcal{F}$ is an ultrafilter in X, then $X \backslash S \in \{A \backslash R \mid A \in \mathcal{F},\ R \subseteq S\} = \mathcal{F}$ by 10.1.16. $\qquad\square$

Theorem 10.1.22

Suppose X is a non-empty set and $\mathcal{F}$ is a filter in X. Then there is an ultrafilter in X which includes $\mathcal{F}$. Moreover, $\mathcal{F} = \bigcap\{\mathcal{U} \mid \mathcal{U}$ is an ultrafilter in $X;\ \mathcal{F} \subseteq \mathcal{U}\}$.

Proof

Let $\mathfrak{F}$ denote the collection of all filters in X which are finer than $\mathcal{F}$. $\mathfrak{F}$ is partially ordered by inclusion. If $\mathfrak{S}$ is any well ordered subset of $\mathfrak{F}$, it follows from 10.1.18 that $\bigcup \mathfrak{S}$ is an upper bound for $\mathfrak{S}$ in $\mathfrak{F}$. By Zorn's Lemma, $\mathfrak{F}$ has a maximal element. Such a maximal element is clearly an ultrafilter in X. Lastly, $\mathcal{F}$ is included in the intersection of all the ultrafilters which include $\mathcal{F}$ and the reverse inclusion follows from 10.1.16: if $S \in \mathcal{P}(X) \backslash \mathcal{F}$, there is a filter $\mathcal{G}$ which includes $\mathcal{F}$ and contains $X \backslash S$; then there is an ultrafilter $\mathcal{U}$ which includes $\mathcal{G}$, yielding $\mathcal{F} \subseteq \mathcal{U}$ and $S \notin \mathcal{U}$. $\qquad\square$

Theorem 10.1.23

Suppose X and Y are non-empty sets and $f \colon X \to Y$. Suppose $\mathcal{U}$ is an ultrafilter in X. Then $f(\mathcal{U})$ is an ultrafilter in $f(X)$ and $[f(\mathcal{U})]$ is an ultrafilter in Y.

Proof

$f(\mathcal{U})$ is a filter in $f(X)$ by 10.1.7. Suppose that $S \subseteq f(X)$. Then either $f^{-1}(S) \in \mathcal{U}$ or $X \backslash f^{-1}(S) \in \mathcal{U}$ by 10.1.21. So either $S = f(f^{-1}(S)) \in f(\mathcal{U})$ or $f(X) \backslash S = f(X \backslash f^{-1}(S)) \in f(\mathcal{U})$, and $f(\mathcal{U})$ is an ultrafilter in $f(X)$ by 10.1.21. Towards the second assertion, suppose $V \subseteq Y$ and $V \notin [f(\mathcal{U})]$. Then $V \cap f(X) \notin f(\mathcal{U})$, so that, since $f(\mathcal{U})$ is an ultrafilter in $f(X)$, we have, by 10.1.21, $f(X) \backslash V \in f(\mathcal{U})$ and hence $Y \backslash V \in [f(\mathcal{U})]$. Since $[f(\mathcal{U})]$ is a filter in Y by 10.1.7, it follows from 10.1.21 that $[f(\mathcal{U})]$ is an ultrafilter in Y. $\qquad\square$

Example 10.1.24
Since quotient maps and natural projections of products onto co-ordinate sets are surjective maps, such maps carry ultrafilters onto ultrafilters.

Example 10.1.25
10.1.23 is the result corresponding to 10.1.7 for ultrafilters rather than filters. There is, however, no result corresponding entirely to 10.1.10 as this example shows. Suppose X and Y are non-empty sets, $y \in Y$ and $\mathcal{U}$ is the point filter at y. Then $\mathcal{U}$ is an ultrafilter in Y. Let $f: X \to Y$ be the function with constant value y. Then $f^{-1}(\mathcal{U}) = \{X\}$ and $[f^{-1}(\mathcal{U})] = \{X\}$, which is not an ultrafilter in X unless X is a singleton set. However, we do have 10.1.26 below.

Theorem 10.1.26
Suppose X and Y are non-empty sets and $\mathcal{U}$ is an ultrafilter in Y. Suppose $f: X \to Y$ is injective and $\varnothing \notin f^{-1}(\mathcal{U})$. Then $f^{-1}(\mathcal{U})$ is an ultrafilter in X.

Proof
Suppose $S \subseteq X$. Then either $f(S) \in \mathcal{U}$ or $Y \backslash f(S) \in \mathcal{U}$. So, because f is injective, $S = f^{-1}(f(S)) \in f^{-1}(\mathcal{U})$ or $X \backslash S = f^{-1}(Y \backslash f(S)) \in f^{-1}(\mathcal{U})$. Since $f^{-1}(\mathcal{U})$ is a filter in X by 10.1.10, $f^{-1}(\mathcal{U})$ is an ultrafilter in X by 10.1.21. □

EXERCISES

Q 10.1.1 Suppose X is a non-empty set and $\mathcal{F}$ is a filter in X. Show that $\mathcal{F} \cup \{\varnothing\}$ is a topology on X. Show also that the converse to this is not in general true by giving an example of a topology $\mathcal{O}$ on a set X such that $\mathcal{O} \backslash \{\varnothing\}$ is not a filter in X.

Q 10.1.2 Suppose X is a non-empty set and $\mathcal{U}$ is a collection of non-empty subsets of X which has the property that, for each subset S of X, we have either $S \in \mathcal{U}$ or $X \backslash S \in \mathcal{U}$. What other conditions on $\mathcal{U}$ are necessary to ensure that $\mathcal{U}$ is an ultrafilter in X?

10.2 Limits

Definition 10.2.1
Suppose X is a topological space and $x \in X$. A collection $\mathcal{P}$ of subsets of X is said to CONVERGE to x in X if and only if every open neighbourhood of x in X includes some non-empty member of $\mathcal{P}$. In this case we write $\mathcal{P} \to x$ and

call x a LIMIT of $\mathcal{P}$. If $\mathcal{P}$ converges to x but to no other point of X, then we call x THE LIMIT of $\mathcal{P}$ and denote it by $\lim \mathcal{P}$. Note that, if Y is a topological space of which X is a subspace, then $\mathcal{P} \to x$ in Y if and only if $\mathcal{P} \to x$ in X.

Example 10.2.2

Suppose X is a set and $\mathcal{O}_s$ and $\mathcal{O}_w$ are topologies on X with $\mathcal{O}_s$ stronger than $\mathcal{O}_w$. If $\mathcal{P}$ is a collection of subsets of X which converges in $(X, \mathcal{O}_s)$, then every collection of subsets of X which includes $\mathcal{P}$ converges in $(X, \mathcal{O}_w)$ to the same limits (though it may converge to more). In particular, with a fixed topology on X, every filter finer than a convergent filter converges to the same limits.

Example 10.2.3

Suppose that X is a topological space and that $x \in X$. Then $\mathrm{Nbd}(x) \to x$ and $\mathrm{nbd}(x) \to x$. And, if $\mathcal{P} \subseteq \mathrm{nbd}(x)$, then $\mathcal{P} \to x$ if and only if $\mathcal{P}$ is a neighbourhood base at x (7.3.1).

Example 10.2.4

Evidently, a filter $\mathcal{F}$ in a topological space X converges to $x \in X$ if and only if $\mathrm{nbd}(x) \subseteq \mathcal{F}$, or, equivalently, $\mathrm{Nbd}(x) \subseteq \mathcal{F}$. In fact, since $\mathcal{F}$ is closed under non-trivial finite intersections and under non-trivial unions, $\mathcal{F} \to x$ if and only if $\mathcal{S} \subseteq \mathcal{F}$, where $\mathcal{S}$ is the collection of subbasic open neighbourhoods of x with respect to some subbase.

Example 10.2.5

Let X be a topological space and $\mathcal{P}$ a non-empty set of non-empty subsets of X. If X has the trivial topology, then $\mathcal{P}$ converges to every point of X. If, on the other hand, X is discrete, then $\mathcal{P}$ converges to $x \in X$ if and only if $\{x\} \in \mathcal{P}$.

Theorem 10.2.6

Suppose X is a topological space. The following are equivalent:

- X is Hausdorff;
- every convergent filter in X has unique limit;
- every convergent ultrafilter in X has unique limit.

Proof

Suppose firstly that X is Hausdorff and $\mathcal{F}$ is a filter in X. Suppose $x, y \in X$ and $\mathcal{F} \to x$ and $\mathcal{F} \to y$. Then $\mathrm{nbd}(x) \subseteq \mathcal{F}$ and $\mathrm{nbd}(y) \subseteq \mathcal{F}$, so that, if $U \in \mathrm{nbd}(x)$ and $V \in \mathrm{nbd}(y)$ then $U \cap V \neq \varnothing$. Since X is Hausdorff, we must conclude that

$x = y$. So convergent filters have unique limits. In particular convergent ultra-filters have unique limits. For the converse, suppose every convergent ultrafilter in X has unique limit. Suppose $x, y \in X$ are such that every open neighbourhood of x has non-empty intersection with every open neighbourhood of y. Then $\mathcal{F} = \{U \cap V \mid U \in \mathrm{Nbd}(x), \ V \in \mathrm{Nbd}(y)\}$ is a filter in X. By 10.1.22, $\mathcal{F}$ is included in an ultrafilter $\mathcal{U}$ of X. But $\mathrm{Nbd}(x) \subseteq \mathcal{U}$ and $\mathrm{Nbd}(y) \subseteq \mathcal{U}$, so that $\mathcal{U} \to x$ and $\mathcal{U} \to y$, whence, by hypothesis, $x = y$. So X is Hausdorff. $\qquad \square$

Convergence of Sequences

Definition 10.2.7

Suppose X and Y are topological spaces and $f: X \rightarrowtail Y$. Suppose $l \in Y$ and $c \in \mathrm{acc}_X(\mathrm{dom}(f))$. We shall say that $f(x)$ CONVERGES to l in Y as x converges to c, and write $f(x) \to l$ as $x \to c$, if and only if $\{f(U \backslash \{c\}) \mid U \in \mathrm{nbd}(c)\}$ converges to l in Y. If l is unique with this property, then we shall call l the LIMIT of $f(x)$ as x converges to c and write $l = \lim_{x \to c} f(x)$.

Suppose X is a topological space. A sequence x in X is a function from $\mathbb{N}$ to X, and the familiar phrase '$x_n \to z$ as $n \to \infty$' is made meaningful as follows. $\mathbb{N}$ with its usual discrete topology is a subspace of $\tilde{\mathbb{N}}$ with its usual topology (8.3.3). Then $x: \tilde{\mathbb{N}} \rightarrowtail X$, and 10.2.7 says $x_n \to z$ as $n \to \infty$ if and only if $\{x(U \backslash \{\infty\}) \mid U \in \mathrm{nbd}(\infty)\}$ converges to z in X. The basic open neighbourhoods of ∞ are $v_n = \{m \in \tilde{\mathbb{N}} \mid n \leq m\}$ for $n \in \mathbb{N}$, and $x(v_n \backslash \{\infty\})$ is the nth tail of x; so '$x_n \to z$ as $n \to \infty$' means that the filter determined by x converges to z. So $x_n \to z$ as $n \to \infty$ if and only if every open neighbourhood of z includes a tail of x; if X is a semimetric space, this occurs if and only if every open ball of X which contains z includes a tail of x. For complex sequences, this is the meaning of '$x_n \to z$' in 4.3.9; and the generalization of 4.3.9 in 10.2.8 below is in this manner consistent with 10.2.7. Just as for complex sequences, limits of sequences in metric spaces are unique, every convergent sequence is bounded, and every subsequence of a convergent sequence converges to the same limit. Note that limits are never unique in non-metric semimetric spaces.

Definition 10.2.8

Suppose X is a topological space, (x_n) is a sequence in X and $z \in X$. We say that (x_n) CONVERGES to z in X if and only if every open neighbourhood of z in X includes a tail of (x_n). If this is the case, we write $x_n \to z$ (or $x_n \to z$ as $n \to \infty$); if z is unique as a limit, we shall denote it by $\lim x_n$. Of course, if Y is a topological space of which X is a subspace, then (x_n) converges to z in X if and only if (x_n) converges to z in Y.

Example 10.2.9

No sequence in a Hausdorff space has more than one limit, by 10.2.6. But there are non-Hausdorff spaces in which all convergent sequences have unique limits: in $\mathbb{R}$ with the countable complement topology, a sequence converges if and only if it is eventually constant; but the space is not T_2 because every pair of non-empty open sets has non-empty intersection. But see Q 10.2.6.

Example 10.2.10

Suppose S is a non-empty set, X is a normed linear space, $f \in B(S, X)$ and (f_n) is a sequence in $B(S, X)$. Convergence $f_n \to f$ in $B(S, X)$ is often described as UNIFORM CONVERGENCE to distinguish it from the usually weaker POINTWISE CONVERGENCE given by $f_n(s) \to f(s)$ for all $s \in S$.

Definition 10.2.11

Suppose X is a normed linear space and $(x_n)_{n \in \mathbb{N}}$ is a sequence in X. If the series $\sum_{n \in \mathbb{N}} x_n$ converges, its limit is called its SUM and may be denoted by $\sum_{n=1}^{\infty} x_n$. If, for every permutation ϕ of $\mathbb{N}$, the series $\sum_{n \in \mathbb{N}} x_{\phi(n)}$ converges to the same sum, then we say that the series $\sum_{n \in \mathbb{N}} x_n$ is UNCONDITIONALLY CONVERGENT. If S is a countable subset of X and $\sum_n x_n$ converges unconditionally for some enumeration (x_n) of S, then we may use loose notation like $\sum_{s \in S} s$ unambiguously for the sum and say that $\sum_{s \in S} s$ converges unconditionally.

Example 10.2.12

Suppose $(x_n)_{n \in \mathbb{N}}$ is a sequence in $\mathbb{R}^{\oplus}$ and $\sum_{n \in \mathbb{N}} x_n$ converges to s. Let $\epsilon \in \mathbb{R}^+$ and $m \in \mathbb{N}$ be such that the mth tail of the series is in $(s - \epsilon, s]$; then, if ϕ permutes $\mathbb{N}$ and $k = \max\{\phi^{-1}(i) \mid 1 \leq i \leq m\}$, it is easy to check that the kth tail of $\sum_{n \in \mathbb{N}} x_{\phi(n)}$ is also in $(s - \epsilon, s]$. So $\sum_{n \in \mathbb{N}} x_n$ converges unconditionally.

Convergence and Closure. Sequential Closure

Closure points are identified as limits of filters. And the Axiom of Choice ensures that filters can be replaced by sequences in first countable spaces.

Theorem 10.2.13

Suppose X is a topological space, $A \subseteq X$ and $x \in X$. These are equivalent:

- $x \in \overline{A}$;
- there exists a filter in A which converges to x in X;
- there exists an ultrafilter in A which converges to x in X.

Proof

If a filter converges to a point then every finer filter converges to the same point; and since every filter is included in an ultrafilter, this ensures that the second and third statements are equivalent. If $x \in \overline{A}$, then each open neighbourhood of x in X has non-empty intersection with A, so that $[\{A \cap U \mid U \in \text{nbd}_X(x)\}]_A$ is a filter in A which converges to x in X. Suppose now that there exists a filter $\mathcal{F}$ in A which converges to x in X. Then each member of $\text{nbd}_X(x)$ includes some member of $\mathcal{F}$, so has non-empty intersection with A, giving $x \in \overline{A}$.   □

Definition 10.2.14

Suppose X is a topological space and $A \subseteq X$. Then A is said to be SEQUEN-TIALLY CLOSED in X if and only if no sequence in A has a limit in $X \backslash A$.

Theorem 10.2.15

Suppose X is a topological space and A is a closed subset of X. Then A is sequentially closed in X.

Proof

Suppose (a_n) is a sequence in A with a limit $z \in X$. Then the filter determined by (a_n) converges to z in X; so, since $A = \overline{A}$, 10.2.13 ensures that $z \in A$.   □

Example 10.2.16

The converse of 10.2.15 is not always true. $\mathbb{R} \backslash \mathbb{Q}$ is a sequentially closed subset of $\mathbb{R}$ with the countable complement topology (7.1.6), but is not closed: every neighbourhood of 0 contains an irrational number, so that $0 \in \overline{\mathbb{R} \backslash \mathbb{Q}}$; whereas, if (x_n) is a sequence in $\mathbb{R} \backslash \mathbb{Q}$, then $\mathbb{R} \backslash \{x_n \mid n \in \mathbb{N}\}$ is an open neighbourhood of every $q \in \mathbb{Q}$, so that (x_n) has no limit in $\mathbb{Q}$.

Theorem 10.2.17

Suppose X is a first countable topological space and $A \subseteq X$. If $x \in X$, then $x \in \overline{A}$ if and only if there exists a sequence in A which converges to x in X. Consequently A is closed in X if and only if A is sequentially closed in X.

Proof

Suppose $x \in \overline{A}$ and let (B_n) be a decreasing sequence whose terms form a local base at x (7.3.1). If $x \in A$, then the constant sequence (x) converges to x. Otherwise $x \in \text{acc}(A) \backslash A$ and, by the Product Theorem, there is a sequence $(a_n)_{n \in \mathbb{N}}$ with $a_n \in A \cap B_n$ for each $n \in \mathbb{N}$. Then, clearly, $a_n \to x$. The converse is in 10.2.13, and the second assertion follows easily.   □

Convergence and Continuity

Continuous functions are identified as those which map convergent filters onto filters which converge to the images of their limits.

Theorem 10.2.18

Suppose X and Y are topological spaces, $x \in X$ and $f \colon X \to Y$. Then the following are equivalent:

- f is continuous at x;
- for every filter $\mathcal{F}$ in X for which $\mathcal{F} \to x$, we have $f(\mathcal{F}) \to f(x)$;
- for every ultrafilter $\mathcal{U}$ in X for which $\mathcal{U} \to x$, we have $f(\mathcal{U}) \to f(x)$;
- $f(\mathrm{Nbd}(x)) \to f(x)$;
- $f(\mathrm{nbd}(x)) \to f(x)$.

Proof

Suppose f is continuous at x; then $\mathrm{nbd}(f(x)) \subseteq [f(\mathrm{nbd}(x))]$ by definition; if $\mathcal{F}$ is a filter in X and $\mathcal{F} \to x$, then $\mathrm{nbd}(x) \subseteq \mathcal{F}$, so that $\mathrm{nbd}(f(x)) \subseteq [f(\mathcal{F})]$, or, equivalently, $f(\mathcal{F}) \to f(x)$. So the first proposition implies the second; that the second implies the third, that the fourth implies the fifth and that the fifth implies the first are more or less obvious. We now assume the third proposition and prove the fourth. Let $\mathfrak{U}$ be the set of ultrafilters in X which include $\mathrm{Nbd}(x)$; for each $\mathcal{U} \in \mathfrak{U}$, we have $\mathcal{U} \to x$ and then, by hypothesis, $f(\mathcal{U}) \to f(x)$. Let $V \in \mathrm{nbd}(f(x))$; by the Product Theorem, there is a family $(A_{\mathcal{U}})_{\mathcal{U} \in \mathfrak{U}}$ such that $A_{\mathcal{U}} \in \mathcal{U}$ and $f(A_{\mathcal{U}}) \subseteq V$ for each $\mathcal{U} \in \mathfrak{U}$. Set $U = \bigcup(A_{\mathcal{U}})$; then $f(U) \subseteq V$ and $U \in \bigcap \mathfrak{U} = \mathrm{Nbd}(x)$ by 10.1.22 and, since V is arbitrary, $f(\mathrm{Nbd}(x)) \to f(x)$. $\square$

Theorem 10.2.19

Suppose $(X_i)_{i \in I}$ is a family of topological spaces and $P = \prod_i X_i$ is endowed with the product topology. Suppose $\mathcal{F}$ is a filter in P and $x \in P$. Then $\mathcal{F} \to x$ if and only if $\pi_j(\mathcal{F}) \to x_j$ for all $j \in I$. (This is expressed by saying that convergence in the product topology is POINTWISE CONVERGENCE.)

Proof

If $\mathcal{F} \to x$ then continuity of the natural projections ensures that $\pi_j(\mathcal{F}) \to x_j$ for all $j \in I$, by 10.2.18. For the converse, suppose V is a subbasic open neighbourhood of x in P; then there exist $j \in I$ and $U \in \mathrm{nbd}(x_j)$ such that $V = \pi_j^{-1}(U)$. By 10.1.7, $\pi_j(\mathcal{F})$ is a filter in X_j. If $\pi_j(\mathcal{F}) \to x_j$, then $U \in \pi_j(\mathcal{F})$, so that $V = \pi_j^{-1}(U) \in \pi_j^{-1}(\pi_j(\mathcal{F})) \subseteq \mathcal{F}$ by 10.1.15. Then, as in 10.2.4, $\mathcal{F} \to x$. $\square$

Sequential Continuity

Functions which map convergent sequences onto convergent sequences with the appropriate limits are said to be *sequentially continuous*. By 10.2.18, every continuous function is sequentially continuous (10.2.21); the converse is not generally true (10.2.22), but the Axiom of Choice ensures that it holds if the domain is first countable (10.2.23, 10.2.24).

Definition 10.2.20

Suppose X and Y are topological spaces and $f: X \to Y$. Then f is said to be SEQUENTIALLY CONTINUOUS at $x \in X$ if and only if, for every sequence (a_n) in X which converges to x in X, the sequence $(f(a_n))$ converges to $f(x)$ in Y. f is said to be SEQUENTIALLY CONTINUOUS if and only if f is sequentially continuous at every point of X.

Theorem 10.2.21

Suppose X and Y are topological spaces, $x \in X$ and $f: X \to Y$. If f is continuous at x, then f is also sequentially continuous at x. If f is continuous, then f is sequentially continuous.

Proof

The first assertion was proved in 10.2.18, and the second is immediate.   □

Example 10.2.22

The converse to 10.2.21 is not true. Let X be the set of real numbers with the countable complement topology. Consider the identity map $\iota: X \to \mathbb{R}$, where $\mathbb{R}$ has its usual topology. If $r \in X$ and $U \in \mathrm{nbd}_X(r)$, then U is not included in $(r - 1, r + 1)$. So ι is not continuous at r. But, if (a_n) is a sequence in X which converges in X to r, then, since the set $\{a_n \mid n \in \mathbb{N}\} \setminus \{r\}$ is countable, its complement is in $\mathrm{nbd}_X(r)$ and therefore includes a tail of (a_n), whence (a_n) is eventually constant with value r. So ι is sequentially continuous at every point of its domain, but continuous at none.

Theorem 10.2.23

Suppose X is a first countable space, Y is a topological space, $f: X \to Y$ and $x \in X$. Then f is continuous at x if and only f is sequentially continuous at x.

Proof

The forward implication is in 10.2.21. For the converse, suppose f is not continuous at x. Then there exists $V \in \mathrm{nbd}(f(x))$ such that no member of $f(\mathrm{nbd}(x))$

is included in V. Let $(U_n)_{n \in \mathbb{N}}$ be a decreasing sequence whose terms form a local base at x (7.3.1). By the Product Theorem, there is a sequence (a_n) in X such that, for each $n \in \mathbb{N}$, $a_n \in U_n$ and $f(a_n) \notin V$. Then $a_n \to x$ and $f(a_n)$ does not converge to $f(x)$, whence f is not sequentially continuous at x.   □

Corollary 10.2.24

Suppose X and Y are topological spaces, X is first countable, and $f: X \to Y$. Then f is continuous if and only if f is sequentially continuous. This applies in particular if X is a semimetric space (7.3.3).

Bolzano–Weierstrass Theorem

Theorem 10.2.25 BOLZANO–WEIERSTRASS THEOREM III

The seminormed linear spaces in which every bounded sequence has a convergent subsequence are precisely those which have a dense finite dimensional subspace.

Proof

Suppose firstly that Z is a seminormed linear space which has no dense finite dimensional subspace; using the Riesz Lemma (6.2.28), the Recursive Choice Theorem ensures that there exists a sequence (x_n) in X such that $\|x_n\| = 1$ and $\mathrm{dist}(x_{n+1}, \langle \{x_i \mid 1 \le i \le n\} \rangle) > 1/2$ for all $n \in \mathbb{N}$. Then (x_n) is bounded and has no convergent subsequence. We prove the converse by induction. A space with a zero-dimensional dense subspace necessarily has the zero seminorm and the trivial topology, so that every sequence converges. Let $d \in \mathbb{N}$ and suppose that every seminormed linear space which has a dense subspace of dimension $d - 1$ has the stated property. Let X be a seminormed linear space which has a dense subspace M of dimension d and suppose (a_n) is a bounded sequence in X. By the Product Theorem, there exists a sequence (b_n) in M such that $\|a_n - b_n\| < 1/n$ for all $n \in \mathbb{N}$; then (b_n) is bounded. Let $z \in M \backslash \{0\}$; then $M/\mathbb{K}z$ has dimension $d - 1$ by 3.2.16 and is seminormed as in 6.2.25; moreover, the sequence $(\pi(b_n))$ is bounded in $M/\mathbb{K}z$ because the quotient map $\pi: M \to M/\mathbb{K}z$ is bounded (6.2.28). By the inductive hypothesis, there exist a subsequence (b_{m_n}) of (b_n) and $w \in M$ such that $\|\pi(b_{m_n}) - \pi(w)\| < 1/n$ for all $n \in \mathbb{N}$, and it follows, again by the Product Theorem, that there exists a sequence (λ_{m_n}) of scalars such that $\|b_{m_n} - w - \lambda_{m_n} z\| < 1/n$ also. But (b_n) is bounded; so either $\|z\| = 0$ or (λ_{m_n}) is bounded; in any case, there exist $\mu \in \mathbb{K}$ and a subsequence (λ_{k_n}) of (λ_{m_n}) such that $|\lambda_{k_n} - \mu| \, \|z\| < 1/n$ for each $n \in \mathbb{N}$ by 4.3.10; then $\|a_{k_n} - w - \mu z\| \le 3/n$ for each $n \in \mathbb{N}$, whence $a_{k_n} \to w + \mu z$. The Principle of Finite Induction now completes the argument.   □

Corollary 10.2.26
Suppose X is a normed linear space and M is a finite dimensional subspace of X. Then M is closed in X.

Proof
By 10.2.25, every bounded sequence in M has a subsequence which converges in M. Since convergent sequences are bounded and limits are unique in metric spaces, this implies that M is sequentially closed; then 10.2.17 implies that M is closed.   □

Corollary 10.2.27
The normed linear spaces in which every bounded sequence has a convergent subsequence are precisely those of finite dimension.

Corollary 10.2.28
Suppose X is finite dimensional linear space. All norms on X are equivalent.

Proof
Let $n = \dim(X)$; if $n = 0$ the result is trivial, so we suppose otherwise. Let $\mathbb{I} = \{i \in \mathbb{N} \mid 1 \le i \le n\}$ and let B be a basis for X enumerated as $(e_i)_{i \in \mathbb{I}}$. Let $\|\cdot\|$ and $\|\|\cdot\|\|$ be norms on X. For each $i \in \mathbb{I}$, let $M_i = \langle B \backslash \{e_i\} \rangle$ and $d_i = \text{dist}_{\|\|\cdot\|\|}(e_i, M_i)$. Let $s = \max\{\|\|e_i\|\| \mid i \in \mathbb{I}\}$ and $t = \min\{d_i \mid i \in \mathbb{I}\}$; then $t > 0$ by 10.2.26. Suppose $(\lambda_i)_{i \in \mathbb{I}}$ is a finite sequence in $\mathbb{K}$, and let $j \in \mathbb{I}$ be such that $|\lambda_j| = \max\{|\lambda_i| \mid i \in \mathbb{I}\}$. Then

$$\left\| \sum_{i=1}^{n} \lambda_i e_i \right\| \le n |\lambda_j| s \le n s t^{-1} |\lambda_j| d_j \le n s t^{-1} \left\|\left\| \sum_{i=1}^{n} \lambda_i e_i \right\|\right\|,$$

and $\|\cdot\|$ is weaker than $\|\|\cdot\|\|$ by 6.2.23. A similar argument establishes that $\|\|\cdot\|\|$ is weaker than $\|\cdot\|$. So the norms are equivalent.   □

Schauder Bases

Hamel bases have only limited use in analysis. A more fruitful concept might emerge if we discover bases which interact with the topology. *Schauder bases*, where they exist, facilitate the analysis of separable normed linear spaces. But Enflo has shown that not every such space admits a Schauder basis; and it is now known that many subspaces of the ℓ_p spaces do not do so either.

Definition 10.2.29

Suppose X is a normed linear space. A sequence $(e_n)_{n \in \mathbb{N}}$ in X is called a BASIC SEQUENCE in X if and only if, for every $x \in \mathrm{Cl}\langle\{e_n \mid n \in \mathbb{N}\}\rangle$, there exists a unique sequence $(\lambda_n)_{n \in \mathbb{N}}$ in $\mathbb{K}$ such that $x = \sum_{n=1}^{\infty} \lambda_n e_n$. And $(e_n)_{n \in \mathbb{N}}$ is called a SCHAUDER BASIS for X if and only if it is a basic sequence and the closure of the linear span of its terms is X.

Example 10.2.30

For $p \in [1, \infty)$, the space ℓ_p has Schauder basis (e_n), where, for each $n \in \mathbb{N}$, e_n is the sequence whose n^{th} term is 1 and whose other terms are all 0. But (e_n) is not a Schauder basis for ℓ_∞ because ℓ_∞ is not separable (7.3.6). In fact, every normed linear space X with a Schauder basis is separable because the countable subset consisting of linear combinations of the Schauder basis elements with coefficients which have rational real and imaginary parts is dense in X.

EXERCISES

Q 10.2.1 Let X be a topological space. Show that every subsequence of a convergent sequence in X converges to the same limits.

Q 10.2.2 Suppose X and Y are topological spaces, $x \in X$ and $f: X \to Y$ is open and injective. Suppose $\mathcal{F}$ is a filter in Y and $\mathcal{F} \to f(x)$. Show that $f^{-1}(\mathcal{F}) \to x$.

Q 10.2.3 Suppose (X, d) is a metric space, $z \in X$ and (x_n) is a sequence in X. Show that $x_n \to z$ in X if and only if for every $\epsilon \in \mathbb{R}^+$, there exists $m \in \mathbb{N}$ such that, for all $n \in \mathbb{N}$ with $m \leq n$, we have $d(x_n, z) < \epsilon$.

Q 10.2.4 Suppose X is an uncountable set endowed with the countable complement topology. Show that every convergent sequence in X is eventually constant.

Q 10.2.5 Suppose X is a topological space, $z \in X$ and $(x_n)_{n \in \mathbb{N}}$ is a sequence in X. Set $x_\infty = z$ and endow $\tilde{\mathbb{N}}$ with its usual topology (8.3.3). Show that $x_n \to z$ if and only if the function $x: \tilde{\mathbb{N}} \to X$ is continuous at ∞.

Q 10.2.6 Suppose X is a first countable space. Show that X is Hausdorff if and only if every convergent sequence in X has a unique limit in X.

Q 10.2.7 Suppose X is a normed linear space, S is a closed subspace of X and F is a finite dimensional subspace of X. Show that $F + S$ is closed in X. Show that the conclusion may fail if S is merely a closed subset of X rather than a closed subspace of X.

11
Compactness

O God, I could be bounded in a nut shell
and count myself a king of infinite space,
were it not that I have bad dreams.   *Hamlet, II,ii.*

Finite sets occupy a special position amongst all sets; it is the *compact* spaces which occupy the corresponding place among topological spaces.

11.1 Compact Topological Spaces

Definition 11.1.1
Suppose X is a topological space and S is a subset of X. A collection $\mathcal{C}$ of open subsets of X is called an OPEN COVER, or simply a COVER, for S in X if and only if $S \subseteq \bigcup \mathcal{C}$. Any subset $\mathcal{D}$ of $\mathcal{C}$ which satisfies $S \subseteq \bigcup \mathcal{D}$ is called a SUBCOVER of $\mathcal{C}$ for S in X. S is called a COMPACT SUBSET of X if and only if every open cover for S in X has a finite subcover. X is called a COMPACT topological space if and only if X is a compact subset of itself.

Example 11.1.2
Every indiscrete space is necessarily compact. Every finite subset of any topological space is a compact subset of that space.

Example 11.1.3
Every compact subset S of a metric space X is closed and bounded. If S were not bounded and $a \in X$, the cover $\{\flat[a\,;n) \mid n \in \mathbb{N}\}$ for S would have no finite subcover. If S were not closed and $c \in \overline{S}\backslash S$, the cover $\{X\backslash\flat[c\,;1/n] \mid n \in \mathbb{N}\}$ for S would have no finite subcover. The converse is not true (11.1.21, 11.2.6).

Example 11.1.4

Suppose X is a topological space. A compact space Y is called a COMPACTIFI-
CATION of X if and only if X is homeomorphic to a dense subspace of Y. Such
compactification is always possible; the simplest example is the ONE-POINT
COMPACTIFICATION. Suppose $(X, \mathcal{O})$ is a non-compact space and $p \notin X$. It is
easy to check that $\mathcal{P} = \mathcal{O} \cup \{A \cup \{p\} \mid A \subseteq X, X \backslash A \text{ closed and compact in } X\}$
is a topology on $\tilde{X} = X \cup \{p\}$ which makes $\tilde{X}$ compact, and that the inclu-
sion map from X to $\tilde{X}$ is a homeomorphism of X onto a dense subspace of
$\tilde{X}$. An example of one-point compactification is the RIEMANN SPHERE, which
is $\tilde{\mathbb{C}} = \mathbb{C} \cup \{\infty\}$ with the topology given above. The standard topology on $\tilde{\mathbb{N}}$
discussed in 8.3.3 is also a one-point compactification. The standard topology
on $\tilde{\mathbb{R}}$ (7.1.2) is a *two point compactification* of $\mathbb{R}$: this is the topology generated
by all open intervals of $\tilde{\mathbb{R}}$; those members which include $\{-\infty, \infty\}$ are the complements of the closed bounded subsets of $\mathbb{R}$, which, as we shall see in 11.1.21,
are precisely the compact subsets of $\mathbb{R}$.

Theorem 11.1.5

Suppose X is a topological space and S is a subset of X. Then S is a compact
subset of X if and only if S with the relative topology is a compact topological
space.

Proof

Suppose firstly that S is a compact space and that $\mathcal{C}$ is an open cover for S in
X. Then $\{U \cap S \mid U \in \mathcal{C}\}$ is an open cover for S in S, so has a finite subcover
$\{U \cap S \mid U \in \mathcal{D}\}$ for some $\mathcal{D} \subseteq \mathcal{C}$; and $\mathcal{D}$ is a finite subcover of $\mathcal{C}$ for S in X.
Towards the converse, suppose that S is a compact subset of X and that $\mathcal{E}$ is
an open cover for S in S. Let $\mathcal{C} = \{U \subseteq X \mid U \text{ open in } X, U \cap S \in \mathcal{E}\}$; since
each $V \in \mathcal{E}$ is of the form $U \cap S$ for some open subset U of X, $\mathcal{C}$ is an open
cover for S in X, so has a finite subcover $\mathcal{F}$; and $\{U \cap S \mid U \in \mathcal{F}\}$ is a finite
subcover of $\mathcal{E}$ for S in S. □

Example 11.1.6

Every closed bounded interval of $\mathbb{R}$ is compact. Suppose $[a, b]$ is such an inter-
val. Let $\mathcal{C}$ be a cover for $[a, b]$ in $\mathbb{R}$ and let $\mathfrak{F} = \{\mathcal{F} \subseteq \mathcal{C} \mid |\mathcal{F}| < \infty\}$. Let
$B = \{r \in \mathbb{R} \mid \exists \mathcal{F} \in \mathfrak{F} : [a, r] \subseteq \bigcup \mathcal{F}\}$. Certainly $a \in B$. Let $c = \sup B$ and suppose $c \in [a, b]$; then there exists $U \in \mathcal{C}$ with $c \in U$; and, since U is open in $\mathbb{R}$,
there exist $s \in U$ and $\mathcal{F} \in \mathfrak{F}$ with $a \leq s \leq c$ and $[a, s] \subseteq \bigcup \mathcal{F}$; and there exists
$t \in U$ with $c < t$, so that $\mathcal{F} \cup \{U\}$ covers $[a, t]$, contradicting the definition
of c. We conclude that $b < c$ and therefore also that $[a, b]$ is covered by some
member of $\mathfrak{F}$.

Convergence and Compactness

Theorem 11.1.7

Let X be a topological space. The following are equivalent:

- X is compact;
- Every non-empty collection of closed subsets of X with the Finite Intersection Property has non-empty intersection;
- Every ultrafilter in X converges in X;
- Every filter in X is included in a filter which converges in X.

Proof

Suppose firstly that X is compact and that S is a non-empty collection of closed subsets of X with the Finite Intersection Property. For each non-empty finite subset G of S, we have $\bigcap G \neq \varnothing$ so that $\{X \backslash F \mid F \in G\}$ is not a cover for X; then, since X is compact, $\{X \backslash F \mid F \in S\}$ is not a cover for X, whence $\bigcap S \neq \varnothing$.

Now suppose that the second condition is satisfied and that $\mathcal{U}$ is an ultrafilter in X. As $X \in \mathcal{U}$, the collection $\mathcal{F}$ of closed members of $\mathcal{U}$ is non-empty; and since $\mathcal{F} \subseteq \mathcal{U}$, $\mathcal{F}$ has the Finite Intersection Property. So, by hypothesis, $\bigcap \mathcal{F} \neq \varnothing$. Let $x \in \bigcap \mathcal{F}$. For each $U \in \mathrm{nbd}(x)$, $X \backslash U$ is closed and $X \backslash U \notin \mathcal{F}$, so that $X \backslash U \notin \mathcal{U}$; and, by 10.1.21, $U \in \mathcal{U}$. Therefore $\mathcal{U} \to x$.

The third condition implies the fourth by 10.1.22. So, lastly, we assume that every filter in X is included in a convergent filter and prove that X is compact. Suppose $\mathcal{C}$ is a collection of open subsets of X which has no finite subset which covers X; then, letting $\mathcal{D}$ denote the collection of non-trivial finite intersections of complements of members of $\mathcal{C}$, we have $\varnothing \notin \mathcal{D}$. So $[\mathcal{D}]$ is a filter by 10.1.3; and it is, by hypothesis, included in a convergent filter $\mathcal{U}$; let $x \in X$ be such that $\mathcal{U} \to x$. For each $U \in \mathrm{nbd}(x)$, we have $U \in \mathcal{U}$, whence $X \backslash U \notin \mathcal{D}$ and $U \notin \mathcal{C}$. So $x \notin \bigcup \mathcal{C}$ and $\mathcal{C}$ is not a cover for X. Therefore X is compact. $\qquad \square$

Continuity and Compactness

Theorem 11.1.8

Every continuous image of a compact space is compact.

Proof

Suppose X and Y are topological spaces, X is compact and $f \colon X \to Y$ is continuous. Suppose $\mathcal{C}$ is an open cover for $f(X)$ in Y. Then, because f is continuous, $f^{-1}(\mathcal{C})$ is an open cover for X, so has a finite subcover $f^{-1}(\mathcal{D})$, where $\mathcal{D} \subseteq \mathcal{C}$. Then $\mathcal{D}$ covers $f(X)$. Therefore $f(X)$ is compact. $\qquad \square$

Example 11.1.9

Continuity on compact subsets of a metric space is uniform. Specifically, if (X, d) and (Y, d') are metric spaces and $f \colon X \to Y$ is continuous at every point of a compact subset C of X, then $f|_C$ is uniformly continuous. Let $\epsilon \in \mathbb{R}^+$. For each $z \in C$, there is $\gamma_z \in \mathbb{R}^+$ such that $d(x, z) < \gamma_z \Rightarrow d'(f(x), f(z)) < \epsilon/2$ for all $x \in X$ by 8.1.14. The cover $\{b[w \,; \gamma_w/2] \mid w \in C\}$ for C has a finite subcover $\mathcal{F}$; let $\delta = \min\{\gamma_w/2 \mid b[w \,; \gamma_w/2) \in \mathcal{F}\}$. For each $a, b \in C$ with $d(a, b) < \delta$, there exists $z \in C$ such that $\delta \leq \gamma_z/2$ and $d(a, z) < \gamma_z/2$, whence $d(b, z) < \gamma_z$. So $d'(f(a), f(z)) < \epsilon/2$ and $d'(f(b), f(z)) < \epsilon/2$, which yields $d'(f(a), f(b)) < \epsilon$.

Example 11.1.10

The EXTREME VALUE THEOREM states that every real continuous function f defined on a closed bounded real interval I is bounded and attains its bounds, in the sense that there exist $a, b \in I$ such that $f(a) = \sup f(I) \in \mathbb{R}$ and $f(b) = \inf f(I) \in \mathbb{R}$. But in fact we know more: if I is a closed bounded interval of $\mathbb{R}$ then I is compact (11.1.6) and connected (9.1.5); so $f(I)$ is compact (11.1.8) and connected (9.1.13) and is therefore itself a closed bounded interval of $\mathbb{R}$ by 9.1.5 and 11.1.3.

Example 11.1.11

Suppose X is a non-empty topological space. Then $\mathcal{C}(X)$ is a normed algebra with isometric involution (8.3.24). Now 11.1.8 allows us to identify significant subalgebras with the same properties. A function $f \colon X \to \mathbb{K}$ is said to VANISH AT INFINITY if and only if, for every $\epsilon \in \mathbb{R}^+$, there exists a compact subset K of X such that $|f(x)| < \epsilon$ for all $x \in X \backslash K$. A continuous function $f \colon X \to \mathbb{K}$ which vanishes at infinity is necessarily bounded and *attains its bound* in the sense that $\sup \operatorname{ran}|f| = \max \operatorname{ran}|f|$; this is obvious if $f = 0$ and otherwise is shown as follows: let $\epsilon \in (0, \sup \operatorname{ran}|f|)$ and K be a compact subset of X for which $|f(x)| \leq \max\{\epsilon, \sup |f|(K)\}$ for all $x \in X$; then continuity of f and of the modulus function, together with 11.1.8 and 11.1.3, imply that $|f|(K)$ is a non-empty closed bounded subset of $\mathbb{R}$ and therefore contains its supremum. The set of continuous functions from X to $\mathbb{K}$ which vanish at infinity is clearly closed under the algebraic operations of $\mathcal{C}(X)$ and under involution and is therefore a normed subalgebra of $\mathcal{C}(X)$ with involution; we denote it by $\mathcal{C}_0(X)$. It may fail to be unital; indeed, it is immediate but striking that it is unital if and only if X is compact, which is both necessary and sufficient to ensure that the function $x \mapsto 1$ vanishes at infinity. A further subalgebra is $\mathcal{C}_c(X)$, which comprises those members of $\mathcal{C}(X)$ which have compact SUPPORT, the support of a function $f \colon X \to \mathbb{K}$ being $\operatorname{Cl}\{x \in X \mid f(x) \neq 0\}$. In general, there is set inclusion $\mathcal{C}_c(X) \subseteq \mathcal{C}_0(X) \subseteq \mathcal{C}(X)$, but the algebras coincide if X is compact.

Compactness of Derived Spaces

Every closed subspace of a compact space is compact. Every quotient of a compact space is compact. And the Axiom of Choice implies that every product of compact spaces is compact.

Theorem 11.1.12
Every closed subset of a compact space is compact.

Proof
Suppose X is a compact space and S is a closed subset of X. Suppose $\mathcal{C}$ is an open cover for S in X. Then $\mathcal{C} \cup \{X \backslash S\}$ is an open cover for X, so has a finite subcover $\mathcal{D}$; and $\mathcal{D} \backslash \{X \backslash S\}$ is a finite subcover of $\mathcal{C}$ for S. □

Example 11.1.13
Arbitrary subsets of compact spaces need not be compact. Consider the interval $(0, 1]$ as a subspace of the compact interval $[0, 1]$ with the relative topology derived from $\mathbb{R}$. This has an open cover $\{(1/n, 1] \mid n \in \mathbb{N}\}$ in $[0, 1]$, which certainly has no finite subcover. So $(0, 1]$ is not compact in $[0, 1]$.

Theorem 11.1.14
Every quotient of a compact space is compact.

Proof
Every continuous image of a compact space is compact (11.1.8) and every quotient map is continuous because it determines the quotient topology. □

Theorem 11.1.15 TYCHONOFF'S THEOREM
Suppose $(X_i)_{i \in I}$ is a family of topological spaces. The product $P = \prod_i X_i$ is compact if and only if each of the co-ordinate spaces is compact.

Proof
Certainly, if P is compact then, by 11.1.8, each of the spaces $X_j = \pi_j(P)$ is compact because the projection π_j is continuous. Towards the converse, suppose that each of the X_j is compact and that $\mathcal{U}$ is an ultrafilter in P. Then, for each $j \in I$, $\pi_j(\mathcal{U})$ is an ultrafilter in X_j by 10.1.23, and, by 11.1.7, $\pi_j(X_j)$ converges. Since limits need not be unique, we must now invoke the Product Theorem to assert that there exists $x \in P$ such that $\pi_j(\mathcal{U}) \to x_j$ for all $j \in I$. Then $\mathcal{U} \to x$, by 10.2.19. So P is compact by 11.1.7. □

Sequential Compactness

There is a concept of *sequential compactness* corresponding to those of sequential closure and sequential continuity, but its relationship to compactness has unexpected subtleties. Neither compactness nor sequential compactness implies the other; there are even first countable sequentially compact spaces which are not compact (11.1.17). But every first countable compact space is sequentially compact (11.1.19) and, happily, the two concepts are identical in metric spaces (11.1.20).

Definition 11.1.16

Suppose X is a topological space. Then X is said to be SEQUENTIALLY COMPACT if and only if every sequence in X has a convergent subsequence.

Example 11.1.17

Here is an example of a first countable sequentially compact space which is not compact. Using 1.4.7, let ω be the smallest uncountable ordinal and endow ω with the topology ω_+ (7.1.8). Then ω is a first countable space because, for each $\alpha \in \omega$, the singleton set $\{\alpha_+\}$ is a local base at α. Also ω is sequentially compact: suppose (x_n) is a sequence in ω; then each term is a countable ordinal; therefore $s = \bigcup (x_n)_{n \in \mathbb{N}}$ is also countable by 2.4.5, and is an ordinal by 1.4.8; so $s \in \omega$, and $x_n \to s$ because every open neighbourhood of s includes $\{x_n \mid n \in \mathbb{N}\}$. But ω is not compact, nor indeed Lindelöf (Q 11.2.1), because ω is an open cover for ω and each countable subset of ω has countable union so does not cover ω.

Example 11.1.18

Here is an example of a compact space which is not sequentially compact. Let I denote the interval $[0, 1)$ with its usual topology as a subspace of $\mathbb{R}$; endow $\{0, 1\}$ with the discrete topology and let $X = \{0, 1\}^I$ with its product topology. Then X is compact by Tychonoff's Theorem. Now define a sequence of functions $f_n: I \to \{0, 1\}$ by setting $f_n(x)$ to be the nth digit in the binary expansion of x (see 4.3.16) for each $x \in I$. Suppose (f_{k_n}) is any subsequence of (f_n). Let $z = \sum_{n=1}^{\infty} 2^{-k_{2n}}$. Then $(f_{k_n}(z))_{n \in \mathbb{N}}$ is an alternating sequence of ones and zeroes, so is not convergent. Therefore (f_{k_n}) is not convergent either (10.2.19). So (f_n) has no convergent subsequence. Therefore X is not sequentially compact.

Theorem 11.1.19

Every first countable compact space is sequentially compact.

Proof

Suppose X is a first countable compact space and (x_n) is a sequence in X. Let $\mathcal{C}$ be the collection of all open subsets of X which contain only a finite number of terms of (x_n). Then, certainly, no finite subset of $\mathcal{C}$ covers X, so that, since X is compact, $\mathcal{C}$ is not a cover for X. Therefore there exists $z \in X$ every one of whose neighbourhoods contains an infinite number of terms of (x_n). Let (V_n) be a sequence of open subsets of X, decreasing with respect to inclusion, whose terms form a local base at z (7.3.1). By the Product Theorem, there exists a subsequence (x_{k_n}) of (x_n) with $x_{k_n} \in V_n$ for each $n \in \mathbb{N}$. Every open neighbourhood of z includes V_m for some $m \in \mathbb{N}$ and therefore includes also the mth tail of (x_{k_n}). So $x_{k_n} \to z$. It follows that X is sequentially compact. $\qquad\square$

Theorem 11.1.20

A metric space X is sequentially compact if and only if it is compact.

Proof

Since every metric space is first countable (7.3.3), compactness of X implies sequential compactness by 11.1.19. Towards the converse, we suppose that X is sequentially compact and that $\mathcal{U}$ is an ultrafilter in X. For each $w \in X$, define $\delta(w) = \sup\{\mathrm{dist}(w,S) \mid S \in \mathcal{U}\}$, and set $\gamma = \inf\{\delta(w) \mid w \in X\}$; we show that $\gamma = 0$. Suppose otherwise and let $\epsilon \in (0, \gamma)$. Let $x_0 \in X$. By the Recursive Choice Theorem, there exist sequences (x_n) in X and (F_n) in $\mathcal{U}$ such that, for each $n \in \mathbb{N}$, we have $x_n \in \bigcap\{F_i \mid 1 \leq i \leq n\}$ and $\mathrm{dist}(x_{n-1}, F_n) \geq \epsilon$. For $i, j \in \mathbb{N}$ with $i \neq j$, we have $\mathrm{dist}(x_i, x_j) \geq \epsilon$, so that (x_n) can have no convergent subsequence, contradicting sequential compactness. This establishes our claim that $\gamma = 0$. By the Product Theorem, there exists a sequence (z_n) in X with $\delta(z_n) < 1/n$ for each $n \in \mathbb{N}$. Since X is sequentially compact, there exists $v \in X$ which is a limit of a convergent subsequence of (z_n). Certainly $\delta(v) = 0$. So, if $V \in \mathrm{nbd}(v)$, then $V \cap U \neq \varnothing$ for each $U \in \mathcal{U}$. In particular, $X \backslash V \notin \mathcal{U}$, so that $V \in \mathcal{U}$ by 10.1.21. Then $\mathcal{U} \to v$, and X is compact by 11.1.7. $\qquad\square$

Corollary 11.1.21 HEINE–BOREL THEOREM

Suppose X is a finite dimensional normed linear space. Then the compact subsets of X are precisely the closed bounded subsets of X.

Proof

Suppose S is a closed bounded subset of X and (x_n) is a sequence in S. By 10.2.25, (x_n) has a subsequence which converges in X; since S is closed in X, its limit is in S by 10.2.13. So S is sequentially compact, and therefore compact by 11.1.20. The converse was proved in 11.1.3. $\qquad\square$

EXERCISES

Q 11.1.1 Let Y be a subspace of a topological space X. If $Z \subseteq Y$, show that Z is a compact subset of Y if and only if Z is a compact subset of X.

Q 11.1.2 Suppose X is a topological space and $\mathcal{C}$ is a collection of compact subsets of X. Show that if $\mathcal{C}$ is finite then $\bigcup \mathcal{C}$ is compact. Show also that if every member of $\mathcal{C}$ is closed, then $\bigcap \mathcal{C}$ is compact.

Q 11.1.3 Let X be a compact metric space. Show that X is second countable.

Q 11.1.4 Show that the Cantor set (Q 9.1.3) is a compact subspace of $\mathbb{R}$.

Q 11.1.5 In contrast to Q 6.2.6, show that the sum of a closed subset and a compact subset of a normed linear space must be closed.

Q 11.1.6 Show how to eliminate the implicit use of Axiom of Choice in 11.1.9.

Q 11.1.7 The following argument is clearly wrong; where is the flaw? The closed unit ball of $\ell_\infty(\mathbb{R})$ is $[0, 1]^{\mathbb{N}}$, which is compact by Tychonoff's Theorem; so ℓ_∞ is finite dimensional by 11.2.6.

11.2 Compact Hausdorff Spaces

Theorem 11.2.1
Suppose X is a Hausdorff space and K and L are disjoint compact subsets of X. There exist disjoint open subsets V and W of X with $L \subseteq V$ and $K \subseteq W$.

Proof
If K or L is empty the result is trivial; we assume otherwise. Suppose $z \in K$. Let $\mathcal{C} = \{U \subseteq X \mid U \text{ open in } X, z \notin \overline{U}\}$. $\mathcal{C}$ is a cover for $X \backslash \{z\}$, and hence for L, because X is Hausdorff. So $\mathcal{C}$ has a finite subcover $\mathcal{D}$ for L. Then $z \in \bigcap\{X \backslash \overline{U} \mid U \in \mathcal{D}\}$, which is open and disjoint from the open set $\bigcup \mathcal{D}$; so its closure is disjoint from L. Since z is arbitrary in K, the set $\mathcal{S} = \{G \subseteq X \mid G \text{ open in } X; \overline{G} \cap L = \varnothing\}$ is a cover for K. So $\mathcal{S}$ has a finite subcover $\mathcal{E}$ for K. Set $V = \bigcap\{X \backslash \overline{G} \mid G \in \mathcal{E}\}$ and $W = \bigcup \mathcal{E}$. $\qquad \square$

Corollary 11.2.2
Every compact Hausdorff space is normal.

Proof
Suppose X is a compact Hausdorff space. Then X is T_1. Also, since closed subsets of X are compact by 11.1.12, X is T_4 by 11.2.1. $\qquad \square$

Corollary 11.2.3

Every compact subset of a Hausdorff space is closed.

Proof

Suppose X is a Hausdorff space and K is a compact subset of X. By 11.2.1, the collection $\mathcal{C}$ of open subsets of X which have empty intersection with K covers $X \backslash K$. So $X \backslash K = \bigcup \mathcal{C}$, which is open. Therefore K is closed. $\qquad \square$

Example 11.2.4

Compact subsets do not have to be closed in general. Consider $X = \{0, 1\}$ with the trivial topology. The set $\{0\}$ is compact but not closed.

Theorem 11.2.5

Suppose X is a Hausdorff space and $\mathcal{A}$ is a non-empty collection of compact subsets of X with the Finite Intersection Property. Then $\bigcap \mathcal{A} \neq \varnothing$.

Proof

Let $K \in \mathcal{A}$. Since X is Hausdorff, the members of $\mathcal{A}$ are closed by 11.2.3. Suppose $\bigcap \mathcal{A} = \varnothing$. Then $\{X \backslash F \mid F \in \mathcal{A}, F \neq K\}$ is an open cover for K and therefore there is a finite subset $\mathcal{C}$ of $\mathcal{A}$ such that $\{X \backslash F \mid F \in \mathcal{C}, F \neq K\}$ is an open cover for K, implying the contradiction that $K \cap \bigcap \mathcal{C} = \varnothing$. $\qquad \square$

Compactness of the Closed Unit Ball

The closed unit ball of a normed linear space is compact in the norm topology if and only if the space is finite dimensional; it is compact in the weak topology if and only if the space is reflexive. The closed unit ball of a dual space is always weak*-compact.

Theorem 11.2.6

Suppose X is a normed linear space. The closed unit ball of X is compact if and only if $\dim(X) < \infty$.

Proof

If $\dim(X) < \infty$, then the closed unit ball of X is compact by 11.1.21. For the converse, suppose it is compact and let (x_n) be a bounded sequence in X. Let $r \in \mathbb{R}^+$ be such that $\|x_n\| < r$ for all $n \in \mathbb{N}$. The sequence (x_n/r), being in the compact unit ball, has a convergent subsequence (x_{m_n}/r); then the sequence (x_{m_n}) also converges, so that X has finite dimension by 10.2.27. $\qquad \square$

Theorem 11.2.7 BANACH–ALAOGLU THEOREM

Suppose X is a normed linear space. Let B^* denote the closed unit ball $\flat_{X^*}[0\,;1]$ of X^* with the relative topology as a subspace of X^* with the weak* topology. Then B^* is a compact Hausdorff space.

Proof

The topology on B^* is Hausdorff (8.4 and 8.3.10). For each $x \in X$, let $B_x = \{z \in \mathbb{C} \mid |z| \le \|x\|\}$. Then $\prod_{x\in X} B_x \subseteq \mathbb{K}^X$ and $B^* = X^* \cap \prod_{x\in X} B_x$. The topology on B^* is that determined by the restrictions of the point evaluation functions to B^* (8.4 and 8.3.5); this is precisely the topology on B^* as a subspace of the product space $\prod_x B_x$ by 8.3.17 and 8.3.5. But $\prod_x B_x$, being a product of closed discs or intervals, is compact by Tychonoff's Theorem. So, by 11.1.12, we need only show that B^* is closed in $\prod_x B_x$. Towards this, suppose $f \in \prod_x B_x$ is in the closure of B^* in $\prod_x B_x$. Suppose $a, b \in X$ and $\epsilon \in \mathbb{R}^+$; then $\hat{a}^{-1}(\flat[f(a)\,;\epsilon/3)) \cap \hat{b}^{-1}(\flat[f(b)\,;\epsilon/3)) \cap \widehat{a+b}^{-1}(\flat[f(a+b)\,;\epsilon/3))$, being an open neighbourhood of f, contains some $g \in B^*$. Since $g(a+b) = g(a) + g(b)$, it follows that $|f(a+b) - f(a) - f(b)| < \epsilon$, and, since ϵ is arbitrary, that f is additive. By a similar calculation, $f(\lambda a) = \lambda f(a)$ for all $\lambda \in \mathbb{K}$, so that f is linear. Since $f \in \prod_x B_x$, f is bounded and $\|f\| \le 1$. $\square$

Lemma 11.2.8

Suppose X is a normed linear space. The copy B' of the closed unit ball $B = \flat_X[0\,;1]$ of X in X^{**} is dense in the closed unit ball $B^{**} = \flat_{X^{**}}[0\,;1]$ of X^{**} with its weak* topology.

Proof

We treat X as a real space. Suppose $\psi \in X^{**}$ is not in the weak*-closure of B'. Since the topology is locally convex, there exists a convex weak*-open neighbourhood N of ψ such that $N \cap B' = \varnothing$. By 5.2.31, there exist $\phi \in X^{**}$ and a maximal subspace M of X^{**} such that $\psi \notin M$, $N \subseteq M + \{\phi\} + \mathbb{R}^+\psi$ and $B' \subseteq M + \{\phi\} + \mathbb{R}^{\ominus}\psi$. Let $h \in \mathcal{L}(X^{**}, \mathbb{R})$ be such that $M = \ker(h)$ and $h(\psi) = 1$. The weak*-open set $N - \{\phi\}$ is included in wedge (h); so h is weak*-continuous by Q 8.4.4, whence $h = \hat{f}$ for some $f \in \mathcal{B}(X, \mathbb{R})$ by 8.4.5. Then, for all $x \in B$, we have $\psi(f) = h(\psi) > h(\phi) \ge h(\hat{x}) = \hat{x}(f) = f(x)$, whence $\psi(f) > h(\phi) \ge \|f\|$. Therefore $\|\psi\| > 1$. $\square$

Theorem 11.2.9

Suppose X is a normed linear space. The closed unit ball $B = \flat_X[0\,;1]$ of X is compact in the weak topology if and only if X is reflexive.

Proof

Recall from 8.4 that B with the weak topology is homeomorphic to its copy B' in X^{**} with its weak* topology. So one is compact if and only if the other is also, by 11.1.8. But the closed unit ball B^{**} of X^{**} is compact in the weak* topology (11.2.7), and B' is weak*-dense in B^{**} (11.2.8); so, since the weak* topology is Hausdorff, B' is weak*-compact if and only if $B' = B^{**}$ by 11.2.3. This last clearly occurs if and only if X is reflexive. $\square$

Approximation

Approximation is an important tool in analysis. Functions which occur naturally in scientific problems may not exhibit the nice properties of linear maps; but a great many of those functions can be likened locally to approximations of linear maps and the properties of linearity used to study them; these *differentiable* functions lie outside the scope of this book. Continuous functions need not be differentiable, but they may still be approximated by more tractable functions; the WEIERSTRASS APPROXIMATION THEOREM establishes that every continuous function $f: [0, 1] \to \mathbb{R}$ can be approximated uniformly with any desired level of accuracy by an easily constructed polynomial function. Stone extended this theorem to more abstract algebras of functions (Q 11.2.6); and the celebrated theorem of Gelfand and Naĭmark (12.4.20) can be viewed as a further extension of Weierstrass's result. *Machado's Lemma*, though not constructive, provides a neat way of proving the existence of such approximations. We adopt the following notation: if X is a topological space, S is a non-empty subset of X, A is a non-empty subset of $\mathcal{C}_0(X)$ (11.1.11), and $f \in \mathcal{C}_0(X)$, then $\|f\|_S = \sup\{|f(x)| \mid x \in S\}$ and $\mathrm{dist}_S(f, A) = \inf\{\|f - g\|_S \mid g \in A\}$.

Theorem 11.2.10 MACHADO'S LEMMA

Suppose X is a non-empty Hausdorff space, A is a closed subalgebra of $\mathcal{C}_0(X)$ and $f \in \mathcal{C}_0(X)$. There exists a closed non-empty subset S of X such that both $(g \in A$ and $g(S) \subseteq \mathbb{R} \Rightarrow g|_S$ constant$)$ and $\mathrm{dist}_S(f, A) = \mathrm{dist}_X(f, A)$.

Proof

Let $\mu = \mathrm{dist}_X(f, A)$. If $\mu = 0$ then $f \in A$, because A is closed, and any singleton subset of X satisfies our requirements. Suppose $\mu > 0$. Let $\mathcal{F}$ be the set of all non-empty closed subsets F of X for which $\mathrm{dist}_F(f, A) = \mu$. This is non-empty since $X \in \mathcal{F}$. We order $\mathcal{F}$ by inclusion and consider any totally ordered subset $\mathcal{B}$ of $\mathcal{F}$. For each $g \in A$ and $F \in \mathcal{B}$, let $N_{g,F} = \{x \in F \mid |f(x) - g(x)| \geq \mu\}$; then $N_{g,F}$ is closed because $f - g$ is continuous; $N_{g,F}$ is non-empty because $f - g$ attains its bound (11.1.11); and, since $f - g$ vanishes at infinity, $N_{g,F}$ is a subset of a compact set and is therefore compact by 11.1.12. So, for each $g \in A$, the collection $\{N_{g,F} \mid F \in \mathcal{B}\}$ is a nest of non-empty compact subsets of the Hausdorff space X; being nested, it has the Finite Intersection Property, so its intersection is non-empty by 11.2.5. But $\bigcup\{\bigcap\{N_{g,F} \mid F \in \mathcal{B}\} \mid g \in A\} \subseteq \bigcap \mathcal{B}$; so $\bigcap \mathcal{B} \neq \varnothing$ and, by construction, $\mathrm{dist}_{\bigcap \mathcal{B}}(f, A) = \mu$; thus $\bigcap \mathcal{B}$, being closed, is in $\mathcal{F}$. Zorn's lemma implies that $\mathcal{F}$ has a minimal element. Let S be such a set. Suppose there exists $p \in A$ such that $p|_S$ is real and not constant. If $p(S) = \{0, \|p\|_S\}$, let $q = p$; otherwise let $q = p^2\|p\|_S - p^3$; now set $h = q/\|q\|_S$; then $h \in A$ and $h|_S$ is real and $\inf h(S) = 0$ and $\sup h(S) = 1$. Let $c \in (0, 1/2)$ and $d \in (1/2, 1)$. Now let $M_1 = \{x \in S \mid h(x) \leq d\}$ and $M_2 = \{x \in S \mid c \leq h(x)\}$; then M_1 and M_2 are proper non-empty subsets of S, closed because h is continuous; and the minimality of S ensures that there exist functions $g_1, g_2 \in A$ satisfying $\|f - g_1\|_{M_1} < \mu$ and $\|f - g_2\|_{M_2} < \mu$. For each $n \in \mathbb{N}$ define

$$h_n = (1 - h^n)^{2^n} \quad \text{and} \quad k_n = g_1 h_n + g_2(1 - h_n).$$

Then $0 \leq h_n \leq 1$ for each $n \in \mathbb{N}$; also $1 - h_n \in A$, so that $g_1 - k_n \in A$ and hence $k_n \in A$. Also $|f - k_n| \leq |f - g_1| h_n + |f - g_2| (1 - h_n)$ for all $n \in \mathbb{N}$; we use this inequality on $S \backslash M_2$, $S \backslash M_1$ and $M_1 \cap M_2$. Firstly, on $S \backslash M_2$, by 4.4.2,

$$h_n = (1 - h^n)^{2^n} \geq 1 - (2h)^n > 1 - (2c)^n \to 1,$$

so that $\limsup \|f - k_n\|_{S \backslash M_2} \leq \|f - g_1\|_{M_1} < \mu$. Secondly, on $S \backslash M_1$,

$$h_n \leq (1 + h^n)^{-2^n} < 1/(2h)^n < 1/(2d)^n \to 0,$$

so that $\limsup \|f - k_n\|_{S \backslash M_1} \leq \|f - g_2\|_{M_2} < \mu$. And lastly, on $M_1 \cap M_2$, we have $|f - k_n| \leq \max\{|f - g_1|, |f - g_2|\} < \mu$ for all $n \in \mathbb{N}$. So $\|f - k_m\|_S < \mu$ for some sufficiently large $m \in \mathbb{N}$, contradicting the definition of S.   □

EXERCISES

Q 11.2.1 A topological space X is called a LINDELÖF SPACE if and only if every cover for X has a countable subcover. Prove LINDELÖF'S THEOREM that every second countable space is a Lindelöf space.

Q 11.2.2 A topological space in which every countable open cover has a finite subcover is said to be COUNTABLY COMPACT. Show that a second countable space is countably compact if and only if it is compact.

Q 11.2.3 Verify that every finite dimensional normed linear space is reflexive.

Q 11.2.4 Show that every bijective continuous mapping from a compact space onto a Hausdorff space is a homeomorphism.

Q 11.2.5 Show that every non-empty compact metric space X is a continuous image of the Cantor set C.

Q 11.2.6 (STONE–WEIERSTRASS THEOREM) Suppose X is a Hausdorff space and A is a closed subalgebra of $\mathcal{C}_0(X)$, closed under conjugation, such that, for each $x, y \in X$, there exist $g, h \in A$ with $g(x) \neq 0$ and $h(x) \neq h(y)$. Show that $A = \mathcal{C}_0(X)$.

11.3 Local Compactness

Compactness can be used to great advantage in spaces such as $\mathbb{R}$ which are not compact but have the related property of *local compactness*.

Definition 11.3.1
A topological space X is said to be LOCALLY COMPACT if and only if every element of X has an open neighbourhood whose closure is compact.

Example 11.3.2
Compact spaces are locally compact. Discrete spaces are locally compact. And a normed linear space is locally compact if and only if it has finite dimension (Q 11.3.5).

Example 11.3.3
A one-point compactification of a topological space X is Hausdorff if and only if X is a locally compact Hausdorff space. Suppose $\tilde{X}$ is a one-point compactification of X and $\tilde{X} \backslash X = \{p\}$. If X is locally compact and $x \in X$, there exists $U \in \mathrm{nbd}_X(x)$ such that $\overline{U}$ is compact, whence $\tilde{X} \backslash \overline{U} \in \mathrm{nbd}_{\tilde{X}}(p)$. Since $U \in \mathrm{nbd}_{\tilde{X}}(x)$, $\tilde{X}$ is Hausdorff if X is. Conversely, if $\tilde{X}$ is Hausdorff and $x \in X$, there exist $U \in \mathrm{nbd}_{\tilde{X}}(x)$ and a closed compact subset K of X with $U \subseteq K$; then $\overline{U}$ is compact by 11.1.12. So X is locally compact; and X is T_2 by 8.3.10.

Example 11.3.4
Subspaces of locally compact spaces need not be locally compact (Q 11.3.1), but all subspaces of locally compact Hausdorff spaces which are either open or closed are so (Q 11.3.2). Moreover, unlike their compact counterparts, locally compact Hausdorff spaces need not be normal (Q 11.3.4); this fact can be used in Q 11.3.4 to give an example of a normal space with a non-normal subspace.

Theorem 11.3.5

Suppose X is a locally compact Hausdorff space, K is a compact subset of X and U is an open subset of X with $K \subseteq U$. Then there exists an open subset V of X with $\overline{V}$ compact such that $K \subseteq V \subseteq \overline{V} \subseteq U$. This applies in particular when K is a singleton set $\{v\}$; in such a case, $V \in \mathrm{nbd}(v)$.

Proof

Let $\tilde{X}$ be a one-point compactification of X. Then $\tilde{X}$ is compact Hausdorff, by 11.3.3, and hence normal by 11.2.2. U is open in $\tilde{X}$; and K, being compact, is closed (11.2.3) in $\tilde{X}$. By 7.4.6, there exists an open subset V of $\tilde{X}$ such that $K \subseteq V \subseteq \mathrm{Cl}_{\tilde{X}}(V) \subseteq U$. Since $U \subseteq X$, both V and $\mathrm{Cl}_{\tilde{X}}(V)$ are subsets of X; so V is open in X and $\mathrm{Cl}_{\tilde{X}}(V)$, being closed in X, equals $\mathrm{Cl}_X(V)$.   □

Theorem 11.3.6 BAIRE'S THEOREM

Every locally compact Hausdorff space is a Baire space.

Proof

Suppose X is a locally compact Hausdorff space. Suppose $\mathcal{U}$ is a non-empty countable collection of open dense subsets of X. Let $(U_n)_{n \in \mathbb{I}}$ be an enumeration of the members of $\mathcal{U}$ and let B_0 be any non-empty open subset of X. By 11.3.5, 7.2.20 and the Recursive Choice Theorem, there exists a decreasing sequence $(B_n)_{n \in \mathbb{I}}$ of non-empty open subsets of X with compact closure such that $\overline{B_n} \subseteq U_n \cap B_{n-1}$ for each $n \in \mathbb{I}$. By 11.2.5, $\bigcap \{\overline{B_n} \mid n \in \mathbb{I}\} \neq \varnothing$. So $B_0 \cap \bigcap \mathcal{U} \neq \varnothing$. Since B_0 is arbitrary, $\bigcap \mathcal{U}$ is dense in X by 7.2.20.   □

EXERCISES

Q 11.3.1 Find a subspace of a compact space which is not locally compact.

Q 11.3.2 Suppose X is a locally compact Hausdorff space and S is a non-empty open subset of X. Show that, if S is either open or closed in X, then S is a locally compact Hausdorff space in the relative topology.

Q 11.3.3 Suppose X is a compact Hausdorff space and $x \in X$. Show that $X \backslash \{x\}$ is locally compact.

Q 11.3.4 Find a locally compact Hausdorff space which is not normal and a normal space with a non-normal subspace.

Q 11.3.5 Show that a normed linear space is locally compact if and only if it is finite dimensional.

12
Completeness

A non-closed subspace of a topological space is singularly defective, particularly if the topology determines geometric structure. We therefore present *completeness* as a property of universal closure, confining our attention to metric spaces. We shall see that extraordinarily powerful theorems are available in a metric space which contains every possible boundary point in every possible metric superspace.

12.1 Complete Metric Spaces

Definition 12.1.1
Suppose X is a set. A metric d on X is said to be COMPLETE if and only if the metric space (X, d) is closed in every metric superspace; in this case (X, d) is called a COMPLETE METRIC SPACE.

Example 12.1.2
The trivial metric space $(\varnothing, \varnothing)$ is complete. Every finite metric space is complete. Indeed, every compact metric space X is complete: suppose Y is a superspace of X; then X is a compact subset of Y by 11.1.5 and so is closed in Y by 11.1.3. But locally compact metric spaces need not be complete: $[0, 1)$, for example, is not closed in $[0, 1]$.

245

Virtual Points

We associate with each metric space X a set $\mathrm{vp}(X)$ of *virtual points* of X and show that $\mathrm{vp}(X)$ is empty if and only if X is complete.

Definition 12.1.3

Suppose X is a metric space. A function $u \colon X \to \mathbb{R}^{\oplus}$ will be called a POINT SIMULATOR for X if and only if $u(a) - u(b) \le d(a,b) \le u(a) + u(b)$ for all $a, b \in X$ and $\inf \mathrm{ran}(u) = 0$; if, moreover, $0 \notin \mathrm{ran}(u)$, then u will be called a VIRTUAL POINT of X. The set of point simulators for X will be denoted by $\mathrm{ps}(X)$ and the set of virtual points of X by $\mathrm{vp}(X)$. Note that point simulators are contractions (8.1.15) and are therefore continuous.

Example 12.1.4

Suppose (X, d) is a metric space; then each $z \in X$ induces a point simulator $x \mapsto d(z, x)$, denoted by δ_z in 8.2.3; then $\delta_z \notin \mathrm{vp}(X)$ because $\delta_z(z) = 0$. Conversely, if $u \in \mathrm{ps}(X) \setminus \mathrm{vp}(X)$, then there exists $z \in X$ such that $u(z) = 0$, which yields $u(x) \le d(z, x) \le u(x)$ for all $x \in X$, and therefore $u = \delta_z$. So $\mathrm{ps}(X) \setminus \mathrm{vp}(X) = \{\delta_z \mid z \in X\}$. The function $x \mapsto (1 - x)$ defined on $[0, 1)$ is a virtual point of $[0, 1)$; defined on $[0, 1]$, it is the point simulator δ_1.

Theorem 12.1.5

Suppose (X, d) is a metric space. Then X is complete if and only if $\mathrm{vp}(X) = \varnothing$.

Proof

Suppose $u \in \mathrm{vp}(X)$ and $p \notin X$. Set $d(p, p) = 0$ and $d(p, x) = u(x) = d(x, p)$ for each $x \in X$, thus extending d to the set $(X \cup \{p\}) \times (X \cup \{p\})$. This extension is certainly symmetric; and, since $d(p, a) - d(p, b) \le d(a, b) \le d(p, a) + d(p, b)$ for all $a, b \in X$ and also $0 \notin \mathrm{ran}(u)$, it is a metric on $X \cup \{p\}$. Moreover $\mathrm{dist}(p, X) = \inf \mathrm{ran}(u) = 0$, so that X is not closed in $X \cup \{p\}$, whence X is not complete. For the converse, if X is not complete and (Y, d) is a superspace of (X, d) in which X is not closed, then $z \in (\mathrm{Cl}_Y X) \setminus X \Rightarrow \delta_z|_X \in \mathrm{vp}(X)$. $\square$

Example 12.1.6

$\mathbb{R}$ and $\mathbb{C}$ are both complete; indeed, every finite dimensional normed linear space X is complete. Suppose $u \in \mathrm{ps}(X)$ and let $D = \{x \in X \mid u(x) \le 1\}$; then D is closed by continuity of u and $D \subseteq \flat[a \,; u(a) + 1]$ for any $a \in X$; so D is compact by 11.1.21. Then $\{X \setminus u^{-1}(\flat[0 \,; r]) \mid r \in \mathbb{R}^{+}\}$ is not a cover for D, as it can have no finite subcover for D because $\inf \mathrm{ran}(u) = 0$. Therefore $0 \in \mathrm{ran}(u)$.

Theorem 12.1.7

Every isometric image of a complete metric space is complete.

Proof

Suppose that X and Y are metric spaces and that $\phi\colon X \to Y$ is an isometry onto Y. Suppose $u \in \operatorname{ps}(Y)$ and let $v = u \circ \phi$. It follows, for all $a, b \in X$, that $v(a) - v(b) = u(\phi(a)) - u(\phi(b)) \leq d(\phi(a), \phi(b)) = d(a, b)$ and, similarly, that $d(a, b) \leq v(a) + v(b)$. Certainly $\inf \operatorname{ran}(v) = 0$; so $v \in \operatorname{ps}(X)$. If X is complete, there exists $a \in X$ with $v(a) = 0$; then $u(\phi(a)) = 0$ and $u \notin \operatorname{vp}(Y)$. □

Theorem 12.1.8

Suppose X is a complete metric space and Z is a subspace of X. Then Z is complete if and only if Z is closed in X.

Proof

If Z is complete, Z is closed by definition. For the converse, suppose $u \in \operatorname{ps}(Z)$; define $\tilde{u}$ on X to be $x \mapsto \inf\{d(x, z) + u(z) \mid z \in Z\}$. For every $a, b \in X$, we have $\tilde{u}(a) \leq d(a, z) + u(z) \leq d(a, b) + d(b, z) + u(z)$ for all $z \in Z$, from which we infer that $\tilde{u}(a) \leq d(a, b) + \tilde{u}(b)$. Furthermore, for all $w, z \in Z$, we have $d(a, b) \leq d(a, w) + d(w, z) + d(b, z) \leq d(a, w) + u(w) + u(z) + d(b, z)$, whence $d(a, b) \leq \tilde{u}(a) + \tilde{u}(b)$. Clearly $\tilde{u}|_Z = u$, so that $\inf \operatorname{ran}(\tilde{u}) = 0$ and $\tilde{u} \in \operatorname{ps}(X)$. Since X is complete, there exists $c \in X$ with $\tilde{u}(c) = 0$; then $\operatorname{dist}(c, Z) = 0$. If Z is closed, then $c \in Z$, so that $u \notin \operatorname{vp}(Z)$; and Z is complete by 12.1.5. □

Completely Ordered Metric Spaces

We have used the term *completeness* in different contexts to refer to orderings and to metrics. A simple criterion links the two ideas; and, although we have already shown that the metric on $\mathbb{R}$ is complete, it is worth noting that it is the relationship (6.1.9) between the ordering and the metric which makes it so.

Theorem 12.1.9

Suppose $(X, <)$ is a completely ordered set and d is a metric on X. Suppose that every bounded non-empty subset S of X is order-bounded and satisfies $\operatorname{dist}(\sup S, S) = 0 = \operatorname{dist}(\inf S, S)$. Then (X, d) is a complete metric space.

Proof

Firstly, note that the infima alluded to in the proposition do exist by Q 1.3.2. If $X = \varnothing$, the result is immediate; so we suppose otherwise. Suppose $u \in \operatorname{ps}(X)$.

Let $\epsilon \in \mathbb{R}^+$ and let $p \in X$ be such that $u(p) < \epsilon$. Also, for each $\alpha \in \mathbb{R}^+$, let $F_\alpha = \{x \in X \mid u(x) \le \alpha\}$ and $\mathcal{F} = \{F_\alpha \mid \alpha \in \mathbb{R}^+\}$; then $F_\alpha \ne \varnothing$ because $\inf \operatorname{ran}(u) = 0$; and, for each $x \in F_\alpha$, $d(x,p) \le u(x) + u(p) < \alpha + \epsilon$, so that $F_\alpha \subseteq \flat[p\,;\alpha + \epsilon)$. By hypothesis, each F_α is order-bounded and, since X is completely ordered, each has a supremum in X. Let $U = \{\sup F_\alpha \mid \alpha \in (0,\epsilon)\}$. By the hypothesis on suprema, $U \subseteq \flat[p\,;2\epsilon]$; then U is order-bounded by hypothesis. Now $\mathcal{F}$ is nested, so that $\{\sup F_\alpha \mid \alpha \in \mathbb{R}^+\}$ is bounded below and has infimum $s = \inf U$ (which is therefore independent of ϵ). By the hypothesis on infima, $s \in \flat[p\,;2\epsilon]$. Then $u(s) \le u(p) + d(p,s) < 3\epsilon$. Since ϵ is arbitrary in $\mathbb{R}^+$, it follows that $u(s) = 0$, so that $u \notin \operatorname{vp}(X)$. Then X is complete by 12.1.5.   □

Completions

Every metric d on a set X can be extended to make of $X \cup \operatorname{vp}(X)$ a complete metric space which has X as a dense subspace.

Definition 12.1.10

Suppose X is a metric space and Y is a complete metric space. Then Y is called a COMPLETION of X if and only if there exists an isometric map $\phi\colon X \to Y$ such that $\phi(X)$ is dense in Y.

Lemma 12.1.11

Suppose (X,d) is a metric space and u and v are point simulators for X. Then $\sup \operatorname{ran}(u - v) = \inf \operatorname{ran}(u + v) = \sup \operatorname{ran}(v - u)$.

Proof

For each $a, b \in X$, $u(a) - u(b) \le d(a,b) \le v(a) + v(b)$; so $u(a) - v(a) \le u(b) + v(b)$ and, since a and b are arbitrary, $\sup \operatorname{ran}(u - v) \le \inf \operatorname{ran}(u + v)$. Let $\epsilon \in \mathbb{R}^+$; then, since $\inf \operatorname{ran}(v) = 0$, there exists $z \in X$ such that $v(z) < \epsilon$, whence $\inf \operatorname{ran}(u + v) - \epsilon \le u(z) \le \sup \operatorname{ran}(u - v) + \epsilon$ and, since ϵ is arbitrary, we get $\inf \operatorname{ran}(u + v) \le \sup \operatorname{ran}(u - v)$. Putting together the two inequalities and reversing the rôles of u and v, we have the result.   □

Theorem 12.1.12

Suppose (X,d) is a metric space. Let s be the function $(u,v) \mapsto \inf \operatorname{ran}(u + v)$ defined on $\operatorname{ps}(X) \times \operatorname{ps}(X)$ Then s is a complete metric on $\operatorname{ps}(X)$ and $(\operatorname{ps}(X), s)$ is a completion of X.

Proof

s is symmetric, non-negative and real; and if $u, v \in \mathrm{ps}(X)$ and $s(u, v) = 0$, then, by 12.1.11, $\mathrm{ran}|u - v| = \{0\}$, whence $u = v$. Also, if $u, v, w \in \mathrm{ps}(X)$, then, for all $x \in X$, using 12.1.11, $s(u, w) \leq u(x) + w(x) \leq s(u, v) + v(x) + w(x)$, whence $s(u, w) \leq s(u, v) + s(v, w)$ because x is arbitrary. So s is a metric on $\mathrm{ps}(X)$. Also $s(\delta_a, \delta_b) \leq \delta_a(a) + \delta_b(a) = d(a, b) = \delta_a(b) - \delta_b(b) \leq s(\delta_a, \delta_b)$ for all $a, b \in X$, so that $\Delta = \{\delta_x \mid x \in X\}$ is an isometric copy of X in $\mathrm{ps}(X)$. And, for $u \in \mathrm{ps}(X)$ and $a \in X$, we have $s(u, \delta_a) \leq u(a)$; so $\mathrm{dist}_{\mathrm{ps}(X)}(u, \Delta) \leq \inf \mathrm{ran}(u) = 0$, whence Δ is dense in $\mathrm{ps}(X)$. It remains to show that $\mathrm{ps}(X)$ is complete. Suppose $h \in \mathrm{ps}(\mathrm{ps}(X))$ and define $\tilde{h} \colon X \to \mathbb{R}^{\oplus}$ to be $x \mapsto h(\delta_x)$; then $\tilde{h} \in \mathrm{ps}(X)$ because $x \mapsto \delta_x$ is isometric. Also $h(\tilde{h}) - h(\delta_x) \leq s(\tilde{h}, \delta_x) = \inf \mathrm{ran}(\tilde{h} + \delta_x) \leq \tilde{h}(x)$ for all $x \in X$; it follows that $h(\tilde{h}) \leq 2 \inf \mathrm{ran}(\tilde{h}) = 0$. So $h \notin \mathrm{vp}(\mathrm{ps}(X))$. Therefore $\mathrm{ps}(X)$ is complete by 12.1.5. □

The suggestion that a point simulator for a metric space X might be a member of X may seem bizarre; it is in fact a consequence of the Axiom of Foundation that it cannot be (see Q 1.2.5). So $X \cap \mathrm{vp}(X) = \varnothing$. It then follows from 12.1.7 and 12.1.12 that the superspace $X \cup \mathrm{vp}(X)$ of X is a completion of X when the metric d is extended by setting $d(x, u) = u(x) = d(u, x)$ and $d(u, v) = \inf \mathrm{ran}(u + v)$ for each $x \in X$ and $u, v \in \mathrm{vp}(X)$. This completion has the advantage over $\mathrm{ps}(X)$ that it is a metric superspace of X.

Theorem 12.1.13

Suppose (X, d) is a metric space. A metric space is a completion of X if and only if it is isometric to $(\mathrm{ps}(X), s)$ where s is $(u, v) \mapsto \inf \mathrm{ran}(u + v)$ defined on $\mathrm{ps}(X) \times \mathrm{ps}(X)$.

Proof

It is a consequence of 12.1.12 and 12.1.7 that every isometric copy of $(\mathrm{ps}(X), s)$ is a completion of X. Towards the converse, suppose that (Y, d') is a completion of X and $\phi \colon X \to Y$ is isometric with $\phi(X)$ dense in Y. For each $y \in Y$, define ψ_y on X to be $a \mapsto d'(y, \phi(a))$. Since $\phi(X)$ is dense in Y, it follows that $\inf \mathrm{ran}(\psi_y) = 0$; moreover, for all $a, b \in X$, we have the inequality $d'(y, \phi(a)) - d'(y, \phi(b)) \leq d'(\phi(a), \phi(b)) \leq d'(y, \phi(a)) + d'(y, \phi(b))$, which yields $\psi_y(a) - \psi_y(b) \leq d(a, b) \leq \psi_y(a) + \psi_y(b)$ because ϕ is an isometric map. So $\psi_y \in \mathrm{ps}(X)$. Now s is a complete metric on $\mathrm{ps}(X)$ by 12.1.12. Suppose that $p, q \in Y$. Let $\epsilon \in \mathbb{R}^+$; then, because $\phi(X)$ is dense in Y, there exists $z \in X$ such that $d'(q, \phi(z)) < \epsilon$. Since ϵ is arbitrary, the calculations $s(\psi_p, \psi_q) = \inf \mathrm{ran}(\psi_p + \psi_q) \leq d'(p, \phi(z)) + d'(q, \phi(z)) \leq d'(p, q)$ and $s(\psi_p, \psi_q) = \sup \mathrm{ran}(\psi_p - \psi_q) \geq d'(p, \phi(z)) - d'(q, \phi(z)) \geq d'(p, q) - 2\epsilon$ imply that $s(\psi_p, \psi_q) = d'(p, q)$. So $\psi\colon Y \to \mathrm{ps}(X)$ is isometric; then $\psi(Y)$ is complete by 12.1.7 and therefore closed in $\mathrm{ps}(X)$. But, for each $x \in X$, $\psi_{\phi(x)} = \delta_x$; and $\{\delta_x \mid x \in X\}$ is dense in $\mathrm{ps}(X)$ by 12.1.12. Therefore $\psi(Y) = \mathrm{ps}(X)$.   □

Cantor's Intersection Theorem

One of the most useful properties of complete spaces is that, under certain conditions, a nest of closed sets has a unique point of intersection.

Theorem 12.1.14 CANTOR'S INTERSECTION THEOREM
Suppose X is a metric space. Then X is complete if and only if, for every non-empty nest $\mathcal{F}$ of non-empty subsets of X with $\inf\{\mathrm{diam}(F) \mid F \in \mathcal{F}\} = 0$, the set $\bigcap\{\overline{F} \mid F \in \mathcal{F}\}$ is a singleton set.

Proof

Suppose X is complete and $\mathcal{F}$ is a non-trivial nest of non-empty subsets of X for which $\inf\{\mathrm{diam}(F) \mid F \in \mathcal{F}\} = 0$. For each $x \in X$, define $u(x)$ to be $\sup\{\mathrm{dist}(x, F) \mid F \in \mathcal{F}\}$. Let $\epsilon \in \mathbb{R}^+$ and let $C \in \mathcal{F}$ be such that $\mathrm{diam}(C) < \epsilon$. Then $u(x) \le \mathrm{dist}(x, C) + \epsilon$ for each $x \in X$, because $\mathcal{F}$ is nested; in particular u is a real function and $u(C) \subseteq [0, \epsilon]$, so that, since ϵ is arbitrary, $\inf \mathrm{ran}(u) = 0$. For $a, b \in X$, we have $d(a, b) \le \mathrm{dist}(a, C) + \epsilon + \mathrm{dist}(b, C) \le u(a) + u(b) + \epsilon$, whence $d(a, b) \le u(a) + u(b)$ because ϵ is arbitrary; and, for each $F \in \mathcal{F}$, $\mathrm{dist}(a, F) \le d(a, b) + u(b)$, whence $u(a) - u(b) \le d(a, b)$. So $u \in \mathrm{ps}(X)$ and, since X is complete, there exists $z \in X$ with $u(z) = 0$. So $z \in \bigcap\{\overline{F} \mid F \in \mathcal{F}\}$ by 6.1.24. Towards the converse, suppose that the condition is satisfied and that $u \in \mathrm{ps}(X)$. Let $\mathcal{F} = \{u^{-1}[0, r) \mid r \in \mathbb{R}^+\}$; $\mathcal{F}$ is clearly a nest of non-empty subsets of X and $\inf\{\mathrm{diam}(F) \mid F \in \mathcal{F}\} = 0$. By hypothesis, there exists $z \in \bigcap\{\overline{F} \mid F \in \mathcal{F}\}$. So $u(z) = 0$ and $u \notin \mathrm{vp}(X)$. X is complete by 12.1.5.   □

Example 12.1.15

It is certainly necessary to have some restriction on the diameters of the subsets in Cantor's Intersection Theorem in order to ensure that the intersection is a singleton. But this condition is used also to ensure that the intersection is non-empty; curiously, larger diameters may yield an empty intersection, as happens for the nest $\{[n, \infty) \mid n \in \mathbb{N}\}$ of closed intervals of $\mathbb{R}$. Even closure and boundedness together are not sufficient to infer non-emptiness; consider the example of the nest of intervals $\{(0, 1/n) \mid n \in \mathbb{N}\}$ in $\mathbb{R}$ with the discrete metric.

Theorem 12.1.16

Suppose (X, d) is a complete metric space and S is a non-empty set. Then $B(S, X)$, endowed, as usual, with the supremum metric (6.1.18) is complete.

Proof

Suppose $\mathcal{N}$ is a nest of non-empty subsets of $B(S, X)$ the infimum of whose diameters is 0. For each $s \in S$, the nest $\{\{f(s) \mid f \in N\} \mid N \in \mathcal{N}\}$ of subsets of X has the same property; so, by Cantor's Intersection Theorem, there exists a unique member of X in $\bigcap \{\mathrm{Cl}\{f(s) \mid f \in N\} \mid N \in \mathcal{N}\}$; label this $g(s)$, thus defining a function $g: S \to X$. We claim $g \in \bigcap \{\overline{N} \mid N \in \mathcal{N}\}$. Suppose $A \in \mathcal{N}$ and $\epsilon \in \mathbb{R}^+$; let $M \in \mathcal{N}$ be such that $\mathrm{diam}(M) < \epsilon/2$ and $M \subseteq A$. Suppose $f \in M$ and $u \in S$. By definition of g, there is $h \in M$ with $d(g(u), h(u)) < \epsilon/2$; but also $d(f(u), h(u)) \leq \epsilon/2$ because $f \in M$, so that $d(f(u), g(u)) < \epsilon$; so $g \in B(S, X)$ and $\mathrm{dist}(g, A) \leq \sup\{d(f(s), g(s)) \mid s \in S\} \leq \epsilon$. So $\mathrm{dist}(g, A) = 0$ and $g \in \overline{A}$. Therefore $g \in \bigcap \{\overline{N} \mid N \in \mathcal{N}\}$, as claimed, and $B(S, X)$ is complete by Cantor's Intersection Theorem. $\qquad \square$

Banach's Fixed Point Theorem

This theorem, known also as the CONTRACTION MAPPING PRINCIPLE is important in the applications of mathematical analysis to the solution of differential and integral equations; and it is at the heart of many algorithmic procedures.

Definition 12.1.17

Suppose X is a set, $z \in X$ and $f: X \to X$. Then z is called a FIXED POINT for f if and only if $f(z) = z$.

Theorem 12.1.18 BANACH'S FIXED POINT THEOREM

Suppose X is a complete metric space and $f: X \to X$ is a strong contraction with Lipschitz constant $k \in [0, 1)$. Then f has a unique fixed point in X; moreover, for each $a \in X$, $(f^n(a))$ converges to this fixed point.

Proof

For each $x \in X$ and $n \in \mathbb{N}$, we have $d(f^n(x), f^{n+1}(x)) \leq k^n d(x, f(x))$ by induction; since $k < 1$, each of the sets $B_n = \{w \in X \mid d(w, f(w)) < 1/n\}$ is non-empty. Then $d(a, b) \leq d(a, f(a)) + d(f(a), f(b)) + d(b, f(b))$ for each $n \in \mathbb{N}$ and $a, b \in B_n$, whence $(1 - k)d(a, b) \leq 2/n$, so that $\mathrm{diam}(B_n) \leq 2/(n - nk)$ and thus $\inf\{\mathrm{diam}(B_n) \mid n \in \mathbb{N}\} = 0$. Since $\{B_n \mid n \in \mathbb{N}\}$ is nested, there is a unique

$z \in X$ such that $d(z, f(z)) = 0$ by Cantor's Intersection Theorem. So $f(z) = z$ and, for each $a \in X$ and $n \in \mathbb{N}$, $d(z, f^n(a)) = d(f^n(z), f^n(a)) \leq k^n d(z, a) \to 0$, whence $f^n(a) \to z$.   □

Baire's Theorem

Complete metric spaces are Baire spaces (7.2.22). This fact, dependent on the Axiom of Choice, is a powerful tool which plays a seminal rôle in Operator Theory. Unlike Cantor's Intersection Theorem, Baire's result holds in some metric spaces which are not complete (Q 12.1.2).

Theorem 12.1.19 BAIRE'S THEOREM
Every complete metric space is a Baire space.

Proof
Suppose (X, d) is a complete metric space and $\mathcal{U}$ is a non-empty countable collection of open dense subsets of X. Let B_0 be any non-empty open subset of X and let $(U_n)_{n \in \mathbb{I}}$ be an enumeration of the members of $\mathcal{U}$. The Recursive Choice Theorem and 7.2.21 ensure that there exists a corresponding sequence $(B_n)_{n \in \mathbb{I}}$ of open balls of X, with a radius of each B_n less than $1/n$, such that $\overline{B_n} \subseteq U_n \cap B_{n-1}$ for each $n \in \mathbb{I}$. Then $\bigcap\{\overline{B_n} \mid n \in \mathbb{I}\} \neq \varnothing$, trivially if $\mathbb{I}$ is finite and, otherwise, since X is complete, by Cantor's Intersection Theorem. So $B_0 \cap \bigcap \mathcal{U} \neq \varnothing$. Since B_0 is arbitrary, $\bigcap \mathcal{U}$ is dense in X by 7.2.20.   □

Sequential Completeness

By weakening slightly the definition of a convergent sequence we achieve the wider concept of a *Cauchy sequence*. It is usual to define a complete metric space to be one in which every Cauchy sequence converges; and this criterion is, in any case, particularly easy to use. We show in 12.1.22 below that this notion of completeness is equivalent to our own in ZFC.

Definition 12.1.20
Suppose (X, d) is a metric space. A sequence (x_n) in X is called a CAUCHY SEQUENCE if and only if, for every $\epsilon \in \mathbb{R}^+$, there exists a ball of X of radius ϵ which includes a tail of (x_n). We shall say that (X, d) is a SEQUENTIALLY COMPLETE space if and only if every Cauchy sequence in X converges in X.

Theorem 12.1.21

Suppose X is a metric space and (x_n) is a Cauchy sequence in X which has a convergent subsequence. Then (x_n) is itself convergent to the same limit.

Proof

Suppose (x_{m_n}) is a subsequence of (x_n) and $x_{m_n} \to z \in X$. Let $\epsilon \in \mathbb{R}^+$. There is a ball B of radius $\epsilon/3$ which includes a tail t_k of (x_n); and $C = \flat[z\,;\epsilon/3]$ includes a tail of (x_{m_n}). So $C \cap B \neq \varnothing$ and $t_k \subseteq C \cup B \subseteq \flat[z\,;\epsilon)$. $\square$

Theorem 12.1.22

A metric space (X, d) is complete if and only if it is sequentially complete.

Proof

Firstly, suppose that X is complete and that (a_n) is a Cauchy sequence in X. For each $n \in \mathbb{N}$, let t_n denote the nth tail of (a_n). Then $\mathcal{F} = \{t_n \mid n \in \mathbb{N}\}$ is a nest and, because (a_n) is Cauchy, $\inf\{\operatorname{diam}(t_n) \mid n \in \mathbb{N}\} = 0$. By Cantor's Intersection Theorem, there exists $z \in \bigcap\{\overline{F} \mid F \in \mathcal{F}\}$; then $d(a_n, z) \leq \operatorname{diam}(t_n)$ for each $n \in \mathbb{N}$, so that $a_n \to z$. Towards the converse, suppose that every Cauchy sequence in X converges in X and let $u \in \operatorname{ps}(X)$. For each $n \in \mathbb{N}$, $\{x \in X \mid u(x) < 1/n\} \neq \varnothing$ because $\inf \operatorname{ran}(u) = 0$. By the Product Theorem, there is a sequence (a_n) in X with $u(a_k) < 1/k$ for each $k \in \mathbb{N}$; then $\flat[a_k\,;2/k]$ includes the kth tail of (a_n). So (a_n) is Cauchy and, by hypothesis, converges to some $z \in X$. Since $u(z) \leq u(a_n) + d(z, a_n)$ for each $n \in \mathbb{N}$, it follows that $u(z) = 0$ and that $u \notin \operatorname{vp}(X)$. So $\operatorname{vp}(X) = \varnothing$ and X is complete by 12.1.5. $\square$

Note that the Axiom of Choice was used in 12.1.22. The assertion that every Cauchy sequence in a metric space X converges in X is equivalent in ZF to the assertion that X is sequentially closed in every metric superspace. It cannot be shown in ZF that a sequentially closed subset of an arbitrary metric space is necessarily closed, simply because it cannot be proved that an arbitrary infinite set has a countable subset. For the same reason, it cannot be proved in ZF that an arbitrary set admits any metric which is not sequentially complete.

Equivalence of Complete Metrics

It is evident from 6.1.30 that equivalent metrics yield the same convergent sequences and the same Cauchy sequences; it follows from 12.1.22 that every metric which is equivalent to a complete metric is complete. The following rather technical result will have implications for normed linear spaces in 12.2.

Theorem 12.1.23

Suppose that X is a set, that d and d' are metrics on X and that d is stronger than d'. Suppose that (X, d) is complete and that there exists $k \in \mathbb{R}^+$ such that $b'[a\,;kr) \subseteq \mathrm{Cl}'(b[a\,;r))$ for all $a \in X$ and $r \in \mathbb{R}^+$, where b' and Cl' refer respectively to balls and closures in (X, d'). Then d is equivalent to d' and (X, d') is complete.

Proof

Note that the inclusion condition yields the implication that for each $a \in X$ and $s \in \mathbb{R}^+$, $v \in b'[a\,;ks) \Rightarrow b[a\,;s) \cap b'[v\,;ks/2) \neq \varnothing$. Suppose $b[x_0\,;r)$ is an arbitrary ball in (X, d) and $v \in b'[x_0\,;kr/4)$. Then, because of the foregoing implication, the Recursive Choice Theorem implies that there exists a sequence (x_n) in X such that $x_n \in b[x_{n-1}\,;r/2^{n+1}) \cap b'[v\,;kr/2^{n+2})$. Clearly (x_n) is Cauchy in (X, d) and converges in (X, d') to v. Since (X, d) is complete and d' is weaker than d, it follows using 6.1.30 that (x_n) converges to v in (X, d) also. Note that $d(x_0, x_n) \leq \sum_{i=1}^{n} d(x_{i-1}, x_i) \leq r/2$, so that $d(x_0, v) < r$. Since v was arbitrarily chosen in $b'[x_0\,;kr/4)$, we then have $b'[x_0\,;kr/4) \subseteq b[x_0\,;r)$. Since $b[x_0\,;r)$ is an arbitrary ball in (X, d), this implies that d' is stronger than d and hence that d and d' are equivalent metrics. $\square$

EXERCISES

Q 12.1.1 Show that every discrete metric is complete.

Q 12.1.2 Find an incomplete metric space which is a Baire space.

Q 12.1.3 Find a continuous image of a complete space which is not complete.

12.2 Banach Spaces

Definition 12.2.1

A normed linear space $(X, \|\cdot\|)$ is called a BANACH SPACE if and only if the metric induced by its norm is complete; in this case, we shall say also that $\|\cdot\|$ is a COMPLETE NORM.

Example 12.2.2

Let X be a non-empty topological space. Since $\mathbb{K}$ is complete, so also is $B(X, \mathbb{K})$ by 12.1.16. Then, since $\mathcal{C}(X)$ is closed in $B(X, \mathbb{K})$ (8.3.24), it too is complete by

12.1.8. We show that $\mathcal{C}_0(X)$ is closed in $\mathcal{C}(X)$ and therefore complete. Suppose $f \in \mathcal{C}(X)$ and $\operatorname{dist}(f, \mathcal{C}_0(X)) = 0$. Let $\epsilon \in \mathbb{R}^+$. There exists $g \in \mathcal{C}_0(X)$ such that $\|g - f\| < \epsilon/2$ and there exists a compact subspace K of X such that $|g(x)| < \epsilon/2$ for all $x \in X \backslash K$. Then $|f(x)| < \epsilon$ for all $x \in X \backslash K$, so that $f \in \mathcal{C}_0(X)$. In contrast, $\mathcal{C}_c(\mathbb{R})$ is not complete (Q 12.2.1).

Example 12.2.3

The usual topology on $\mathbb{N}$ is discrete, so that every sequence in a topological space is automatically continuous. Consequently, ℓ_∞ is $\mathcal{C}(\mathbb{N})$ (Q 8.3.4) and c_0 (6.2.10) is $\mathcal{C}_0(\mathbb{N})$, whence ℓ_∞ and c_0 are both complete by 12.2.2. Also, because of the identification (Q 10.2.5) between continuity at ∞ and convergence, the map $(x_n)_{n \in \mathbb{N}} \mapsto (x_n)_{n \in \mathbb{N}} \cup \{(\infty, \lim x_n)\}$ is an isometry from the space c of convergent scalar sequences onto $\mathcal{C}(\check{\mathbb{N}})$. So c is complete by 12.2.2 and 12.1.7.

Example 12.2.4

Suppose X and Y are normed linear spaces and Y is complete. Then, by 12.1.16, $B(\flat_X[0;1), Y)$ is complete. Therefore $\{T|_{\flat_X[0;1)} \mid T \in \mathcal{B}(X,Y)\}$, being a closed subset (6.2.16) of $B(\flat_X[0;1), Y)$, is complete, and $\mathcal{B}(X,Y)$, being an isometric copy of $\{T|_{\flat_X[0;1)} \mid T \in \mathcal{B}(X,Y)\}$, is complete by 12.1.7.

Example 12.2.5

Since $\mathbb{K}$ is complete, 12.2.4 ensures that every dual space is complete and then, by 12.1.7, that every reflexive space is complete; thus ℓ_1, being isometric to c_0^* (Q 8.4.5) is complete and, for $p \in (1, \infty)$, the spaces ℓ_p are complete by 8.4.7.

Suppose X is a normed linear space. Then X is isometric to a subspace of the complete space X^{**} (8.4); the closure of this subspace in X^{**} is also a subspace of X^{**} and is complete by 12.1.8; and this closure is a completion of X. Therefore X has a completion which is a Banach space. It follows that every completion of X, being an isometric copy of this completion, can be endowed with algebraic operations and with an extension of the norm which make it into a Banach space (6.2.37). But some care is required here, because the algebraic operations may conflict with naturally occurring ones; $\operatorname{ps}(X)$, for example, is a completion of X whose members are real functions which can be added and multiplied, but $\operatorname{ps}(X)$ is not generally even closed under these operations, and they certainly cannot be used to make $\operatorname{ps}(X)$ into a Banach space completion of X. Nonetheless, our observations here imply that a normed linear space is complete if and only if it is closed in every normed linear superspace.

Example 12.2.6

A series $\sum_{n\in\mathbb{N}} x_n$ in a normed linear space X is said to be ABSOLUTELY CON-VERGENT if and only if the real series $\sum_{n\in\mathbb{N}}\|x_n\|$ converges. In contrast to 4.3.20, absolute convergence need not imply convergence. In fact it necessarily does so if and only if X is complete. If X is complete and $\sum_{n\in\mathbb{N}}\|x_n\|$ converges, then $\sum_{n\in\mathbb{N}} x_n$ is a Cauchy sequence, therefore converges. On the other hand, if X is not complete, there exists a Cauchy sequence (w_n) in X which does not converge; and, for each $n \in \mathbb{N}$, there exists a ball B_n of X of radius less than $1/2^{n+2}$ which includes a tail of (w_n). By the Recursive Choice Theorem, (w_n) has a subsequence (w_{k_n}) with $w_{k_n} \in \bigcap\{B_i \mid 1 \leq i \leq n\}$ for each $n \in \mathbb{N}$. Let $k_0 = w_0 = 0$ and, for each $n \in \mathbb{N}$, set $x_n = w_{k_n} - w_{k_{n-1}}$. The series $\sum_{n\in\mathbb{N}} x_n$ is precisely the sequence (w_{k_n}), so does not converge by 12.1.21. But, for $n \in \mathbb{N}$, $\sum_{i=1}^{n}\|x_i\| = \sum_{i=1}^{n}\|w_{k_i} - w_{k_{i-1}}\| \leq \sum_{i=1}^{n} 1/2^i$. So $\sum_{n\in\mathbb{N}}\|x_n\|$ converges.

Theorem 12.2.7

Suppose X is a Banach space and M is a closed linear subspace of X. Then X/M is a Banach space.

Proof

Suppose (x_n/M) is a Cauchy sequence in X/M. By the Product Theorem, there exists a sequence (z_n) in M such that $(x_n - z_n)$ is Cauchy in X. Since X is complete, there exists $w \in X$ such that $x_n - z_n \to w$, and it follows that $x_n/M \to w/M$ in X/M.   □

Equivalence of Complete Norms

Baire's Theorem ensures that complete norms on a single linear space are either equivalent or not comparable. It yields other important theorems, notably the Open Mapping Theorem (12.2.10), the Closed Graph Theorem (Q 12.2.3) and the Banach–Steinhaus Uniform Boundedness Principle (Q 12.2.4).

Theorem 12.2.8

Suppose X is a linear space and $\|\cdot\|$ and $\|\cdot\|'$ are norms on X, the latter weaker than the former. Suppose $(X, \|\cdot\|)$ is a Banach space and $(X, \|\cdot\|')$ is a Baire space. Then the norms are equivalent and $\|\cdot\|'$ is complete.

Proof

$\{X\backslash\mathrm{Cl}'(\flat[0\,;n)) \mid n \in \mathbb{N}\}$ is a nest of open subsets of $(X, \|\cdot\|')$ with empty intersection. By Baire' Theorem, there exists $m \in \mathbb{N}$ such that $X\backslash\mathrm{Cl}'(\flat[0\,;m))$ is not dense in $(X, \|\cdot\|')$. So there exists $r \in \mathbb{R}^+$ such that $b'[0\,;r) \subseteq \mathrm{Cl}'(b[0\,;m))$. Using 6.2.12, $b'[a\,;r) \subseteq \mathrm{Cl}'(b[a\,;m))$ for all $a \in X$, and we invoke 12.1.23.   □

Theorem 12.2.9

Suppose X and Y are Banach spaces and $T \in \mathcal{B}(X,Y)$. If T is bijective then $T^{-1} \in \mathcal{B}(Y,X)$.

Proof

T^{-1} is certainly linear. Denote by $\|\cdot\|'$ the norm $x \mapsto \|Tx\|$ induced by T on X (6.2.9). $T\colon X \to Y$ is an isometry when X is endowed with this norm, so that $\|\cdot\|'$ is complete by 12.1.7. Also, for each $x \in X$, $\|x\|' = \|Tx\| \le \|T\| \|x\|$, so that $\|\cdot\|'$ is weaker than the given norm on X and so equivalent to it by 12.2.8. Therefore there exists $k \in \mathbb{R}^+$ such that, for all $y \in Y$, $\|T^{-1}y\| \le k\|T^{-1}y\|' = k\|y\|$.   □

Theorem 12.2.10   OPEN MAPPING THEOREM

Suppose X and Y are Banach spaces and $T \in \mathcal{B}(X,Y)$. If T is surjective then T is open.

Proof

Because T is bounded, $\ker(T)$ is closed (6.2.17); so $X/\ker(T)$ is complete by 12.2.7. The map $x/\ker(T) \mapsto Tx$ is well defined on X; we denote it by $\tilde{T}$; it is clearly bijective and bounded with $\|\tilde{T}\| = \|T\|$. By 12.2.9, $\tilde{T}$ is open. But the quotient map $\pi\colon X \to X/\ker(T)$ is also open (8.3.34); so $T = \tilde{T} \circ \pi$ is open.   □

EXERCISES

Q 12.2.1 Show that $\mathcal{C}_c(\mathbb{R})$ (11.1.11) is not complete.

Q 12.2.2 (DINI'S THEOREM) Suppose X is a compact space and (f_n) is an increasing sequence of real functions in $\mathcal{C}(X)$ which converges pointwise to $f \in \mathcal{C}(X)$. Show that $f_n \to f$ in $\mathcal{C}(X)$.

Q 12.2.3 (CLOSED GRAPH THEOREM) Suppose X and Y are Banach spaces and $T \in \mathcal{L}(X,Y)$. Show that T is bounded if and only if its graph $\Gamma(T)$ is closed in $X \times Y$ with the product topology.

Q 12.2.4 (UNIFORM BOUNDEDNESS PRINCIPLE) Suppose X is a Banach space and $\mathcal{B}$ is a non-empty set of bounded linear maps from X into normed linear spaces. Suppose that, for each $x \in X$, $\{\|Tx\| \mid T \in \mathcal{B}\}$ is bounded in $\mathbb{R}$. Show that $\{\|T\| \mid T \in \mathcal{B}\}$ is also bounded in $\mathbb{R}$.

12.3 Hilbert Spaces

Definition 12.3.1
A complex inner product space is called a HILBERT SPACE if and only if the
metric determined by the inner product is complete.

If H is an inner product space and $\tilde{H}$ is a completion of H, then certainly the
algebraic operations of H can be reproduced and extended to $\tilde{H}$ to make it into
a linear space; the inner product can also be similarly extended to $\tilde{H}$ so that it
determines the extended metric and makes $\tilde{H}$ into a Hilbert space (Q 12.3.1).
It follows as for Banach spaces that an inner product space is complete if and
only if it is closed in every inner product superspace.

Example 12.3.2
The map $(x, y) \mapsto \sum_{n=1}^{\infty} x_n \overline{y_n}$ is an inner product on ℓ_2 (6.3.10); it determines
the norm $\|\cdot\|_2$, which is complete (12.2.5). So ℓ_2 is a Hilbert space. And ℓ_2^n is a
Hilbert space for each $n \in \mathbb{N}$; in fact every complex linear space of dimension
n can be made into a Hilbert space using an isomorphism from ℓ_2^n (3.4.22).

Convexity in Hilbert Spaces

Closed convex subsets of a Hilbert space have a special property (12.3.3), not
shared by arbitrary Banach spaces, which determines much of the extraordinary
structure of Hilbert spaces. It ensures, for example, that Hilbert spaces, unlike
all other Banach spaces, have the property that, for every closed subspace M,
there is a closed complementary subspace; in fact $M^\perp$ is such a complement.

Theorem 12.3.3 SMALLEST VECTOR THEOREM
Suppose H is a Hilbert space and C is a closed non-empty convex subset of H.
The C contains a unique vector of smallest norm.

Proof
Let $d = \inf\{\|x\| \mid x \in C\}$. For each $r \in \mathbb{R}^+$, let $B_r = C \cap \flat[0\,;d+r]$. For
$a, b \in B_r$, convexity of C ensures that $(a + b)/2 \in C$, so that $\|a + b\| \geq 2d$;
then $\|a - b\|^2 = 2\|a\|^2 + 2\|b\|^2 - \|a + b\|^2 \leq 4(d + r)^2 - 4d^2 = 4r(2d + r)$ by
the Parallelogram Law, whence $\inf\{\operatorname{diam}(B_r) \mid r \in \mathbb{R}^+\} = 0$. So, by Cantor's
Intersection Theorem, the nest $\{B_r \mid r \in \mathbb{R}^+\}$ of closed subsets of H has a
unique point z of intersection. Then $z \in C$ and, clearly, $\|z\| = d$. $\square$

Theorem 12.3.4

Suppose H is a Hilbert space and M is a closed subspace of H. For each $x \in H$, there exists a unique $y \in M$ with $\|x - y\| = \text{dist}(x, M)$; moreover $x - y \in M^\perp$. Consequently, $H = M \oplus M^\perp$.

Proof

Let $x \in H$; then $x + M$ is a closed convex subset of H. By 12.3.3, there exists a unique $z \in x + M$ with $\|z\| = \text{dist}(x, M)$. Set $y = x - z$ and the first assertion follows. Moreover $\langle z, 0 \rangle = 0$, and if $m \in M \backslash \{0\}$, then, for each $\alpha \in \mathbb{C}$, we have $y + \alpha m \in M$ so that $\|z\|^2 \le \|x - y - \alpha m)\|^2 = \|z\|^2 - 2\Re\langle z, \alpha m \rangle + \|\alpha m\|^2$, from which it follows that $|\alpha|^2 \|m\|^2 \ge 2\Re\langle z, \alpha m \rangle$ which is contradicted by $\alpha = \langle z, m \rangle / \|m\|^2$ unless $\langle z, m \rangle = 0$. So $z \in M^\perp$ and $x = y + z \in M \oplus M^\perp$. $\square$

Representation of Functionals

It was shown in 6.3.11 that each vector g in an inner product space H determines a unique functional $x \mapsto \langle x, g \rangle$, denoted there by ϕ_g. It was shown also that the map $g \mapsto \phi_g$ is an isometric conjugate-linear mapping from H to H^*. If H is complete, then this mapping is bijective. This result effectively tells us that the dual of a Hilbert space is a mirror image of the space itself and, consequently, that every Hilbert space is reflexive.

Theorem 12.3.5 RIESZ REPRESENTATION THEOREM

Suppose H is a Hilbert space. With respect to the notation of 6.3.11, the map $g \mapsto \phi_g$ is a bijective isometric conjugate-linear mapping from H onto H^*.

Proof

The map $g \mapsto \phi_g$ is an isometric conjugate-linear map from H to H^* by 6.3.11; being isometric, it is injective; we show that it is surjective. Certainly $0 = \phi_0$. Suppose $f \in H^* \backslash \{0\}$; then $\ker(f)$ is a proper closed subspace of H (6.2.41). By 12.3.4, there exists $z \in (\ker(f))^\perp \backslash \{0\}$. Set $w = \overline{f(z)}z/\|z\|^2$. For $y \in \ker(f)$ and $\alpha \in \mathbb{C}$, $\phi_w(y + \alpha z) = \left\langle y + \alpha z, \overline{f(z)}z/\|z\|^2 \right\rangle = \alpha f(z) = f(y + \alpha z)$. Since $H = \ker(f) \oplus \mathbb{C}z$, we therefore have $f = \phi_w$. $\square$

Orthonormal Bases

Every separable Hilbert space has a Schauder basis with extra properties. Moreover, something very like a Schauder basis bearing the same special traits exists in each non-separable Hilbert space.

Definition 12.3.6

Suppose H is a Hilbert space and $E \subseteq H$. Then E is called an ORTHONORMAL SET in H if and only if, for every $a, b \in E$, $\langle a, b \rangle$ is 1 if $a = b$ and $\langle a, b \rangle = 0$ otherwise. If E is maximal with this property, E is called an ORTHONORMAL BASIS for H.

Theorem 12.3.7

Every Hilbert space has an orthonormal basis.

Proof

Suppose H is a Hilbert space. Order the collection of orthonormal subsets of H by inclusion. The union of any totally ordered collection of orthonormal subsets of H is orthonormal and is an upper bound for this collection. Zorn's lemma ensures that H has a maximal orthonormal subset. $\qquad\square$

Example 12.3.8

ℓ_2 has orthonormal basis $\{e_n \mid n \in \mathbb{N}\}$, where, for each $n \in \mathbb{N}$, e_n is the sequence whose n^{th} term is 1 and whose other terms are all 0.

Theorem 12.3.9

Suppose H is a Hilbert space, $x \in H$, and E is an orthonormal basis for H. Then the set $S = \{e \in E \mid \langle x, e \rangle \neq 0\}$ is countable, $\sum_{e \in S} \langle x, e \rangle e$ converges unconditionally to x, and $\sum_{e \in S} |\langle x, e \rangle|^2$ converges unconditionally to $\|x\|^2$.

Proof

If $x = 0$, the result certainly holds. Suppose $x \neq 0$ and F is any finite subset of E; then $\left\| \sum_{e \in F} \langle x, e \rangle e \right\|^2 = \langle \sum_{e \in F} \langle x, e \rangle e, \sum_{e \in F} \langle x, e \rangle e \rangle = \sum_{e \in F} |\langle x, e \rangle|^2$. Since $\langle x, \sum_{e \in F} \langle x, e \rangle e \rangle = \sum_{e \in F} |\langle x, e \rangle|^2$, it follows that $\left\| x - \sum_{e \in F} \langle x, e \rangle e \right\|^2$ equals $\|x\|^2 - \sum_{e \in F} |\langle x, e \rangle|^2$ and that $\sum_{e \in F} |\langle x, e \rangle|^2 \leq \|x\|^2$. Therefore, for each $n \in \mathbb{N}$, the set $C_n = \{e \in E \mid |\langle x, e \rangle| > \|x\| /n\}$ is finite and $S = \bigcup(C_n)_{n \in \mathbb{N}}$ is countable. Also, since the terms $|\langle x, e \rangle|^2$ are real and positive and every finite sum of them is bounded by $\|x\|^2$, 4.3.5 and 10.2.12 ensure that $\sum_{e \in S} |\langle x, e \rangle|^2$ converges unconditionally. Let $(b_i)_{i \in \mathbb{N}}$ be an enumeration of S, extended if S is finite by setting $b_j = 0$ for all $j > |S|$. For each $n \in \mathbb{N}$, set $z_n = \sum_{i=1}^{n} \langle x, b_i \rangle b_i$. Then $\|z_n - z_m\|^2 = \left\| \sum_{i=n+1}^{m} \langle x, b_i \rangle b_i \right\|^2 = \sum_{i=n+1}^{m} |\langle x, b_i \rangle|^2$ for all $m, n \in \mathbb{N}$ with $n < m$, so that (z_n) is a Cauchy sequence in H; since H is complete, $z_n \to w$ for some $w \in H$. Note that, for each $j, n \in \mathbb{N}$ with $n > j$, we have $\langle x - z_n, b_j \rangle = 0$. It follows from the continuity of the inner product (Q 8.3.3) that $\langle x - w, b_j \rangle = 0$ and also that $\langle x - w, e \rangle = 0$ for all $e \in E \backslash S$. Since E is a maximal orthonormal subset of H, we have $x - w = 0$ and therefore $w = x$. Then also $\|x\|^2$ is the unconditional limit of $\sum_{e \in S} |\langle x, e \rangle|^2$. $\square$

If H is a Hilbert space, E is an orthonormal basis for H and $x \in H$, the expression $x = \sum_E \langle x, e \rangle e$ can now be used unambiguously to indicate that x is the unconditional limit of any series $\sum_S \langle x, e \rangle e$ where S is countable and $\langle x, e \rangle = 0$ for all $e \in E \backslash S$. Moreover, it is easy to check that the coefficients are uniquely determined; that, if $x = \sum_{e \in E} \lambda_e e$ is a similar representation of x, then $\lambda_e = \langle x, e \rangle$ for all $e \in E$. It follows that any enumeration of a countable orthonormal basis is a Schauder basis.

Separable Hilbert Spaces

It is easy to show that all orthonormal bases for a Hilbert space have the same cardinality, which we call the ORTHOGONAL DIMENSION of the space. Though it clearly coincides with the algebraic dimension for finite dimensional spaces, it differs otherwise. This difference is illustrated forcibly by the fact that the Hilbert space ℓ_2 has orthogonal dimension ∞, whereas there is no Hilbert space with linear dimension ∞ (Q 12.3.3).

Theorem 12.3.10
Let H be a Hilbert space. Then H is separable if and only if its orthogonal dimension does not exceed ∞.

Proof
Certainly, if H has a countable orthonormal basis, then H is separable (10.2.30). Towards the converse, suppose E is an orthonormal basis of H and D is a dense subset of H. For each $e \in E$, the ball $B_e = \flat[e; 1/\sqrt{2})$ has non-empty intersection with D; but, for $e, f \in E$ with $f \neq e$, we have $\|e - f\| = \sqrt{2}$, so that $B_e \cap B_f = \varnothing$. We thus have $|E| \leq |D|$. $\square$

Suppose H is an infinite dimensional separable Hilbert space. Then the orthogonal dimension of H is ∞ by 12.3.10. Let $(e_n)_{n \in \mathbb{N}}$ be an enumeration of an orthonormal basis of H. It is easy to verify that $x \mapsto (\langle x, e_n \rangle)_{n \in \mathbb{N}}$ is an isometric isomorphism from H onto ℓ_2. Consequently, every infinite dimensional separable Hilbert space is isometrically isomorphic to ℓ_2. However, it is far from being the case that the only separable Hilbert space worth studying is ℓ_2, since there are many separable Hilbert spaces which differ greatly in other structure.

EXERCISES

Q 12.3.1 Suppose X is an inner product space. Show that every completion of X can be endowed with an inner product in such a way that X is isometrically isomorphic to a dense inner product subspace of the completion. Deduce that an inner product space is complete if and only if it is closed in every inner product superspace.

Q 12.3.2 Suppose H is a Hilbert space and M and N are orthogonal closed subspaces of H. Show that $M \oplus N$ is a closed subspace of H. Show that the conclusion might fail if M is not orthogonal to N.

Q 12.3.3 Show that no Hilbert space has algebraic dimension ∞.

12.4 Banach Algebras

Our finale is merely a glimpse into the great playground of complex Banach algebras and C^*-algebras, where the richness of structure guarantees the success of many beautiful results. We confine our attention to complex algebras.

Definition 12.4.1
A complete complex normed algebra is called a BANACH ALGEBRA.

Example 12.4.2
The simplest Banach algebra is, of course, $\mathbb{C}$. Of the various Banach spaces presented in 12.2, those which are also normed algebras, that is those which are closed under a multiplicative operation and have submultiplicative norm, are Banach algebras. In particular ℓ_∞ is a Banach algebra, and, more generally, $\mathcal{C}(X)$ is a Banach algebra for every topological space X. If X is a Banach space, then $\mathcal{B}(X)$ is a Banach algebra in which the multiplicative operation is composition.

Spectral Radius Formula

The most fundamental fact concerning Banach algebras is that every member of a unital Banach algebra has non-empty spectrum; indeed, its spectrum is compact and there is a neat formula for the largest modulus of its members (3.3.5). This result of Gelfand and Beurling can be viewed as an extension of the Fundamental Theorem of Algebra.

Definition 12.4.3

Suppose A is a unital algebra and $a \in A$. We define the SPECTRAL RADIUS of a in A to be $\sup\{|\lambda| \mid \lambda \in \sigma(a)\}$ if $\sigma(A) \neq \varnothing$ and to be $-\infty$ otherwise. It will be denoted by $\rho(a)$.

Theorem 12.4.4

Suppose A is a unital Banach algebra. Then $\mathrm{inv}(A)$ is open in A and, for each $a \in A$, $\rho(a) \leq \|a\|$.

Proof

Suppose $a \in A$ and $\lambda \in \mathbb{C}$ with $|\lambda| > \|a\|$. Then the series $\sum_{n=0}^{\infty} a^n/\lambda^{n+1}$ is Cauchy, so converges to some $z \in A$. Continuity of the algebraic operations implies that $\lambda - a$ is inverse to z, so that $\lambda \notin \sigma(a)$ and $\rho(a) \leq \|a\|$. Now suppose $v \in \mathrm{inv}(A)$ and $w \in A$ with $\|w\| < 1/\|v^{-1}\|$; then $1 > \|v^{-1}w\| \geq \rho(v^{-1}w)$, whence $1 - v^{-1}w \in \mathrm{inv}(A)$ and $v - w = v(1 - v^{-1}w) \in \mathrm{inv}(A)$.   □

Theorem 12.4.5 GELFAND–BEURLING THEOREM

Suppose A is a unital Banach algebra and $a \in A$. Then $\sigma(a)$ is a non-empty compact subset of $\mathbb{C}$ and $\rho(a) = \inf\{\|a^n\|^{1/n} \mid n \in \mathbb{N}\} = \lim\|a^n\|^{1/n}$. This formula for $\rho(a)$ is known as the SPECTRAL RADIUS FORMULA.

Proof

Suppose $k \in \mathbb{N}$. For each $n > k$, use 4.1.10 to write $n = d_n k + r_n$ where $0 \leq r_n < k$. Then $\|a^n\|^{1/n} = \|a^{d_n k} a^{r_n}\|^{1/n} \leq \|a^k\|^{d_n/n} \|a\|^{r_n/n} \to \|a^k\|^{1/k}$, from which we infer that $\limsup\|a^n\|^{1/n} \leq \inf\{\|a^n\|^{1/n} \mid n \in \mathbb{N}\}$ and hence that $\inf\{\|a^n\|^{1/n} \mid n \in \mathbb{N}\} = \lim\|a^n\|^{1/n} \leq \|a\|$; we denote this limit by s. Either $\rho(a) = -\infty < s$ or, for each $n \in \mathbb{N}$, $\rho(a) = (\rho(a^n))^{1/n}$ by 4.3.29; and $\rho(a^n) \leq \|a^n\|$ by 12.4.4, so that $\rho(a) \leq \|a^n\|^{1/n}$ and hence $\rho(a) \leq s$. We suppose $\rho(a) < s$ and achieve a contradiction. Recall the Average Inverse Theorem (4.3.29) and, for each $n \in \mathbb{N}$, let f_n denote the function $\lambda \mapsto \lambda^n(\lambda^n - a^n)^{-1}$ defined on $\{\mu \in \mathbb{C} \mid |\mu| > \rho(a)\}$. Continuity of the algebraic operations and of the inverse function (8.1.29) ensure that these functions are continuous, and therefore uniformly continuous on compact subsets of the domain (11.1.9). Let $\epsilon \in \mathbb{R}^+$. Then there exists $t \in (s, \infty)$ such that $\|f_1(\eta t) - f_1(\eta s)\| < \epsilon$ for all $\eta \in \mathbb{T}$. Invoking 4.3.29, for all $n \in \mathbb{N}$, there exists $\omega \in \mathbb{T}$ such that

$$f_n(t) - f_n(s) = \frac{1}{n} \sum_{i=1}^{n} (f_1(\omega^i t) - f_1(\omega^i s)),$$

whence $\|f_n(t) - f_n(s)\| < \epsilon$. Now $\lim\|a^n/t^n\|^{1/n} = s/t < 1$, whence $a^n/t^n \to 0$ and $f_n(t) = (1 - t^{-n}a^n)^{-1} \to 1$ by continuity of the operations and of the inverse function. So, for all sufficiently large $n \in \mathbb{N}$, we have $\|1 - f_n(s)\| < 2\epsilon$. Since ϵ is arbitrary, $s \neq 0$ and $(1 - s^{-n}a^n)^{-1} = f_n(s) \to 1$. Again by continuity, $a^n/s^n \to 0$. So, for some $k \in \mathbb{N}$, $\|a^k\|^{1/k} < s$, which contradicts the definition of s. Therefore $\rho(a) = s \in \mathbb{R}^\oplus$ and $\sigma(a) \neq \varnothing$. The openness of A^{-1} in A (12.4.4) ensures that $\sigma(a)$ is closed; being bounded, it is compact by 11.1.21. $\qquad \Box$

Corollary 12.4.6 GELFAND–MAZUR THEOREM
Suppose A is a unital Banach algebra which is also a division algebra. Then A consists simply of the scalar multiples of the identity.

Proof
Let $a \in A$. Then, by 12.4.5, there exists $\lambda \in \sigma(a)$. Since A is a division algebra, it follows that $\lambda - a = 0$. $\qquad \Box$

Example 12.4.7
For each $m \in \mathbb{N}$, $\mathcal{M}_{m \times m}(\mathbb{C})$ is an algebra which is finite dimensional as a linear space. It can be equipped with various complete norms, all of which are equivalent (10.2.28). For each $A \in \mathcal{M}_{m \times m}(\mathbb{C})$ and for any such norm, $\lim\|A^n\|^{1/n}$ is independent of the norm and is the largest modulus of an eigenvalue of A.

Characters and Maximal Ideals

The kernel of a non-zero linear functional on an algebra A is certainly a maximal subspace of A; if the functional preserves multiplication, then its kernel is necessarily a maximal algebra ideal. In general, there is no guarantee that a maximal ideal is a maximal subspace of A; but in a commutative unital Banach algebra, this must be so (12.4.12).

Theorem 12.4.8
Suppose A is a Banach algebra and $\phi: A \to \mathbb{C}$ is a non-zero algebra homomorphism. Then $\phi \in A^*$ and $\|\phi\| \leq 1$. If A is unital then $\phi(1) = 1$ and $\|\phi\| = 1$.

Proof
Suppose $z \in \flat_A[0\,;1)$. Then $\sum_{n \in \mathbb{N}} z^n$ is Cauchy, so converges to some $u \in A$. Continuity of the algebraic operations implies that $z = u - zu$, whence also $\phi(z) = \phi(u)(1 - \phi(z))$, yielding $\phi(z) \neq 1$. So, if $w \in A\backslash\ker(\phi)$, we have

$w/\phi(w) \notin \flat_A[0\,;1)$, and then $\phi(w) \le \|w\|$. Thus $\|\phi\| \le 1$. If A is unital, then $\phi(1) \ne 0$ because $\phi \ne 0$, and $\phi(1)^2 = \phi(1)$, yielding $\phi(1) = 1$ and $\|\phi\| = 1$. $\quad\square$

Definition 12.4.9

Suppose A is a Banach algebra. A non-zero algebra homomorphism $\phi\colon A \to \mathbb{C}$ is called a CHARACTER. The set of characters on A endowed with the weak* topology inherited from A^* is called the CHARACTER SPACE or SPECTRUM of A and will be denoted by $\Omega(A)$. If $\Omega(A) \ne \varnothing$, then, for each $a \in A$, the point evaluation function $\hat{a}$ with domain $\Omega(A)$ is called the GELFAND TRANSFORM of a. Being a restriction of a continuous function, $\hat{a}$ is continuous; it is also in $B(\Omega(A)\,,\mathbb{C})$ because $|\hat{a}(\phi)| \le \|a\|$ for all $\phi \in \Omega(A)$ by 12.4.8. So $\{\hat{a} \mid a \in A\}$ is a subspace of $\mathcal{C}(\Omega(A))$; it will be denoted by $\hat{A}$. The map $a \mapsto \hat{a}$ from A to $\hat{A}$ is called the GELFAND MAPPING.

Theorem 12.4.10

Suppose A is a Banach algebra. Then $\Omega(A)$ is a compact Hausdorff space.

Proof

$B^* = \{f \in A^* \mid \|f\| \le 1\}$ with the weak* topology is a compact Hausdorff space by the Banach–Alaoglu Theorem (11.2.7); and $\Omega(A)$ is a subspace of B^* by 12.4.8; so, by 11.1.12, it is sufficient to show that $\Omega(A)$ is closed in B^*. Suppose that ϕ is in the closure of $\Omega(A)$ in B^*, that $a, b \in A$ and that $\epsilon \in \mathbb{R}^+$. Since $\hat{a}^{-1}(\flat[\phi(a)\,;\epsilon)) \cap \hat{b}^{-1}(\flat[\phi(b)\,;\epsilon)) \cap \widehat{ab}^{-1}(\flat[\phi(ab)\,;\epsilon))$ is a neighbourhood of ϕ, it contains some $\psi \in \Omega(A)$. So $\psi(ab) = \psi(a)\psi(b)$; then $\phi(ab) - \phi(a)\phi(b)$ equals $\phi(ab) - \psi(ab) + \phi(a)(\psi(b) - \phi(b)) + \phi(b)(\psi(a) - \phi(a)) + (\psi(b) - \phi(b))(\psi(a) - \phi(a))$. So $|\phi(ab) - \phi(a)\phi(b)| \le \epsilon(1 + |\phi(a)| + |\phi(b)| + \epsilon)$. Since ϵ is arbitrary, it follows that $\phi(ab) = \phi(a)\phi(b)$ and that $\phi \in \Omega(A)$. $\quad\square$

Theorem 12.4.11

Suppose A is a unital Banach algebra. Then every maximal ideal of A is closed.

Proof

Suppose M is a maximal ideal of A. Then $M \cap \text{inv}(A) = \varnothing$ because M is proper, and $1 \notin \overline{M}$ because $\text{inv}(A)$ is open in A by 12.4.4. So the ideal $\overline{M}$ (Q 7.2.9) is proper. Maximality of M ensures that $\overline{M} = M$. $\quad\square$

Theorem 12.4.12

Suppose A is a commutative unital Banach algebra. Then $\Omega(A) \ne \varnothing$ and the map $\phi \mapsto \ker(\phi)$ is a bijection from $\Omega(A)$ onto the set of maximal ideals of A.

Proof

The kernel of a character is certainly a maximal ideal; and if $\phi, \psi \in \Omega(A)$ are distinct, then, since their kernels are maximal subspaces and $\phi(1) = 1 = \psi(1)$, we must have $\ker(\phi) \neq \ker(\psi)$. Conversely, suppose M is a maximal ideal of A. Then M is closed in A by 12.4.11. So A/M is a Banach algebra (12.2.7); it is certainly simple, and, being commutative, is a division algebra (Q 3.2.4). By the Gelfand–Mazur Theorem, there exists an algebra isomorphism $\psi \colon A/M \to \mathbb{C}$ onto $\mathbb{C}$; then $a \mapsto \psi(a/M)$ is a character on A with kernel M. So the mapping $\phi \mapsto \ker(\phi)$ is bijective, and, since A admits maximal ideals (3.1.37), $\Omega(A) \neq \varnothing$. $\qquad \square$

Theorem 12.4.13 GELFAND REPRESENTATION THEOREM

Suppose A is a unital commutative Banach algebra. Then, for each $a \in A$, $\sigma(a) = \hat{a}(\Omega(A))$; and the Gelfand mapping is a norm-decreasing algebra homomorphism with $\|\hat{a}\| = \rho(a)$ for each $a \in A$.

Proof

For each $a \in A$ and $\phi \in \Omega(A)$, we have $\phi(\phi(a) - a) = 0$, so that, since $\phi \neq 0$, $\phi(a) - a \notin \mathrm{inv}(A)$ and $\hat{a}(\phi) = \phi(a) \in \sigma(a)$. Conversely, if $\lambda \in \sigma(a)$, then $(\lambda - a)A$ is a proper ideal of A which is therefore included in some maximal ideal M of A. Let ϕ be the character on A whose kernel is M; then $\lambda = \hat{a}(\phi) \in \hat{a}(\Omega(A))$. So $\hat{a}(\Omega(A)) = \sigma(a)$ and $\|\hat{a}\| = \sup |\hat{a}|(\Omega(A)) = \sup\{|\lambda| \mid \lambda \in \sigma(a)\} = \rho(a) \leq \|a\|$. Lastly, for each $a, c \in A$ and $\phi \in \Omega(A)$, $\widehat{ac}(\phi) = \phi(ac) = \phi(a)\phi(c) = \hat{a}(\phi)\hat{c}(\phi)$, and it follows that the Gelfand map is an algebra homomorphism. $\qquad \square$

C^*-Algebras

There is an immense literature on these algebras. But our intention here is quite modest: we shall define unital commutative C^*-algebras as mathematical structures with the basic properties we associate with $\mathbb{C}$; we shall then unfold the remarkably beautiful discovery of Gelfand and Naĭmark that such objects have an entirely different topological description, thus uniting in a single insight many of the concepts we have presented in the foregoing pages.

Definition 12.4.14

A Banach algebra A with involution * is called a C^*-ALGEBRA if and only if $\|a\|^2 \leq \|a^*a\|$ for all $a \in A$. This inequality implies that $\|a\|^2 \leq \|a^*\|\,\|a\|$ and thus $\|a\| \leq \|a^*\|$ and, since $a^{**} = a$, that $\|a\| = \|a^*\|$; it then yields also $\|a^*a\|^2 \leq \|a^*aa^*a\| \leq \|a\|^4$ and therefore $\|a^*a\| = \|a\|^2$.

Example 12.4.15

$\mathbb{C}$ is the archetypal C^*-algebra, conjugation being the involution. More generally, for any non-empty set S, $B(S,\mathbb{C})$, being a complete (12.1.16) normed algebra with conjugation (6.2.32) is a unital commutative C^*-algebra with the uniform norm (6.2.5). If X is a topological space, then, since conjugation is continuous, $\mathcal{C}(X)$ is a commutative C^*-subalgebra of $B(X,\mathbb{C})$. The content of the Gelfand–Naĭmark Theorem (12.4.20) is that every unital commutative C^*-algebra is isometrically *-isomorphic to $\mathcal{C}(X)$ for some compact Hausdorff space X; the compact Hausdorff space we have in mind is the character space.

Example 12.4.16

Suppose H is a Hilbert space. It is not difficult to show that, for each $T \in \mathcal{B}(H)$, there is an operator $T^* \in \mathcal{B}(H)$ uniquely determined by the equations $\langle x, T^*y \rangle = \langle Tx, y \rangle$ for all $x, y \in H$. Then $T \mapsto T^*$ is an involution on $\mathcal{B}(H)$ and, since $\langle x, T^*Tx \rangle = \|Tx\|^2$ for all $x \in H$, it follows from the Schwarz Inequality that $\|T\|^2 \leq \|T^*T\|$. So $\mathcal{B}(H)$ is a C^*-algebra. Then, of course, every closed subalgebra of $\mathcal{B}(H)$ which is closed under involution is also a C^*-algebra. In fact, every C^*-algebra is isometrically *-isomorphic to such an algebra, though we shall not prove it here.

Lemma 12.4.17

Suppose A is a unital C^*-algebra and $a \in A$. If $a = a^*$, then $\sigma(a) \subseteq \mathbb{R}$.

Proof

Suppose $\mu \in \sigma(a)$ and $t \in \mathbb{R}$; then $\mu + it \in \sigma(a + it)$, and it follows that $|\mu + it|^2 \leq \|a + it\|^2 = \|(a - it)(a + it)\|^2 = \|a^2 + t^2\| \leq \|a\|^2 + t^2$. Now we write $\mu = r + is$ where $r, s \in \mathbb{R}$; then $r^2 + (s + t)^2 = |\mu + it|^2 \leq \|a\|^2 + t^2$, which yields $r^2 + s^2 + 2st \leq \|a\|^2$. Since t is arbitrary, we must have $s = 0$. □

Lemma 12.4.18

Suppose A is a unital C^*-algebra, $a \in A$ and $\phi \in \Omega(A)$. Then $\phi(a^*) = \overline{\phi(a)}$ and $\widehat{a^*} = \overline{\hat{a}}$.

Proof

Write $a = u + iv$, where $u = (a + a^*)/2$ and $v = (a - a^*)/2i$. Then $u = u^*$ and $v = v^*$, so that $\phi(u) \in \sigma(u) \in \mathbb{R}$ and $\phi(v) \in \sigma(v) \in \mathbb{R}$ by 12.4.13 and 12.4.17. Also $a^* = u - iv$. So $\phi(a^*) = \phi(u) - i\phi(v) = \overline{\phi(u) + i\phi(v)} = \overline{\phi(a)}$. Now, for each $\psi \in \Omega(A)$, $\widehat{a^*}(\psi) = \psi(a^*) = \overline{\psi(a)} = \overline{\hat{a}}(\psi)$ so that $\widehat{a^*} = \overline{\hat{a}}$. □

Lemma 12.4.19

Suppose A is a unital C^*-algebra and $a \in A$ with $a^*a = aa^*$. Then $\rho(a) = \|a\|$.

Proof

$\left\|a^2\right\|^2 = \left\|(a^2)^*a^2\right\| = \|(a^*a)^*(a^*a)\| = \|a^*a\|^2 = \|a\|^4$, so that $\left\|a^2\right\| = \|a\|^2$. It follows by induction that, $\left\|a^{2^n}\right\| = \|a\|^{2^n}$ for each $n \in \mathbb{N}$. So the spectral radius formula (12.4.5) gives $\rho(a) = \|a\|$. $\qquad\square$

It follows from 12.4.19 that, given a unital complex algebra A with involution, there is at most one function on A which can be a norm which makes A into a C^*-algebra, namely $a \mapsto \sqrt{\rho(a^*a)}$—so the geometric structure of a C^*-algebra is entirely determined by its algebraic structure (and, of course, the structure of $\mathbb{C}$ itself). Given a non-unital C^*-algebra, it is a tricky exercise to show that the function $a \mapsto \sqrt{\rho(a^*a)}$ is a norm on its unitization. But our task now is simply to give a topological description of unital C^*-algebras.

Theorem 12.4.20 GELFAND–NAĬMARK THEOREM

Suppose A is a commutative unital C^*-algebra. Then the Gelfand map is an isometric *-isomorphism from A onto $\mathcal{C}(\Omega(A))$.

Proof

12.4.18 shows that the Gelfand map preserves involution. Also, we know from the Gelfand Representation Theorem (12.4.13) that it is a norm-decreasing homomorphism. That the norm is actually preserved by the Gelfand map follows from the following calculation which uses 12.4.18, 12.4.13 and 12.4.19:

$$\|\hat{a}\|^2 = \left\|\overline{\hat{a}}\hat{a}\right\| = \left\|\widehat{a^*}\hat{a}\right\| = \left\|\widehat{a^*a}\right\| = \rho(a^*a) = \|a^*a\| = \|a\|^2 .$$

So $\hat{A}$ is complete by 12.1.7 and therefore closed in $\mathcal{C}(\Omega(A))$; moreover, being isometric, the Gelfand map is injective. It remains to show that $\hat{A} = \mathcal{C}(\Omega(A))$. Let $f \in \mathcal{C}(\Omega(A))$. By Machado's Lemma (11.2.10), there exists a non-empty closed subset S of $\Omega(A)$ such that $\hat{a}|_S$ has constant modulus for all $a \in A$ and $\text{dist}_S(f, \hat{A}) = \text{dist}_{\Omega(A)}(f, \hat{A})$. Suppose $\phi, \psi \in S$. Then $|\hat{a}(\phi)| = |\hat{a}(\psi)|$, that is $|\phi(a)| = |\psi(a)|$, for all $a \in A$. So $|\psi| = |\phi|$ and therefore $\ker(\psi) = \ker(\phi)$, which implies $\psi = \phi$ by 12.4.12. So $S = \{\phi\}$. Let $g \in \mathcal{C}(\Omega(A))$ be the constant function whose value is $f(\phi)$. Then $g \in \hat{A}$ and $\|f - g\|_S = 0$, and it follows that $\text{dist}_{\Omega(A)}(f, \hat{A}) = \text{dist}_S(f, \hat{A}) = 0$. Finally, since $\hat{A}$ is closed in $\mathcal{C}(\Omega(A))$, we have $f \in \hat{A}$, as required. $\qquad\square$

Solutions

Sigh no more, ladies, sigh no more,
Men were deceivers ever,
One foot in sea and one on shore,
To one thing constant never.
Then sigh not so, but let them go,
And be you blithe and bonny;
Converting all your sounds of woe
Into Hey nonny nonny.   *Much Ado about Nothing, II,iii.*

1. Sets

Q 1.1.1 There are sets $\{a, b\}$ and $\{c, d\}$ by 1.1.5; and there is a set $\{\{a, b\}, \{c, d\}\}$ by 1.1.5. By Axiom IV, there is a set $\{a, b, c, d\}$.

Q 1.1.2 If $z \in a\backslash\bigcup b$, then $z \in a$ and $(y \in b \Rightarrow z \notin y)$, whence $y \in b \Rightarrow z \in a\backslash y$; then $z \in \bigcap\{a\backslash y \mid y \in b\}$. Conversely, if $z \in \bigcap\{a\backslash y \mid y \in b\}$, then $y \in b \Rightarrow z \in a\backslash y$, whence $z \in a$ and $z \notin \bigcap b$; then $z \in a\backslash\bigcap b$.

Q 1.2.1 • $x \in f^{-1}(\bigcup \mathcal{V}) \Leftrightarrow f(x) \in \bigcup \mathcal{V} \Leftrightarrow \exists V \in \mathcal{V} : x \in f^{-1}(V) \Leftrightarrow x \in \bigcup f^{-1}(\mathcal{V})$.
• $x \in f^{-1}(\bigcap \mathcal{V}) \Leftrightarrow f(x) \in \bigcap \mathcal{V} \Leftrightarrow \forall V \in \mathcal{V} : x \in f^{-1}(V) \Leftrightarrow x \in \bigcap f^{-1}(\mathcal{V})$.
• $y \in f(\bigcup \mathcal{U}) \Leftrightarrow \exists U \in \mathcal{U} : y \in f(U) \Leftrightarrow y \in \bigcup f(\mathcal{U})$.
• $y \in f(\bigcap \mathcal{U}) \Leftrightarrow \exists x \in \bigcap \mathcal{U} : y = f(x) \Rightarrow \forall U \in \mathcal{U}, y \in f(U) \Leftrightarrow y \in \bigcap f(\mathcal{U})$; and the penultimate implication can be reversed if f is injective.

Q 1.2.2 $x \in A \Rightarrow f(x) \in f(A) \Leftrightarrow x \in f^{-1}(f(A))$, with reversal also if f is injective.

Q 1.2.3 $x \in f^{-1}(B) \Leftrightarrow f(x) \in B \Leftrightarrow f(x) \in B \cap f(X)$. So $f(f^{-1}(B)) = B \cap f(X)$.

Q 1.2.4 For each $x \in X$, $((g^{-1} \circ f^{-1}) \circ (f \circ g))(x) = g^{-1}(f^{-1}(f(g(x)))) = x$ and, for each $z \in Z$, $((f \circ g) \circ (g^{-1} \circ f^{-1}))(z) = f(g(g^{-1}(f^{-1}(z)))) = z$ and it follows that $f \circ g$ is bijective and invertible with inverse $g^{-1} \circ f^{-1}$.

Q 1.2.5 Second part; first is similar. Suppose x, r, s, t are sets and $x \in r \in s \in t \in x$; let $A = \{x, r, s, t\}$; then $x \in r \cap A$, $r \in s \cap A$, $s \in t \cap A$ and $t \in x \cap A$; and A violates Axiom VI. So no such chain of membership is possible. Suppose $z \in a$ and $S \in \mathcal{P}(b^a)$. If $f \in S$, then $z \in \{z\} \in \{\{z\}, \{z, f(z)\}\} = (z, f(z)) \in f \in S$, whence $z \neq S$.

Q 1.2.6 $(b, a) \in r \Rightarrow (a, b) \in \text{ran}(r) \times \text{dom}(r)$; so the inverse of r is a set, which we label r^{-1}; since its members are ordered pairs, r^{-1} is a relation. If r is a function, r^{-1} is certainly injective and $r = (r^{-1})^{-1}$; conversely, if r is injective, then, for each $a \in \text{ran}(r)$, $\exists! b \in \text{dom}(r) : (b, a) \in r$, whence r^{-1} is a function.

Q 1.3.1 Suppose u and v are maximal in S; then $\acute{u} = \varnothing = \acute{v}$; so $u \notin \acute{v}$ and $v \notin \acute{u}$; since the ordering is total, $u = v$. The other arguments are similar.

Q 1.3.2 Suppose $A \subseteq S$ and $A \neq \varnothing$; suppose the set B of lower bounds for A is non-empty. B is bounded above by each member of A, so has a supremum $b \in S$. Then $b \in B$, for otherwise there would be some $a \in A$ with $a < b$; and, since a is an upper bound for B, this would contradict the definition of b. So $b = \max B = \inf A$.

269

Q 1.3.3 Suppose $(S, <)$ is a well ordered set. Suppose $a, b \in S$ and neither $a < b$ nor $b < a$; since $\{a, b\}$ has a minimum element, $a = b$; so S is totally ordered. Suppose A is a non-empty subset of S bounded above in S; then the set of upper bounds for A in S has a minimum element s, and $s = \sup A$.

Q 1.3.4 It assumes that there exists b such that $a \sim b$.

Q 1.3.5 Suppose $s \colon A \to B$ is a similarity map. Then s^{-1} is bijective; for $w, z \in B$, $s^{-1}(z) \leq s^{-1}(w) \Rightarrow z = s(s^{-1}(z)) \leq s(s^{-1}(w)) = w$; so $w < z \Rightarrow s^{-1}(w) < s^{-1}(z)$.

Q 1.3.6 $s(x_+) > s(x)$; and, if $y > s(x)$, then, by Q 1.3.5, $s^{-1}(y) > s^{-1}(s(x)) = x$, whence $s^{-1}(y) \geq x_+$ and $y = s(s^{-1}(y)) \geq s(x_+)$. The second part is similar.

Q 1.4.1 If α_- exists, then it is the least ordinal which is not a member of a member of α, so is $\bigcup \alpha$ (1.4.8). Otherwise, $\bigcup \alpha \subseteq \alpha$ (1.4.2) and, for $\beta \in \alpha$, $\exists \gamma$ with $\beta \in \gamma \in \alpha$, whence $\beta \in \bigcup \alpha$.

Q 1.4.2 If α is an ordinal, then $\bigcup \alpha \subseteq \alpha$ by 1.4.2. Conversely, if $\bigcup \alpha \subseteq \alpha$ and $z \in \beta \in \alpha$, then $z \in \alpha$, yielding $\beta \subseteq \alpha$, whence α, well ordered by $\in$, is an ordinal.

Q 1.5.1 Suppose f is a function. Then $\mathcal{C} = \{f^{-1}\{b\} \mid b \in \mathrm{ran}(f)\}$ is a collection of mutually disjoint non-empty sets. By the Singleton Intersection Theorem, there exists a subset A of $\bigcup \mathcal{C}$ which has singleton intersection with each member of $\mathcal{C}$. Then $f|_A$ is injective and has the same range as f.

Q 1.5.2 By the Product Theorem. See 1.2.25.

Q 1.5.3 If the condition is satisfied, then certainly $<$ is inductive. Suppose $<$ is inductive and S is a totally ordered subset of X which has no upper bound in X. By 1.4.13, there is an ordinal ω which is not equinumerous with any member of $\mathcal{P}(S)$. Let $r = \{(g, s) \mid s \in S, \ g \colon \omega \rightarrowtail S, \ \mathrm{ran}(g) \subseteq \grave{s}\}$. If $\alpha \in \omega$ and $f \in S^\alpha$ is strictly increasing, then $\mathrm{ran}(f)$ is well ordered, so has upper bound $z \in X$; since z is not an upper bound for S, $\exists s \in S$ with $\mathrm{ran}(f) \subseteq \grave{s}$; so $f \in \mathrm{dom}(r)$. By the Recursive Choice Theorem, there is an injective function from ω to S, contradicting the definition of ω.

Q 1.5.4 That the Recursive Choice Theorem implies the first form of the Lemma has been proved, and that that form implies the second is easy. We show that the second form implies the Well Ordering Principle. Suppose X is a set and, invoking 1.4.13, let ω be an ordinal which is not in one-to-one correspondence with any member of $\mathcal{P}(X)$. Let $\mathcal{A} = \{f \subseteq X \times \omega \mid f \text{ is an injective function}, \mathrm{ran}(f) \in \omega\}$. Then $\mathcal{A}$ is partially ordered by inclusion. If $\mathcal{C}$ is a nested subset of $\mathcal{A}$, then $\bigcup \mathcal{C}$ is an injective function out of X into ω; its range, being a union of ordinals in ω, is, by 1.4.8, a member of ω_+; and it is not ω, by definition. So $\bigcup \mathcal{C} \in \mathcal{A}$. By Zorn's Lemma II, there is a maximal member g of $\mathcal{A}$. $\mathrm{dom}(g) = X$, for, if $x \in X \backslash \mathrm{dom}(g)$, the function $g \cup \{(x, \mathrm{ran}(g))\}$ contradicts maximality of g. Finally, $\{(a, b) \in X \times X \mid g(a) < g(b)\}$ well orders X.

2. Counting

Q 2.1.1 Intersection must be performed on a set; and there is no set of all sets which satisfy the axiom.

Q 2.1.2 By 2.1.5 and 2.1.6.

Q 2.2.1 Let $h \colon |X| \to X$ be a bijection. $\alpha \subseteq |X|$, $\mathrm{ran}(h|_\alpha) \subseteq X$ and $|\mathrm{ran}(h|_\alpha)| = \alpha$.

Q 2.2.2 Let $h \colon S \to |S|$ be a bijection; well order S by $\{(a, b) \in S \times S \mid h(a) \in h(b)\}$; then h is a similarity mapping.

Q 2.3.1 For each $a, b \in S$, $\acute{a} \cap \grave{b}$, being a subset of S, is finite by 2.2.6; so the order is enumerative. There is a minimum element by 2.3.5; and S is well ordered by 2.3.8.

Q 2.3.2 Suppose there is a non-empty subset V of S with no least element. Then, for each $v \in V$, $\grave{v} \neq \varnothing$; by the Recursive Choice Theorem, there is a strictly decreasing function $f: \infty \to V$. Then $f(\infty)$ is a countable subset of S with no least element.

Q 2.3.3 No finite set has this property by 2.3.4. Suppose S is an infinite set. Let $f: \infty \to S$ be injective. Define $g: S \to S$ by $g(s) = f((f^{-1}(s))_+)$ if $s \in \mathrm{ran}(f)$ and $g(s) = s$ otherwise. Then g is a bijective map from S onto $S \backslash f(0)$.

Q 2.3.4 If $|\acute{w}| = \infty = |\grave{w}|$, apply 2.3.18 to $\{w\} \cup \acute{w}$; note that the reverse of an enumerative ordering is enumerative and do the same with $\grave{w}$ in place of $\acute{w}$ and r^{-1} in place of r; put together the two sequences. Simplify if either segment is finite.

Q 2.4.1 If S is enumeratively well ordered and $b = \max S$, then $S = \{a, b\} \cup (\acute{a} \cap \grave{b})$, where $a = \min S$; and S is finite by 2.4.1. Conversely, if S is finite, the similarity map $s: |S| \to S$ determines an ordering of the required type on S.

Q 2.4.2 By the Product Theorem, $\prod_i X_i$ has a subset $\prod_i S_i$ where $|S_j| = 2$ for all $j \in I$; clearly, this subset is equinumerous with $2^{|I|}$, so that $|\prod_i X_i| \geq |2^{|I|}|$. Moreover, $0 < |\bigcup(X_i)| \leq |I|$ by 2.4.5, so that $|I \times \bigcup(X_i)| = |I|$ by 2.4.6. By 2.2.8, $|\mathcal{P}(I \times \bigcup(X_i))| = |2^{|I|}|$. But $\prod_i X_i \subseteq \mathcal{P}(I \times \bigcup(X_i))$; so we invoke 2.2.6 to finish.

Q 2.4.3 It follows easily from 2.4.6 that if all except a finite number of the sets are singleton sets, then the product is countable. The rest is covered by Q 2.4.2.

Q 2.4.4 This is an instance of Q 2.4.2.

3. Algebraic Structure

Q 3.1.1 The assertion is true by definition if $|\mathcal{C}| \leq 1$. For disjoint sets A and B, $|A \cup B| = |(A \times \{0\}) \cup (B \times \{1\})| = |(|A| \times \{0\}) \cup (|B| \times \{1\})| = |A| + |B|$, and the result follows by finite induction.

Q 3.1.2 The map $r \mapsto xr$ is not injective by 2.2.4. So $\exists a, b \in R$ with $a \neq b$ and $xa = xb$; then $x(a - b) = 0$.

Q 3.1.3 Suppose $u(z - ab) = 1$. Then, since $az = za$, also $z^{-1}(1 + bua)(z - ba) = 1$.

Q 3.1.4 Suppose R is a field and I is an ideal of R. If $x \in I \backslash \{0\}$, then, for each $y \in R$, $y = xx^{-1}y \in xR \subseteq I$, so $I = R$. Conversely, if R has no proper non-trivial ideal, then, for $x \in R \backslash \{0\}$, $x \in xR$; so $xR = R$ and $\exists r \in R$ with $xr = 1$.

Q 3.2.1 If S is an infinite linearly independent subset of V, then the collection of finite subsets of S has cardinality $|S|$ (2.4.7); the linear span of each of these has cardinality $|F|$ (2.4.6); so $|\langle S \rangle| = |S| \, |F|$, which equals $|V|$ only if $|S| = |V|$.

Q 3.2.2 Let A and B be complementary subspaces in S and T respectively of $S \cap T$. $\dim(S + T) + \dim(S \cap T) = \dim(A) + 2\dim(S \cap T) + \dim(B) = \dim(S) + \dim(T)$.

Q 3.2.3 For $a, b \in R$, each of $f \circ g$ and $g \circ f$ satisfies the additive property. For the other, $(f \circ g)(ab) = f(g(a)b + ag(b)) = f(g(a)b) + f(ag(b)) = f(g(a))b + g(a)f(b) + f(a)g(b) + af(g(b))$; similarly $(g \circ f)(ab) = g(f(a))b + f(a)g(b) + g(a)f(b) + ag(f(b))$. By subtraction, $((f \circ g) - (g \circ f))(ab) = ((f \circ g) - (g \circ f))(a)b + a((f \circ g) - (g \circ f))(b)$.

Q 3.2.4 $\mathcal{M}_{n \times n}(F)$ is a unital ring (3.1.28). For each r, c in the indexing set J, let $E_{r,c} \in \mathcal{M}_{n \times n}(F)$ be the matrix whose entry in the rth row and c^{th} column is 1 and whose other entries are 0; $E_{r,c}$ is clearly not invertible, so that $\mathcal{M}_{n \times n}(F)$ is not a division ring. Suppose $\mathcal{I}$ is a non-trivial ideal of $\mathcal{M}_{n \times n}(F)$ and $A = (a_{i,j}) \in \mathcal{I} \backslash \{0\}$; then $a_{p,q} \neq 0$ for some $p, q \in J$. For each $i, j \in J$, $E_{i,p} A E_{q,j} = a_{p,q} E_{i,j}$; so $E_{i,j} \in \mathcal{I}$. Then $\mathcal{M}_{n \times n}(F) = \langle \{E_{i,j} \mid i, j \in J\} \rangle \subseteq \mathcal{I}$.

Q 3.3.1 $+_4$ and $\times_4$ of 3.1.33.

Q 3.3.2 This follows from Q 3.1.3.

Q 3.4.1 $k \mapsto \mathrm{rem}_4(k)$ (see 3.1.33).

Q 3.4.2 For each $a \in A$, define $\phi_a \colon A \to A$ by $\phi_a(x) = ax$ for each $x \in A$. It is easy to verify that this LEFT MULTIPLICATION OPERATOR is a vector space homomorphism. Moreover the map $a \mapsto \phi_a$ is clearly an algebra homomorphism from A into $\mathcal{L}(A)$.

Q 3.4.3 Scalar multiples of I certainly commute with every member of $\mathcal{L}(X)$. Suppose $T \in \mathcal{L}(X)$ is not a scalar multiple of the identity. If, for some linearly independent $x, y \in X$ and scalars λ, μ, we have $Tx = \lambda x$ and $Ty = \mu y$, then either $\lambda = \mu$ or $T(x+y)$ is not a scalar multiple of $x + y$. In any case, $\exists z \in X$ such that z and Tz are linearly independent. Let B be a basis for X with $\{z, Tz\} \subseteq B$. Set $Sb = 0$ for $b \in B \backslash \{z\}$ and $Sz = z$ and extend S linearly to X. Then $S \in \mathcal{L}(X)$, $STz = 0$ and $TSz = Tz \neq 0$.

Q 3.4.4 • If $ST = I$, then $\forall z \in X \backslash \{0\}$, $STz = z \neq 0$; so $Tz \neq 0$. Conversely, if $Tw = 0$ then $\forall S \in \mathcal{L}(X)$, $STw = 0$; so $ST \neq I$.
• If $TS = I$, then $X = TS(X) \subseteq T(X)$. Conversely, if $T(X) = X$, let W be such that $W \oplus \ker(T) = X$; then $T|_W \colon W \to X$ is bijective; define $S \in \mathcal{L}(X)$ to be $x \mapsto T|_W^{-1} x$; then $TS = I$.

Q 3.4.5 By induction, $Tp(T) = p(T)T$. Then $p(T)z = 0 \Rightarrow p(T)Tz = Tp(T)z = 0$. And $a = p(T)b \Rightarrow Ta = Tp(T)b = p(T)Tb$.

Q 3.4.6 Using 3.2.12, we see that the decreasing sequence $(\ker(T) \cap T^n(X))_{n \in \mathbb{N}}$ of finite dimensional spaces is eventually constant and equal to $\ker(T) \cap \mathcal{R}(T)$. Let B be such that $\ker(T) = B \oplus (\ker(T) \cap \mathcal{R}(T))$. For sufficiently large $k \in \mathbb{N}$, we have $\ker(T) + T^k(X) = B \oplus T^k(X)$; therefore $\bigcap(\ker(T) + T^n(X)) = \bigcap(B + T^n(X))$. It is easy to check that $T^{-1}(\mathcal{R}(T)) = \bigcap(\ker(T) + T^n(X))$ and $\bigcap(B + T^n(X)) = B \oplus \mathcal{R}(T)$. So $T^{-1}(\mathcal{R}(T)) = B \oplus \mathcal{R}(T)$ and the result follows because $T(B) = \{0\}$,

Q 3.4.7 Let $\mathcal{S}$ be the set of ordered pairs (M, Q) where M is a subspace of X invariant under T, $Q = Q^2 \in \mathcal{L}(M)$, $(T - Q)M = M$ and $T|_M - Q$ is injective. $\mathcal{S} \neq \varnothing$ because $(\{0\}, 0) \in \mathcal{S}$. Then $\{((M, Q), (N, R)) \in \mathcal{S} \times \mathcal{S} \mid M \subset N, Q = R|_M\}$ is a partial ordering on $\mathcal{S}$. A straightforward application of Zorn's Lemma shows that $\mathcal{S}$ has a maximal element (Y, P). We claim that $Y = X$. First, if $\exists x \in X \backslash Y$ such that $Tx \in Y$, extend P linearly to $Y \oplus Fx$ with $Px = x$ and check that $(Y \oplus Fx, P) \in \mathcal{S}$, contradicting maximality. Second, if $\exists x \in X \backslash Y$ and $p \in \mathrm{poly}(\mathcal{L}(X))$ with non-zero constant term such that $p(T)x \in Y$, consider p and x to satisfy these criteria with $\deg p$ minimum for such a pair; let $W = \langle \{T^k x \mid 0 \leq k < \deg p\} \rangle$ and check $W \cap Y = \{0\}$; extend P linearly to $Y \oplus W$ with $P(W) = \{0\}$ and check that $(Y \oplus W, P) \in \mathcal{S}$, contradicting maximality. Third, suppose $\exists x \in X \backslash Y$ such that $p(T)x \in X \backslash Y$ for all $p \in \mathrm{poly}(\mathcal{L}(X)) \backslash \{0\}$. Then $(T^k x)_{k \in \infty}$ is a linearly independent sequence of vectors; let $V = \langle \{T^k x \mid k \in \infty\} \rangle$ and extend P linearly to $Y \oplus V$ with $PT^{2n} x = T^{2n} x$ and $PT^{2n+1} x = T^{2n+2} x - T^{2n} x$ for all $n \in \infty$. Check that $(Y \oplus V, P) \in \mathcal{S}$, again contradicting maximality. Conclude that $Y = X$ and therefore that $T - P$ is bijective.

4. Analytic Structure

Q 4.1.1 The ordering of 2.3.10 has both minimum and maximum; the standard ordering has neither; the ordering $<'$ has minimum but no maximum.

Q 4.1.2 If $a, b \in R$ with $ab = 0$, then also $(-a)(-b) = a(-b) = (-a)b = 0$. So none of $\{a, b\}$, $\{-a, -b\}$, $\{a, -b\}$ and $\{-a, b\}$ is a subset of R^+. Therefore $a = 0$ or $b = 0$.

Q 4.2.1 Every subfield of $\mathbb{R}$ must contain 1 and then, using induction, must include $\mathbb{Z}$ and then all inverses of integers and so also $\{m/n \mid m \in \mathbb{Z}, n \in \mathbb{N}\}$, which is $\mathbb{Q}$.

Q 4.2.2 By 4.1.11.

Q 4.3.1 Using 2.4.6, $|\mathbb{C}| = |\mathbb{R} \times (\mathbb{R}\backslash\{0\})| = |\mathbb{R}| = |2^\infty|$.

Q 4.3.2 Let $a = \inf I$, $b = \sup I$ and $c \in (0, b - a)$. $\mathbb{R} = \bigcup\{(a + nc, b + nc) \mid n \in \mathbb{Z}\}$ is a countable union of intervals each equinumerous with I. So $|I| = |\mathbb{R}|$ by 2.4.5. For the rest, $\mathbb{T} \subset \mathbb{D} \subset \mathbb{C}$ and the map $z \mapsto \Re z$ is surjective from $\mathbb{T}$ onto $[-1, 1]$; so $|2^\infty| = |[-1, 1]| \le |\mathbb{T}| \le |\mathbb{D}| \le |\mathbb{C}| = |2^\infty|$.

Q 4.3.3 By Q 4.3.2, $|[0, 1]| = |\mathbb{R}| = |2^\infty|$. Then this is an instance of Q 2.4.2.

Q 4.3.4 Let $m \in \mathbb{N}$ be such that the mth tail of (x_n) is included in $D[z\,;1)$. Let $r = |z| + 1 + \max\{|x_i| \mid 1 \le i < m\}$. Then $x_n \in D[0\,;r)$ for all $n \in \mathbb{N}$. And certainly every disc centred at z includes a tail of every subsequence of (x_n). Lastly, the sequence $(n)_{n\in\mathbb{N}}$ is bounded below but has no convergent subsequence; note that its limit inferior is ∞, so there is no contradiction to 4.3.8.

Q 4.3.5 Suppose S is a proper non-empty open subset of $\mathbb{C}$. For each $s \in S$, let $\rho(s) = \sup\{r \in \mathbb{R}^+ \mid D[s\,;r) \subseteq S\}$. Let $z \in S$; check that $\rho(z) < \infty$ and that, for each $r \in \mathbb{R}^+$, $\exists y \in D[z\,;\rho(z))$ such that $\rho(y) < r$. By the Product Theorem, $\exists$ a sequence (x_n) in $D[z\,;\rho(z))$ such that $\rho(x_n) < 1/n$. By 4.3.10, this has a subsequence which converges to some $w \in \mathbb{C}$. For each $r \in \mathbb{R}^+$, there exists $k \in \mathbb{N}$ with $1/k < r/2$ such that $x_k \in D[w\,;r/2)$, whence $D[w\,;r) \cap S \ne \varnothing$ and $D[w\,;r) \cap (\mathbb{C}\backslash S) \ne \varnothing$. So $w \in \mathbb{C}\backslash S$, because S is open, and then $\mathbb{C}\backslash S$ is not open because no $D[w\,;r)$ is included in $\mathbb{C}\backslash S$.

Q 4.3.6 Suppose that S is closed in $\mathbb{C}$ and that $v \in \mathbb{C}\backslash f(S)$. For each $n \in \mathbb{N}$, let $D_n = D[v\,;n^{-1})$. We claim that there exists $k \in \mathbb{N}$ such that $S \cap f^{-1}(D_k) = \varnothing$. Suppose otherwise; then, by the Product Theorem, there is a sequence (a_n) with $a_n \in S \cap f^{-1}(D_n)$ for each $n \in \mathbb{N}$. Notice that $\{a_n \mid n \in \mathbb{N}\} \subseteq f^{-1}(D_1)$, which is bounded by hypothesis. By 4.3.10, (a_n) has a convergent subsequence; let w be the limit of some such sequence. Now $w \in S$, for otherwise, because $\mathbb{C}\backslash S$ is open, there would be an open disc of $\mathbb{C}$ containing w but no point of S, and therefore no term of (a_n). Since $v \notin f(S)$, we have $f(w) \ne v$. Let $m \in \mathbb{N}$ be such that $2m^{-1} < |f(w) - v|$; then $D[f(w)\,;m^{-1})$ is disjoint from D_m, whence $f^{-1}D[f(w)\,;m^{-1})$ is disjoint from the mth tail of (a_n). But $f^{-1}D[f(w)\,;m^{-1})$, being open by hypothesis, includes an open disc which contains w, and we have the contradiction that this disc contains at most $m - 1$ terms of (a_n). So our claim that there exists $k \in \mathbb{N}$ with $S \cap f^{-1}(D_k) = \varnothing$ is true. Then D_k is an open disc containing v which is disjoint from $f(S)$. Since v is arbitrary in $\mathbb{C}\backslash f(S)$, this implies that $\mathbb{C}\backslash f(S)$ is open in $\mathbb{C}$ and then that $f(S)$ is closed in $\mathbb{C}$.

Q 4.3.7 The polynomial function $z^n - 1$ can be written as $\prod_{i=1}^n (z - \alpha_i)$, and this has value 0 only at the α_i. So there are at most n roots. Suppose $\omega \ne 1$ is one which has the property that the number of roots which are powers of ω is the maximum possible. Each member of the set $S = \{\omega^i \mid n \in \mathbb{N} : 0 \le i < n\}$ is a root. If $|S| < n$, then the numbers ω^i with $0 \le i < |S|$ are distinct and by 4.1.10, there exist $q \in \mathbb{N}$ and $r \in \mathbb{Z}$ with $0 \le r < |S|$ such that $n = q|S| + r$, from which we get $\omega^r = 1$, and therefore $r = 0$ because the ω^i are distinct for $0 \le i < |S|$. So $n = q|S|$. Then $q = 1$, for otherwise $\exists \alpha \in \mathbb{C}$ such that $\alpha^q = \omega$, and it would follow that α is an nth root with a greater number of powers as roots than ω.

Q 4.4.1 For $p \in \mathbb{R}^+$, $\|a\|_\infty^p \le \sum|a_i|^p \le n\|a\|_\infty^p$; so $\|a\|_\infty \le \|a\|_p \le n^{1/p}\|a\|_\infty$. The result follows because $\inf\{n^{1/p} \mid p \in \mathbb{R}^+\} = 1$.

Q 4.4.2 For each $(y, z) \in S_x$, let $S_{yz} = \{x \mid (x, y, z) \in A\}$. So $|A| = \sum_{(y,z) \in S_x} |S_{yz}|$, and $|A|^2 = (\sum_{(y,z) \in S_x} 1 \cdot |S_{yz}|)^2 \leq |S_x| \sum_{(y,z) \in S_x} |S_{yz}|^2$ by Cauchy's inequality (4.4.6). For each $(y, z) \in S_x$, let f_{yz} be $(a, b) \mapsto ((z, a), (b, y))$ defined on $S_{yz} \times S_{yz}$. The f_{yz} have disjoint ranges and are injective, so $\sum_{(y,z) \in S_x} |S_{yz}|^2 \leq |S_y| \, |S_z|$, and therefore $|A|^2 \leq |S_x| \, |S_y| \, |S_z|$.

Q 4.4.3 The proposition is true if $|A| = 1$. If it holds for $|A| = m - 1 \geq 1$, then, putting $p = m$ and $q = m/(m - 1)$ in Young's inequality, $(\prod_{i=1}^m a_i)^{1/m}$ equals $a_1^{1/m} (\prod_{i=2}^m a_i)^{1/m} \leq a_1^{1/m} (\sum_{i=2}^m a_i/(m - 1))^{(m-1)/m} \leq (a_1 + \sum_{i=2}^m a_i)/m$, which equals $\sum_{i=1}^m a_i/m$. The result follows by induction.

5. Linear Structure

Q 5.1.1 Since $\mathcal{M} = \mathcal{M}_{n \times n}(\mathbb{C})$ has finite dimension, the set $\{A^n \mid n \in \mathbb{N}\}$ is linearly dependent. So $\exists p \in \text{poly}(\mathcal{M}) \backslash \{0\}$ such that $p(A) = 0$. But p is reducible (5.1.6); so $0 = p(A) = c \prod_{i \in \mathbb{I}} (\lambda_i - A)$ for some finite sequence $(\lambda_i)_{i \in \mathbb{I}}$ in $\mathbb{C}$; so $\lambda_i \in \sigma(A)$, and hence is an eigenvalue of A (as in 3.4.28) for each $i \in \mathbb{I}$. A real 2×2 matrix whose diagonal terms are 0 and other terms 1 and -1 has eigenvalues i and $-i$ only.

Q 5.1.2 No. The real algebra is not a subalgebra of the complex one; the fields differ.

Q 5.1.3 Invoke Q 3.2.1 to show that it is $|2^\infty|$.

Q 5.2.1 If W is a wedge of X, then certainly W includes the specified union; and for each $t \in [0, 1]$ and $a, b \in W$, $ta \in W$ and $(1 - t)b \in W$, whence $ta + (1 - t)b \in W$, yielding convexity of W. Conversely, if W is convex and $W = \bigcup \{\mathbb{R}^\oplus v \mid v \in W\}$, then $\mathbb{R}^\oplus W = W$ and, for each $a, b \in W$, we have $a/2 + b/2 \in W$ and hence $a + b \in W$.

Q 5.2.2 Suppose $w \in W \backslash \{0\}$. If W is a cone, then $-w \notin W$, so that $\mathbb{R}w$ is not included in W. Conversely, if W is a wedge and $\mathbb{R}w$ is not included in W, then, since $w \in W$ and $\mathbb{R}^\oplus W \subseteq W$, we have $-w \notin W$.

Q 5.2.3 Suppose M and N are maximal subspaces of X and $a, b \in X$ satisfy the equation $M + \{a\} = N + \{b\}$. Then $a - b \in N$, so that $N = N + \{b - a\} = M$.

Q 5.2.4 Clearly $C - \{z\} \subseteq \langle C \rangle$. If $0 \in C$, then $z/2 \in C$ and, for each $x \in C$, $x = (x - z) - 2(z/2 - z) \in \langle C - \{z\} \rangle$. So $\langle C \rangle = \langle C - \{z\} \rangle$.

Q 5.2.5 • Firstly, suppose $a, b \in X$, $a \neq b$ and $e \in (a, b)$; then $e \in L_{a,b}$ and, if e is extreme, e is an endpoint of $C \cap L_{a,b}$; so $(e + a)/2$ or $(e + b)/2$ is in $(a, b) \backslash C$. So $(a, b) \not\subseteq C$; and C includes no open line segment which contains an extreme point.
• Secondly, suppose $x, y \in C \backslash \{e\}$; then $[x, y] \subseteq C$. If $e \notin (x, y)$, then $[x, y] \subseteq C \backslash \{e\}$. So, if C includes no open line segment which contains e, then $C \backslash \{e\}$ is convex.
• Thirdly, suppose L is a line of X with $e \in L$; if e is not an endpoint of $C \cap L$, then there exist $a, b \in C \cap L \backslash \{e\}$ such that $e \in [a, b]$, yielding $C \backslash \{e\}$ not convex. So, if $C \backslash \{e\}$ is convex, then e is an extreme point of C.

Q 5.2.6 If e is extreme and $x \in C \backslash \{e\}$, then e is an endpoint of $C \cap L_{x,e}$; so $e \not\leq x$. For the converse, suppose the condition on C holds and e is maximal. Suppose L is a line of X with $e \in L$; if e were not an endpoint of $C \cap L$, then there would be an endpoint $z \neq e$, yielding the contradiction $e < z$.

Q 5.3.1 Write $\mathbb{R}$ as a vector space over $\mathbb{Q}$. By 3.2.9, this space has a basis B. Let $b \in B$. For each $r \in \mathbb{R}$ define $f(r) = q$ where $r = x + qb$ with $x \in \langle B \backslash \{b\} \rangle$ and $q \in \mathbb{Q}$. Then f is additive; but $\sqrt{2} f(b) = \sqrt{2} \notin \mathbb{Q}$, whereas $f(\sqrt{2} b) \in \mathbb{Q}$.

Q 5.3.2 Let $V = \{p \in \text{poly}(\mathbb{R}) \mid p = \sum_{i=0}^{n} \alpha_n z^n : n \in \mathbb{N} \cup \{0\}, \alpha_n > 0\}$. Suppose W is a wedge of $\text{poly}(\mathbb{R})$ which includes V. If W is proper, then $\{-z^n \mid n \in \mathbb{N} \cup \{0\}\}$ is not included in W. Let $m \in \mathbb{N} \cup \{0\}$ be minimal such that $-z^m \notin W$. Then, since $\{z^{m+1}\} + \mathbb{R}^\ominus z^m \subseteq V$, we have $(\{-z^{m+1}\} + \mathbb{R}^\oplus z^m) \cap W = \varnothing$, whence $-z^{m+1} \notin W + \mathbb{R}^\ominus z^m$. So $W + \mathbb{R}^\ominus z^m$ is a proper wedge; it includes W.

Q 5.3.3 $A = \{z \in \mathbb{R}^2 \mid z_2 \in \mathbb{R}^+\} \cup \{(0,0)\}$ and $U = \{z \in \mathbb{R}^2 \mid z_2 \in \mathbb{R}^-\} \cup \{(1,0)\}$.

Q 5.3.4 $U = \{z \in \mathbb{R}^2 \mid z_2 = 0\}$ and $A = \{z \in \mathbb{R}^2 \mid z_1 = 0, z_2 > 0\}$.

6. Geometric Structure

Q 6.1.1 For the triangle inequality, suppose $a, b, c \in X_1 \times X_2$. Note firstly that $e(a,c) \leq ((d_1(a_1,b_1) + d_1(b_1,c_1))^p + (d_2(a_2,b_2) + d_2(b_2,c_2))^p)^{1/p}$. Now, for $x, y \in \mathbb{R}^2$, Minkowski's inequality gives $\|x+y\|_p \leq \|x\|_p + \|y\|_p$. Rewriting this inequality with $x = (d_1(a_1,b_1), d_2(a_2,b_2))$ and $y = (d_1(b_1,c_1), d_2(b_2,c_2))$ yields the inequality $((d_1(a_1,b_1) + d_1(b_1,c_1))^p + (d_2(a_2,b_2) + d_2(b_2,c_2))^p)^{1/p} \leq e(a,b) + e(b,c)$, so that $e(a,c) \leq e(a,b) + e(a,c)$. The other metric properties are easily checked.

Q 6.1.2 For each $v \in V$, let $\delta_v = \text{dist}(v, Y \backslash V)$. Set $U = \bigcup \{\flat_X[v; \delta_v] \mid v \in V\}$.

Q 6.1.3 $Z \backslash F$ is open in Z, so, by Q 6.1.2, there exists an open subset U of X such that $Z \backslash F = U \cap Z$; but $X \backslash Z$ is open in X, so that $X \backslash F = U \cup (X \backslash Z)$ is open in X.

Q 6.1.4 Suppose $a \in X$ and $s \in \mathbb{R}^+$. The ball $\flat[a; s]$ of (X, d) includes the ball $\flat'[a; rs]$ of (X, d'). So d' is stronger than d by 6.1.30.

Q 6.1.5 Closed sets are complements of open sets. By De Morgan's laws (1.1.13), an intersection of closed sets is the complement of a union of open sets and a union of closed sets is the complement of an intersection of open sets. Invoke 6.1.27.

Q 6.1.6 Let $a \in \mathbb{R}^2$ and $r \in \mathbb{R}^+$. Then $\flat_1[a; r] \subseteq \flat_2[a; r] \subseteq \flat_\infty[a; r] \subseteq \flat_1[a; 2r]$, where the suffices 1, 2 and ∞ identify the three metrics in the order given in 6.1.7.

Q 6.2.1 For $x, y \in X$ and $t \in \mathbb{R}$, $q(tx) = p(tx) + p(-tx) = |t|\,(p(x) + p(-x)) = |t|\,q(x)$ and $q(x+y) = p(x+y) + p(-x-y) \leq p(x) + p(y) + p(-x) + p(-y) = q(x) + q(y)$.

Q 6.2.2 The sequence (1) is in $\ell_\infty \backslash \ell_s$. The sequence $(n^{-1/r})$ is in $\ell_s \backslash \ell_r$ (4.3.18).

Q 6.2.3 If $\|\cdot\|$ is not a norm and $a \in X$, then there exists $z \in X \backslash \{a\}$ with $\text{dist}(z, \{a\}) = \|z - a\| = 0$, whence $\{a\}$ not closed in X by 6.1.24. Conversely, if singleton sets are closed, then $X \backslash \{0\}$ is open, so that, for each $x \in X \backslash \{0\}$, $\exists r \in \mathbb{R}^+$ such that $\flat[x; r] \subseteq X \backslash \{0\}$; then $\|x\| \geq r$.

Q 6.2.4 Endow $\text{poly}(\mathbb{C})$ with the norm of 6.2.4 with respect to its usual basis $\{z^n \mid n \in \mathbb{N} \cup \{0\}\}$. For each $n \in \mathbb{N}$, set $Tz^n = nz^n$ and extend T linearly to $\text{poly}(\mathbb{C})$. Then T is injective and unbounded.

Q 6.2.5 It follows from 6.1.23 and 6.2.12 that all translates of A are open. Then $A + B = \bigcup \{(A + \{b\}) \mid b \in B\}$.

Q 6.2.6 For each $n \in \mathbb{N}$, let e_n denote the sequence whose n^{th} term is 1 and whose other terms are all 0. Let $S = \{(1 + 1/n)e_n \mid n \in \mathbb{N}\}$. Note that, for $m, n \in \mathbb{N}$, $\|(1 + 1/n)e_n - (1 + 1/m)e_m\| > \sqrt{2}$. So, if $z \in \ell_2$ and $\text{dist}(z, S) = 0$, then $z \in S$. So S is closed. Also, $0 \notin B + S$ and, for each $n \in \mathbb{N}$, $e_n/n \in B + S$ and $\|e_n/n\| = 1/n$, whence $\text{dist}(0, B + S) = 0$, so that $B + S$ is not closed and $\text{dist}(0, S) = 1$

Q 6.2.7 For each b in an open ball B, $\mathbb{R}^+(B - \{b\}) = X$; so, if $B \subseteq W$, then $-B \cap W = \varnothing$. Check that $\{x \in X \mid \text{dist}(x, W) = 0\}$ is a wedge; it is closed by 6.1.24; it is proper because it does not contain the centre of $-B$; it includes the maximal wedge W, therefore equals W. Conversely, if W is closed, write $W = M + \mathbb{R}^\ominus a$ where M is a maximal subspace of X and $a \in X \backslash W$ (5.2.26). Then $d = \text{dist}(a, W) > 0$ and $\flat[-a\,;d) \subseteq W$.

Q 6.2.8 Considering X as a real space, 5.2.31 ensures that there exist a maximal subspace M of X and $w \in X$ such that the hyperplane $M + \{w\}$ separates G from A. Let $z \in G - \{w\}$. Then $M + \mathbb{R}^\oplus z$ is a maximal wedge of real X; since it includes $G - \{w\}$, it is closed by Q 6.2.7; also $M = W \cap (-W)$ is closed in X by 6.1.27. Let $u \in \mathcal{L}(X, \mathbb{R})$ be such that wedge$(u) = M + \mathbb{R}^\oplus z$ (5.3.4). Since $\ker(u) = M$ is closed, u is bounded (6.2.41); then so is the $f \in \mathcal{L}(X, \mathbb{K})$ for which $u = \Re f$ (5.3.2, 6.2.38). Let $t = u(w)$. Then, for $g \in G$ and $a \in A$, we have $g - w \in M + \mathbb{R}^+ z$ and $a - w \in M + \mathbb{R}^\ominus z$, so that $u(g - w) > 0 \geq u(a - w)$, yielding $u(g) > t \geq u(a)$.

Q 6.3.1 $(a, b) \mapsto a_1 b_1 + a_1 b_2 - a_2 b_1 + a_2 b_2$ defined on $\mathbb{R}^2$.

Q 6.3.2 Suppose $s \in S$, $x, y \in S^\perp$ and $\lambda \in \mathbb{K}$. Then $\langle s, x + y \rangle = \langle s, x \rangle + \langle s, y \rangle = 0$ and $\langle s, \lambda x \rangle = \overline{\lambda} \langle s, x \rangle = 0$; so $S^\perp$ is a linear subspace of X. Let $\epsilon \in \mathbb{R}^+$. If $z \in X$ and $\text{dist}(z, S^\perp) = 0$, then there exists $v \in S^\perp$ with $\|z - v\| < \epsilon$. For each $s \in S$, $|\langle z, s \rangle| = |\langle z - v, s \rangle| < \epsilon \|s\|$ by 6.3.7. Since ϵ is arbitrary in $\mathbb{R}^+$, we have $\langle z, s \rangle = 0$.

Q 6.3.3 Define $\langle \cdot, \cdot \rangle$ on $X \times X$ to be $(x, y) \mapsto \sum_{k=1}^{4} i^k \|x + i^k y\|^2 /4$. An easy calculation establishes that $\langle x, x \rangle = \|x\|^2$ for all $x \in X$. We show that $\langle \cdot, \cdot \rangle$ is sesquilinear. For all $\alpha \in \mathbb{C}$ and $a, b \in X$, $\langle \alpha a, b \rangle + \langle a, \alpha b \rangle = \sum_{k=1}^{4} i^k (\|\alpha a + i^k b\|^2 + \|a + i^k \alpha b\|^2)/4 = \sum_{k=1}^{4} i^k (\|(\alpha + 1)(a + i^k b)\|^2 + \|(\alpha - 1)(a - i^k b)\|^2)/8$, which, swapping order, equals $\sum_{k=1}^{4} i^k (\|(\alpha + 1)(a + i^k b)\|^2 - \|(\alpha - 1)(a + i^k b)\|^2)/8 = (|\alpha + 1|^2 - |\alpha - 1|^2)\langle a, b \rangle /2 = (\alpha + \overline{\alpha})\langle a, b \rangle$. We also have $i\langle x, y \rangle = \langle ix, y \rangle = -\langle x, iy \rangle$ for all $x, y \in X$; then $\langle \alpha a, b \rangle - \langle a, \alpha b \rangle = -i(\langle i\alpha a, b \rangle + \langle a, i\alpha b \rangle) = -i(i\alpha + \overline{i\alpha})\langle a, b \rangle = (\alpha - \overline{\alpha})\langle a, b \rangle$; so $\langle \alpha a, b \rangle = \alpha \langle a, b \rangle$ and $\langle a, \alpha b \rangle = \overline{\alpha} \langle a, b \rangle$. Then, for each $a, b, c \in X$, $\langle a, b \rangle + \langle a, c \rangle = \sum_{k=1}^{4} i^k (\|a + i^k b\|^2 + \|a + i^k c\|^2)/4 = \sum_{k=1}^{4} i^k (\|2a + i^k(b + c)\|^2 + \|b - c\|^2)/8 = \sum_{k=1}^{4} i^k \|2a + i^k(b + c)\|^2 /8 = \langle 2a, b + c \rangle /2 = \langle a, b + c \rangle$. A similar calculation yields $\langle a + b, c \rangle = \langle a, c \rangle + \langle b, c \rangle$.

7. Topological Structure

Q 7.1.1 The collection of lower and upper segments of $\mathbb{R}$ is the standard subbase for the order topology (7.1.20) on $\mathbb{R}$; and each open interval (r, s) can be expressed as $\grave{s} \cap \acute{r}$. Conversely each upper or lower segment is the union of the bounded open intervals it includes.

Q 7.1.2 Let $X = \{1, 2, 3, 4\}$. Then $\mathcal{O} = \{\varnothing, \{1\}, X\}$ and $\mathcal{P} = \{\varnothing, \{2\}, X\}$ are topologies on X but $\mathcal{O} \cup \mathcal{P}$ is not.

Q 7.1.3 In $\mathbb{R}$, $\bigcap\{(-1/n, 1/n) \mid n \in \mathbb{N}\} = \{0\}$.

Q 7.1.4 Consider $\mathbb{N}$ with the finite complement topology. Each singleton set is closed, but the set $\{2n \mid n \in \mathbb{N}\}$ of even integers is not closed.

Q 7.1.5 Let $\mathfrak{P}$ be the set of all topologies on X which are stronger than every $\mathcal{O} \in \mathfrak{T}$. Then $\mathfrak{P} \neq \varnothing$ because the discrete topology belongs to $\mathfrak{P}$. By 7.1.15, $\bigcap \mathfrak{P}$ is a topology on X; clearly $\bigcap \mathfrak{P}$ is stronger than each $\mathcal{O} \in \mathfrak{T}$, and since every topology which has this property is a member of $\mathfrak{P}$, it must be that $\bigcap \mathfrak{P}$ is the smallest such topology.

Q 7.1.6 No. Let $\mathcal{O}$ denote the usual topology on $\mathbb{R}$ and, for each $n \in \mathbb{N}$, let $\mathbb{I}_n = \{k \in \mathbb{N} \mid k \leq n\}$ and $\mathcal{U}_n = \{U \cup S \mid U \in \mathcal{O}, S \subseteq \mathbb{I}_n\}$. Then $\{\mathcal{U}_n \mid n \in \mathbb{N}\}$ is a nest of topologies on $\mathbb{R}$. $\mathbb{N}$ is a union of members of $\bigcup \mathcal{U}_n$ but is not a member of $\bigcup \mathcal{U}_n$.

Q 7.2.1 • If $\partial B \subseteq A \subseteq B$ and $z \in \partial B$ and $U \in \mathrm{nbd}(z)$, then $z \in A \cap U$ and $\varnothing \neq (X \backslash B) \cap U \subseteq (X \backslash A) \cap U$, so that $z \in \partial A$.
• If $A = \mathbb{D} \backslash \{0\}$ then $0 \in \partial A \backslash \partial(\overline{A})$.
• If $x \in \mathrm{acc}(A)$ and $U \in \mathrm{nbd}(x)$, then $\varnothing \neq U \cap A \backslash \{x\} \subseteq U \cap (A \cup B) \backslash \{x\}$, so that $x \in \mathrm{acc}(A \cup B)$. So $\mathrm{acc}(A) \subseteq \mathrm{acc}(A \cup B)$; similarly, $\mathrm{acc}(B) \subseteq \mathrm{acc}(A \cup B)$. Conversely, suppose $x \in \mathrm{acc}(A \cup B) \backslash \mathrm{acc}(A)$. $\exists V \in \mathrm{nbd}(x)$ such that $V \cap A \subseteq \{x\}$. Let $U \in \mathrm{nbd}(x)$. Then $(U \cap V) \cap (A \cup B) \backslash \{x\} \neq \varnothing$; so $U \cap B \backslash \{x\} \neq \varnothing$. Hence $x \in \mathrm{acc}(B)$. Therefore $\mathrm{acc}(A \cup B) \backslash \mathrm{acc}(A) \subseteq \mathrm{acc}(B)$, whence $\mathrm{acc}(A \cup B) = \mathrm{acc}(A) \cup \mathrm{acc}(B)$.
• $\overline{A \cup B} = A \cup B \cup \mathrm{acc}(A \cup B) = A \cup B \cup \mathrm{acc}(A) \cup \mathrm{acc}(B) = \overline{A} \cup \overline{B}$, using 7.2.27.
• $\overline{A} \cap \overline{B}$ is closed and includes $A \cap B$; it therefore also includes $\overline{A \cap B}$ by definition.
• $\partial A = X \backslash (A^\circ \cup A^{c\circ}) = (X \backslash A^\circ) \cap (X \backslash A^{c\circ}) = (X \backslash (X \backslash A)^{c\circ}) \cap (X \backslash A^{c\circ}) = \overline{X \backslash A} \cap \overline{A}$, using 7.2.7.
• $\partial(A \cup B) = \overline{A \cup B} \cap \overline{X \backslash (A \cup B)} = (\overline{A} \cup \overline{B}) \cap \overline{A^c \cap B^c} \subseteq (\overline{A} \cup \overline{B}) \cap \overline{A^c} \cap \overline{B^c} \subseteq (\overline{A} \cap \overline{A^c}) \cup (\overline{B} \cap \overline{B^c}) = \partial A \cup \partial B$.
• $\partial(A \cap B) = \partial(X \backslash (A^c \cup B^c)) = \partial(A^c \cup B^c) \subseteq \partial A^c \cup \partial B^c = \partial A \cup \partial B$.
• The required proper inclusions are all got with $A = [-1, 0)$ and $B = [0, 1]$.

Q 7.2.2 • $A = \varnothing : A^\circ = \varnothing$, $\overline{A} = \varnothing$, $\mathrm{acc}(A) = \varnothing$.
• $A = \{1\} : A^\circ = \{1\}$, $\overline{A} = X$, $\mathrm{acc}(A) = \{2, 3\}$.
• $A = \{2\} : A^\circ = \varnothing$, $\overline{A} = \{2, 3\}$, $\mathrm{acc}(A) = \{3\}$.
• $A = \{3\} : A^\circ = \varnothing$, $\overline{A} = \{3\}$, $\mathrm{acc}(A) = \varnothing$.
• $A = \{2, 3\} : A^\circ = \varnothing$, $\overline{A} = \{2, 3\}$, $\mathrm{acc}(A) = \{3\}$.
• $A = \{3, 1\} : A^\circ = \{1\}$, $\overline{A} = X$, $\mathrm{acc}(A) = \{2, 3\}$.
• $A = \{1, 2\} : A^\circ = \{1, 2\}$, $\overline{A} = X$, $\mathrm{acc}(A) = \{2, 3\}$.
• $A = X : A^\circ = X$, $\overline{A} = X$, $\mathrm{acc}(A) = \{2, 3\}$.

Q 7.2.3 $\mathrm{iso}(A) = \varnothing \Leftrightarrow \mathrm{acc}(A) = \overline{A}$ by 7.2.27; and also $\overline{A} = A \cup \mathrm{acc}(A)$ by 7.2.27.

Q 7.2.4 $(\overline{F})^\circ = \varnothing \Leftrightarrow F^\circ = \varnothing \Leftrightarrow (X \backslash F)^{c\circ} = \varnothing \Leftrightarrow X = (X \backslash F)^\circ \cup \partial(X \backslash F) = \overline{X \backslash F}$.

Q 7.2.5 $X = \{1, 2\}$ with the trivial topology and $S = \{1\}$.

Q 7.2.6 • Suppose the condition is satisfied and $\mathcal{U}$ is a countable collection of dense open subsets of X. For each $U \in \mathcal{U}$, $X \backslash U$ is closed, therefore nowhere dense by Q 7.2.4. By hypothesis $\bigcup \{X \backslash U \mid U \in \mathcal{U}\}$ is nowhere dense, so certainly includes no non-empty open subset of X. Its complement $\bigcap \mathcal{U}$ is therefore dense in X.
• $\mathbb{R}$ is a Baire space (12.1.19); $\mathcal{F} = \{\{q\} \mid q \in \mathbb{Q}\}$ is a countable collection of nowhere dense closed subsets of $\mathbb{R}$; and $\bigcup \mathcal{F} = \mathbb{R}$.

Q 7.2.7 Let $\mathcal{U} = \{X \backslash \overline{F} \mid F \in \mathcal{F}\}$. Each $U \in \mathcal{U}$ is open and dense in X, so that $\bigcap \mathcal{U} \neq \varnothing$, whence $\bigcup \mathcal{F} \subseteq X \backslash \bigcap \mathcal{U} \subset X$.

Q 7.2.8 $\{\mathbb{R} \backslash \{r\} \mid r \in \mathbb{R}\}$.

Q 7.2.9 Let $\epsilon \in \mathbb{R}^+$. • Suppose S is convex, $x, y \in \overline{S}$ and $t \in [0, 1]$. $\exists a, b \in S$ with $t \|x - a\| < \epsilon/2$ and $(1 - t) \|y - b\| < \epsilon/2$. So $\|(tx + (1 - t)y) - (ta + (1 - t)b)\| < \epsilon$, and, since ϵ is arbitrary in $\mathbb{R}^+$, $\mathrm{dist}(tx + (1 - t)y, S) = 0$. So $tx + (1 - t)y \in \overline{S}$.
• Suppose S is balanced, $x \in \overline{S}$ and $\eta \in \mathbb{K}$ with $|\eta| \leq 1$. Then $\exists a \in S$ such that $|\eta| \|x - a\| < \epsilon$, whence $\|\eta x - \eta a\| < \epsilon$; since $\eta a \in S$, $\mathrm{dist}(\eta x, S) = 0$ and $\eta x \in \overline{S}$.
• Suppose S is a wedge, $x \in \overline{S}$ and $t \in \mathbb{R}^\oplus$. Then $\exists a \in S$ with $t \|x - a\| < \epsilon$, whence $\|tx - ta\| < \epsilon$; since $ta \in S$, $\mathrm{dist}(tx, S) = 0$ and $tx \in \overline{S}$. The result follows because $\overline{S}$ is convex from above.

- Suppose S is an algebra ideal. $\overline{S}$ is a subspace of X (7.2.17) . If $x \in X$ and $s \in \overline{S}$, then $\exists a \in S$ with $\|x\| \|a - s\| < \epsilon$, whence $\|xa - xs\| < \epsilon$ and $\|ax - as\| < \epsilon$. Since $xa, ax \in S$ and ϵ is arbitrary, we have $\text{dist}(xs, S) = 0 = \text{dist}(sx, S)$ and $xs, sx \in \overline{S}$.
- The closure of $\{z \in \mathbb{R}^2 \mid z_2 > 0 \text{ or } z = 0\}$ includes $\mathbb{R} \times \{0\}$.

Q 7.3.1 If X is countable, then the set of finite subsets of X is also countable by 2.4.7; so the finite complement topology is countable. Conversely, suppose $|X| > \infty$ and $\mathcal{B}$ is a countable collection of subsets of X each with finite complement in X. $\bigcup\{X \backslash B \mid B \in \mathcal{B}\}$ is countable by 2.4.5, so is not equal to X. Therefore $\bigcap \mathcal{B} \neq \varnothing$; and, for $x \in \bigcap \mathcal{B}$, the set $X \backslash \{x\}$, cannot be expressed as a union of members of $\mathcal{B}$. So $\mathcal{B}$ is not a base for the finite complement topology on X.

Q 7.3.2 Suppose S is a countable subbase for the topology on X. The collection $\mathfrak{F}$ of finite members of $\mathcal{P}(S)$ is countable by 2.4.7. Every member of the base generated by S is $\bigcap \mathcal{F}$ for some $\mathcal{F} \in \mathfrak{F}$. So this base is also countable.

Q 7.3.3 If $\mathcal{B}$ is a base for the topology, then $\{\{x\} \mid x \in \text{iso}(X)\} \subseteq \mathcal{B}$.

Q 7.4.1 By 7.4.4, X is T_1 if and only if the topology includes $\{X \backslash \{x\} \mid x \in X\}$. But this is a subbase for the finite complement topology on X.

Q 7.4.2 That $\tilde{\mathbb{R}}$ is T_1 follows from 7.4.4. Suppose A and B are disjoint closed subsets of $\tilde{\mathbb{R}}$. Then $A \backslash \{-\infty, \infty\}$ and $B \backslash \{-\infty, \infty\}$ are closed in $\mathbb{R}$ and, because $\mathbb{R}$ is normal (7.4.3), there exist disjoint open subsets U, V of $\mathbb{R}$ such that $A \subseteq U \cup \{-\infty, \infty\}$ and $B \subseteq V \cup \{-\infty, \infty\}$. If $\infty \in A$, then $\infty \notin B$ and there exists $r \in \mathbb{R}$ such that $B \cap (r, \infty] = \varnothing$; in this case, let $U' = (r, \infty]$, otherwise let $U' = \varnothing$. If $\infty \in B$, let $V' = (r, \infty]$ where $r \in \mathbb{R}$ is such that $A \cap (r, \infty] = \varnothing$; otherwise let $V' = \varnothing$. If $-\infty \in A$, let $U'' = [-\infty, r)$ where $r \in \mathbb{R}$ is such that $B \cap [-\infty, r) = \varnothing$; otherwise let $U'' = \varnothing$. If $-\infty \in B$, let $V'' = [-\infty, r)$ where $r \in \mathbb{R}$ is such that $A \cap [-\infty, r) = \varnothing$; otherwise let $V'' = \varnothing$. Then $U \cup U' \cup U''$ and $V \cup V' \cup V''$ are open subsets of $\tilde{\mathbb{R}}$ which separate A and B.

8. Continuity and Openness

Q 8.1.1 • Suppose f is continuous and F is closed in Y; then $Y \backslash F$ is open in Y; so $X \backslash f^{-1}(F) = f^{-1}(Y \backslash F)$ is open in X and $f^{-1}(F)$ is closed in X.
- Suppose the second assertion. Let $A \subseteq X$. Then $A \subseteq f^{-1}(f(A)) \subseteq f^{-1}(\overline{f(A)})$, which is closed by hypothesis; so $\overline{A} \subseteq f^{-1}(\overline{f(A)})$. Therefore $f(\overline{A}) \subseteq f(f^{-1}(\overline{f(A)})) \subseteq \overline{f(A)}$.
- Suppose the third assertion. Let $B \subseteq Y$. Then $f(f^{-1}(B)) \subseteq B$ and $\overline{f(f^{-1}(B))} \subseteq \overline{B}$. So, by hypothesis, $f(\overline{f^{-1}(B)}) \subseteq \overline{B}$. Therefore $\overline{f^{-1}(B)} \subseteq f^{-1}(\overline{B})$.
- Suppose the fourth assertion. Let V be open in Y. Then $Y \backslash V = \overline{Y \backslash V}$, so the hypothesis gives $\overline{f^{-1}(Y \backslash V)} \subseteq f^{-1}(Y \backslash V)$. Then $f^{-1}(Y \backslash V)$ is closed and $f^{-1}(V)$, its complement in X, is open. So f is continuous.

Q 8.1.2 If f is open, the condition is certainly satisfied. Conversely, if the condition is satisfied and $\mathcal{A}$ is a finite subset of S, then, since f is injective, $f(\bigcap \mathcal{A}) = \bigcap f(\mathcal{A})$ by Q 1.2.1, so that $f(\bigcap \mathcal{A})$ is open. It follows from 8.1.23 that f is open.

Q 8.1.3 If V is open in Y and either f is continuous or f^{-1} is open, then $f^{-1}(V)$ is open in X, yielding both f continuous and f^{-1} open. The rest follows as easily.

Q 8.2.1 The final topology is the strongest for which f is continuous, so is stronger than $\mathcal{P}$. Towards the converse, suppose V is a member of the final topology. Then $f^{-1}(V) \in \mathcal{O}$. Surjectivity and openness of f imply $V = f(f^{-1}(V)) \in \mathcal{P}$.

Q 8.2.2 The initial topology is the weakest for which f is continuous, so is weaker than $\mathcal{O}$. Towards the converse, suppose $U \in \mathcal{O}$. Since f is open, $f(U) \in \mathcal{P}$. So $f^{-1}(f(U)$ is a member of the initial topology. But $f^{-1}(f(U)) = U$ by injectivity.

Q 8.2.3 f is continuous as f determines a topology; f is open by 8.2.10 or 8.2.12.

Q 8.2.4 • Suppose $x \in X$ and $U \in \mathrm{nbd}(x)$. Then there exist a finite sequence $(f_i)_{i \in \mathbb{I}}$ in S and a corresponding sequence $(W_i)_{i \in \mathbb{I}}$ of open subsets of M such that $x \in \bigcap\{f_i^{-1}(W_i) \mid i \in \mathbb{I}\} \subseteq U$. For each $i \in \mathbb{I}$, $W_i \in \mathrm{nbd}(f_i(x))$, so that, if M is regular, there exists a sequence $(G_i)_{i \in \mathbb{I}}$ of open subsets of M such that $f_i(x) \in G_i \subseteq \overline{G_i} \subseteq W_i$ for each $i \in \mathbb{I}$. Since each f_i is continuous, $f_i^{-1}(G_i)$ is open in X and $f_i^{-1}(\overline{G_i})$ is closed in X (Q 8.1.1). Set $V = \bigcap\{f_i^{-1}(G_i) \mid i \in \mathbb{I}\}$; then $x \in V \subseteq \overline{V} \subseteq U$.
• Suppose $x, y \in X$ and $x \neq y$. If $f(x) = f(y)$ for all $f \in S$, then y is a member of every subbasic open neighbourhood of x; so X is not Hausdorff. On the other hand, if $\exists f \in S$ such that $f(x) \neq f(y)$, then, if M is Hausdorff, there exist $U \in \mathrm{nbd}(f(x))$ and $V \in \mathrm{nbd}(f(y))$ with $U \cap V = \varnothing$, whence $f^{-1}(U) \in \mathrm{nbd}(x)$ and $f^{-1}(V) \in \mathrm{nbd}(y)$ and $f^{-1}(U) \cap f^{-1}(V) = \varnothing$.

Q 8.2.5 Each member of the topology is either a union of open annuli centred at 0 (sets of the form $\{x \mid a < \|x\| < b, a, b \in \mathbb{R}^{\oplus}\}$) or such a union together with an open ball centred at 0.

Q 8.3.1 Immediate from 8.3.25 and Q 8.2.1.

Q 8.3.2 The topology on Y is that generated by the open balls, each of which is clearly the intersection of Y with an open ball of X. It follows that Y is a topological subspace of X. A similar argument applies to the quotient spaces.

Q 8.3.3 Suppose $a, b \in X$ and $\epsilon \in \mathbb{R}^+$; let $\delta \in \mathbb{R}^+$ be such that $\delta(\|a\| + \|b\| + \delta) < \epsilon$. For $x, y \in X$, $\langle x, y \rangle - \langle a, b \rangle = \langle x - a, b \rangle + \langle a, y - b \rangle + \langle x - a, y - b \rangle$ and the Schwarz inequality gives $\langle x, y \rangle \in \flat[\langle a, b \rangle; \epsilon)$. Since ϵ is arbitrary, $\langle \cdot, \cdot \rangle$ is continuous at (a, b).

Q 8.3.4 Give $\mathbb{N}$ its usual discrete topology. Then all functions from $\mathbb{N}$ to $\mathbb{K}$ are continuous. So $\mathcal{C}(\mathbb{N})$ is the set ℓ_∞ of all bounded sequences of scalars. This space is the direct product of a countable number of copies of the scalar field.

Q 8.3.5 For each $E \in \mathcal{P}(F)$, E and $F \backslash E$ are closed in F and are therefore closed in X. Since X is T_4, there is an open subset U of X with $F \cap \overline{U} = E$. But $(X \backslash \overline{U \cap D}) \cap U$ is open and disjoint from D; so it is empty because D is dense in X. Then $U \subseteq \overline{U \cap D}$ and $\overline{U} = \overline{U \cap D}$, whence $F \cap \overline{U \cap D} = E$. So the injective relation $\{(A, B) \in \mathcal{P}(F) \times \mathcal{P}(D) \mid F \cap \overline{B} = A\}$ has domain $\mathcal{P}(F)$. By the Included Function Theorem, $\exists j \colon \mathcal{P}(F) \to \mathcal{P}(D)$ which is injective. So $|F| \leq |D|$ by 2.2.9.

Q 8.3.6 • Suppose X is a T_1 space, Y is a subspace of X and $a, b \in Y$. Then there exist $U \in \mathrm{nbd}_X(a)$ and $V \in \mathrm{nbd}_X(b)$ such that $a \notin V$ and $b \notin U$. Then $Y \cap U$ and $Y \cap V$ are open neighbourhoods in Y of a and b respectively which have the property that $a \notin Y \cap V$ and $b \notin Y \cap U$. So Y is also T_1.
• We use the criterion of 7.4.4. Suppose X is a topological space equipped with an equivalence relation $\sim$, and let π denote the corresponding quotient map. Then π is continuous and, for each $x \in X$, we have $x/\!\!\sim = \pi^{-1}\{\pi(x)\}$. If the singleton set $\{\pi(x)\}$ is closed in $\pi(X)$, then the equivalence class $x/\!\!\sim$ is closed in X by Q 8.1.1. Conversely, if the equivalence class $x/\!\!\sim$ is closed in X, its complement $\pi^{-1}(\pi(X) \backslash \{\pi(x)\})$ is open and, since the quotient topology is the strong topology determined π, it follows that $\pi(X) \backslash \{\pi(x)\}$ is open and hence that $\{\pi(x)\}$ is closed in the quotient space.

Q 8.3.7 Let U be an open subset of $\mathbb{R}$. Then $V = \{r \mid r^{-1} \in U\}$ is open (8.1.29). Because f is continuous, $(1/f)^{-1}(U) = f^{-1}(V)$ is open in X and therefore also in N.

Q 8.3.8 The subset $\{(x, 1/x) \mid x > 0\}$ is closed in $\mathbb{R}^2$, but its projection onto the x-axis, $\{x \in \mathbb{R} \mid x > 0\}$ is not closed in $\mathbb{R}$. The projection map is open (8.3.25).

Q 8.3.9 Suppose X is separable, A is a countable dense subset of X and $f: X \to Y$ is continuous. Certainly $f(A) \cap (Y \backslash \overline{f(A)}) = \varnothing$, so that $A \cap f^{-1}(Y \backslash \overline{f(A)}) = \varnothing$. But $f^{-1}(Y \backslash \overline{f(A)})$ is open in X, by continuity of f, and A is dense in X; it follows that $f^{-1}(Y \backslash \overline{f(A)}) = \varnothing$ and hence that $f(X) \subseteq \overline{f(A)}$. For the second part, see 8.3.30.

Q 8.4.1 Let $z \in X \backslash Y$ and W be a subspace of X satisfying $X = Y \oplus \mathbb{K}z \oplus W$. For each $x \in X$, we have $x = y + \lambda z + w$ for some $y \in Y$, $\lambda \in \mathbb{K}$ and $w \in W$. If $\lambda \neq 0$, then $(z - x/|\lambda|, z + x/|\lambda|) \subseteq X \backslash Y$; otherwise $z + \mathbb{R}x \subseteq X \backslash Y$. So $X \backslash Y$ is $\mathcal{SL}$-open.

Q 8.4.2 The $\mathcal{SL}$-closure of $\{z \in \mathbb{R}^2 \mid z_2 > 0 \text{ or } z = 0\}$ is $\{z \in \mathbb{R}^2 \mid z_2 \geq 0\}$.

Q 8.4.3 We establish that addition is continuous; a similar argument can be used to prove the continuity of scalar multiplication. Suppose $a, b \in X$. Let U be a subbasic open neighbourhood of $a + b$; then there exist $f \in \mathcal{F}$ and an open subset V of $\text{codom}(f)$ such that $U = f^{-1}(V)$. Then $f(a) + f(b) = f(a + b) \in V$ and, since addition on $\text{codom}(f)$ is continuous, there exist $A \in \text{nbd}(f(a))$ and $B \in \text{nbd}(f(b))$ such that $A + B \subseteq V$. Since f is continuous, we have $f^{-1}(A) \in \text{nbd}(a)$ and $f^{-1}(B) \in \text{nbd}(b)$, so that $f^{-1}(A) \times f^{-1}(B) \in \text{nbd}(a, b)$. The image of this neighbourhood under addition is the set $f^{-1}(A) + f^{-1}(B)$. Moreover, $f(f^{-1}(A) + f^{-1}(B)) \subseteq A + B \subseteq V$, whence $f^{-1}(A) + f^{-1}(B) \subseteq f^{-1}(V) = U$. So addition is continuous at (a, b).

Q 8.4.4 If $f = 0$, the result is trivial; suppose otherwise. If f is continuous, then $f^{-1}(0, \infty)$ is a non-empty open subset of X included in wedge (f). Conversely, if U is open in X and $z \in U \subseteq$ wedge (f), then $z \notin \ker(f)$ and there is a maximal subspace M of X such that wedge $(f) = M + \mathbb{R}^{\oplus}z$. For each real open interval (a, b), $f^{-1}(a, b) = M + (az/f(z), bz/f(z))$ which is open in X. So f is continuous.

Q 8.4.5 Most of the argument is sound. The equation beginning with $\|v_n\|_p^p$ can be modified to $\sum_{i=1}^{n} |g(e_i)| = \sum_{i=1}^{n} u_i g(e_i) = g(v_n) \leq \|g\| \|v_n\|_\infty$, which still yields $\|w\|_1 \leq \|g\|$ because $\|v_n\|_\infty$ is either 1 or 0. But the 'easy matter' of checking that $g = \psi_w$ depends crucially on the domains of these functions lying in c_0.

Q 8.4.6 ℓ_1.

9. Connectedness

Q 9.1.1 Suppose S is a connected subset of $\mathbb{R}$ with the specified topology. Let $z \in \mathbb{R}$. The intervals $(-\infty, z]$ and (z, ∞) are each unions of members of $\mathcal{C}$, so they are both open sets. Since $\{S \cap (-\infty, z], S \cap (z, \infty)\}$ is a partition of S, one of the sets is empty. Since z is arbitrary in $\mathbb{R}$, it follows that S has at most one member.

Q 9.1.2 No, because every non-empty open set contains the particular point.

Q 9.1.3 Suppose $a, b \in C$ with $a < b$; then $\exists n \in \mathbb{N}$ such that $|a - b| > 1/3^n$. Since I_n is a union of disjoint closed intervals each of length $1/3^n$, $\exists r \in (a, b) \backslash I_n$. In particular, a and b do not belong to the same connected component of C.

Q 9.1.4 It does not follow from the arbitrariness of j that $U = \varnothing$ or $U = P$.

Q 9.1.5 If X is connected and $f: X \to \{0, 1\}$ is continuous, then $f(X)$ is connected by 9.1.13, so that f is constant. Conversely, if X is not connected and $\{U, V\}$ is a disconnection of X, define $f: X \to \{0, 1\}$ by $f(x) = 1$ if $x \in U$ and $f(x) = 0$ if $x \in V$; then f is non-constant and continuous.

Q 9.2.1 The subset $S = \{z \in \mathbb{R}^2 \mid z_2 = 1\} \cup \{z \in \mathbb{R}^2 \mid z_1^{-1} \in \mathbb{N}\}$ of $\mathbb{R}^2$ is clearly polygonally connected, therefore connected (9.2.5). $(0,0) \in \overline{S}$. So $S' = S \cup \{0,0\}$ is also connected (9.1.7). Suppose $f: [0,1] \to S'$ is continuous with $f(0) = (0,0)$; let $t = \sup f^{-1}\{(0,0)\}$ and let B denote the open unit ball of $\mathbb{R}^2$. Because f is continuous, we have $f(t) = (0,0)$ and there is an open interval I of $\mathbb{R}$ with $t \in I$ and $f(I) \subseteq B \cap S'$. But $I \cap [0,1]$ is connected (9.1.5); so $f(I)$ is also connected (9.1.13). But, for each $n \in \mathbb{N}$, the line segment $B \cap \{z \in \mathbb{R}^2 \mid z_1 = 1/n\}$ is clearly a component of $B \cap S'$, so that $\{(0,0)\}$ is also a component of $B \cap S'$. So $f(I) = \{(0,0)\}$ and $t = 1$. Then $f(1) = (0,0)$ and f is not a path from $(0,0)$ to a point of S.

Q 9.2.2 T.

10. Convergence

Q 10.1.1 Being a filter, $\mathcal{F}$ is closed under non-trivial finite intersections; and, since it contains all supersets of its members, it is closed under arbitrary unions and has X as a member. So $\mathcal{F} \cup \{\varnothing\}$ is a topology on X. On the other hand, any set of three or more elements can be endowed with a topology in such a way that removing the empty set does not produce a filter: if $X = \{a, b, c\}$, then $\{\varnothing, \{a\}, X\}$ is a topology on X, but $\{\{a\}, X\}$ is not a filter.

Q 10.1.2 $\mathcal{U}$ being closed under intersection will suffice. Suppose $\mathcal{U}$ is so, and $A \in [\mathcal{U}]$; then $\exists B \in \mathcal{U}$ with $B \subseteq A$. Since $(X \backslash A) \cap B = \varnothing \notin \mathcal{U}$, $X \backslash A \notin \mathcal{U}$. So $A \in \mathcal{U}$.

Q 10.2.1 Every tail of a sequence includes a tail of every subsequence.

Q 10.2.2 Suppose $U \in \mathrm{nbd}(x)$. Since f is open, $f(U) \in \mathrm{nbd}(f(x))$; but $\mathcal{F} \to f(x)$, so $f(U) \in \mathcal{F}$. Since f is injective, we then have $U = f^{-1}(f(U)) \in f^{-1}(\mathcal{F})$.

Q 10.2.3 If U is an open subset of X with $z \in U$, then $\flat[z\,;\epsilon] \subseteq U$ for some $\epsilon \in \mathbb{R}^+$; if $\exists m \in \mathbb{N}$ such that for all $n \in \mathbb{N}$ with $m \leq n$, we have $d(x_n, z) < \epsilon$, then the mth tail of (x_n) is included in U. Conversely, if every open neighbourhood of z includes a tail of (x_n), then, in particular, for each $\epsilon \in \mathbb{R}^+$, the ball $\flat[z\,;\epsilon]$ has this property.

Q 10.2.4 Suppose (x_n) is a sequence in X which converges to $z \in X$. Then $(X \backslash \{x_n \mid n \in \mathbb{N}\}) \cup \{z\} \in \mathrm{nbd}(z)$; so $\{z\}$ is a tail of (x_n).

Q 10.2.5 x is continuous at ∞ if and only if every neighbourhood of x_∞ includes the image under x of some neighbourhood of ∞; such an image includes a tail of (x_n) by definition.

Q 10.2.6 In Hausdorff spaces, sequences have unique limits by 10.2.6. For the converse, suppose that $a, b \in X$ and that, for all $A \in \mathrm{nbd}(a)$ and $B \in \mathrm{nbd}(b)$, $A \cap B \neq \varnothing$. Suppose (U_n) and (V_n) are sequences whose terms form nested local bases at a and b respectively. By the Product Theorem, $\exists (x_n)$, a sequence in X such that $x_n \in U_n \cap V_n$ for each $n \in \mathbb{N}$. Then $x_n \to a$ and $x_n \to b$. If limits are unique, then $a = b$. It follows that distinct points of X have at least one pair of disjoint neighbourhoods; so X is Hausdorff.

Q 10.2.7 • Let $\pi: X \to X/S$ be the quotient map. Since S is closed, X/S is normed (6.2.25), so that $\pi(F)$, being a finite dimensional subspace of X/S, is closed in X/S by 10.2.26. Since π is continuous, $F + S = \pi^{-1}(\pi(F))$ is closed in X by Q 8.1.1.
• $\{z \in \mathbb{R}^2 \mid z_1 > 0, z_1 z_2 = 1\}$ is a closed subset of $\mathbb{R}^2$ and $\{z \in \mathbb{R}^2 \mid z_2 = 0\}$ is a one dimensional subspace of $\mathbb{R}^2$. But their sum is $\{z \in \mathbb{R}^2 \mid z_2 > 0\}$, and this is not closed in $\mathbb{R}^2$.

11. Compactness

Q 11.1.1 The relative topology on Z as a subspace of X is identical to the relative topology on Z as a subspace of Y (8.3.5). The result follows from 11.1.5.

Q 11.1.2 Suppose firstly that $|\mathcal{C}| < \infty$ and let $\mathcal{U}$ be a cover for $\mathcal{C}$; for each $A \in \mathcal{C}$, there is a finite subcover $\mathcal{F}_A$ of $\mathcal{U}$ for A; then $\bigcup\{\mathcal{F}_A \mid A \in \mathcal{C}\}$ is a finite subcover of $\mathcal{U}$ for $\bigcup\mathcal{C}$. For the second assertion: if each member of $\mathcal{C}$ is closed, then $\bigcap\mathcal{C}$ is a closed subset of a member of $\mathcal{C}$, so is compact (11.1.12).

Q 11.1.3 By the Product Theorem, $\exists(F_n)$, a sequence of finite subsets of X such that, for each $n \in \mathbb{N}$, $\{\flat[x\,;1/n] \mid x \in F_n\}$ covers X. Then $\bigcup(F_n)$ is dense in X. $|\bigcup(F_n)| \leq \infty$; so X is separable, then second countable by 7.3.3.

Q 11.1.4 The Cantor set is, by definition, an intersection of finite unions of closed intervals in $[0, 1]$, so is closed in the compact space $[0, 1]$, therefore compact (11.1.12).

Q 11.1.5 Suppose X is a normed linear space, F is a closed subset of X and K is a compact subset of X. Suppose $z \in \overline{F + K}$. There exist sequences (f_n) and (k_n) in F and K respectively such that $f_n + k_n \to z$. Then (k_n) has a subsequence which converges to some $w \in K$ (11.1.19). It follows that the corresponding subsequence of (f_n) converges to $z - w$. Since F is closed, $z - w \in F$ (10.2.17), whence $z \in F + K$.

Q 11.1.6 For each $z \in C$, set $\gamma_z = 1$ if $f(X) \subseteq \flat'[f(z)\,;\epsilon/2)$; otherwise set $\gamma_z = \text{dist}\big(z\,, X\backslash f^{-1}(\flat'[f(z)\,;\epsilon/2))\big)$.

Q 11.1.7 $[0, 1]^\mathbb{N}$ is compact in the product topology, but not in the topology of ℓ_∞.

Q 11.2.1 Suppose X is a second countable space. Let $\mathcal{B}$ be a countable base for the topology. Suppose $\mathcal{C}$ is an open cover for X. Let $\mathcal{S} = \{(B, C) \in \mathcal{B} \times \mathcal{C} \mid B \subseteq C\}$. Then $\text{dom}(\mathcal{S}) = \mathcal{B}$ and, by the Included Function Theorem, $\exists f: \mathcal{B} \to \mathcal{C}$ such that $B \subseteq f(B)$ for each $B \in \mathcal{B}$. Since $\mathcal{B}$ covers X, so does $f(\mathcal{B})$; but $f(\mathcal{B}) \subseteq \mathcal{C}$ and $|f(\mathcal{B})| \leq |\mathcal{B}| \leq \infty$.

Q 11.2.2 Suppose X is second countable. Then X is Lindelöf by Q 11.2.1; so, if X is countably compact, it is compact. The converse is trivial.

Q 11.2.3 There are easier proofs, but this follows from 11.2.6 and 11.2.9.

Q 11.2.4 Suppose K is a compact space, X is a Hausdorff space and $f: K \to X$ is a bijective continuous mapping. Let U be an open subset of K. Then $K\backslash U$, being closed, is compact by 11.1.12. So $f(K\backslash U)$ is compact by 11.1.8 and closed by 11.2.3. Since f is bijective, $f(U) = X\backslash f(K\backslash U)$; so $f(U)$ is open. Thus f is an open map.

Q 11.2.5 X is second countable by Q 11.1.3. Let $(B_n)_{n\in\mathbb{N}}$ be a sequence whose terms form a base for the topology of X. For each $z \in [0, 1]$ and $n \in \mathbb{N}$, let $\alpha_{z,n} = 2$ if $z = 1$ and $\alpha_{z,n}$ be the nth digit in the unique ternary expansion of z given in 4.3.16 otherwise; thus $z = \sum_{n\in\mathbb{N}} \alpha_{z,n}/3^n$. Check that $C = \{z \in [0, 1] \mid \forall n \in \mathbb{N}, \alpha_{z,n} \in \{0, 2\}\}$. For each $z \in C$, set $\phi_0(z) = X$ and then define recursively, for each $n \in \mathbb{N}$, $\phi_n(z) = \overline{B_n}$ if both $\alpha_{z,n} = 2$ and $\overline{B_n} \cap \bigcap\{\phi_i(z) \mid 0 \leq i < n\} \neq \varnothing$, and $\phi_n(z) = X\backslash B_n$ otherwise. Since each $\phi_n(z)$ is closed in X and therefore compact Hausdorff, 11.2.5 ensures that $\bigcap(\phi_n(z))$ is non-empty; and, since X is Hausdorff, it is a singleton set. Define $f: C \to X$ by setting $f(z)$ to be the unique member of $\bigcap(\phi_n(z))$ for each $z \in C$. Next we show that f is surjective: for each $x \in X$ and $n \in \mathbb{N}$, let $\beta_n = 0$ if $x \notin B_n$ and $\beta_n = 2$ otherwise; let $z = \sum \beta_n 3^{-n}$; then $z \in C$ and $x \in \bigcap(\phi_n(z))$, so that $f(z) = x$. Now we show that f is continuous. Suppose $z \in C$ and $U \in \text{nbd}(f(z))$. Then $(X\backslash U) \cap \bigcap(\phi_n(z)) = \varnothing$ and, by 11.2.5, $\exists k \in \mathbb{N}$ such that $(X\backslash U) \cap \bigcap\{\phi_n(z) \mid n \in \mathbb{N}, 1 \leq n \leq k\} = \varnothing$. If $z = 1$, let $m = k + 1$; otherwise, by the definition of the ternary expansion, $\exists m \in \mathbb{N}$ with $k < m$ such that $\alpha_{z,m} = 0$. Then, if $y \in C \cap (z - 3^{-m}, z + 3^{-m})$, we have $\alpha_{y,i} = \alpha_{z,i}$ when $1 \le i \le k$, so that $f(y) \in U$, establishing continuity of f at z. Since z is arbitrary, f is a continuous function.

Q 11.2.6 Suppose $f \in \mathcal{C}_0(X)$ and let S be as in Machado's Lemma. Let $p, q \in S$; then $|u(p)| = |u(q)|$ for all $u \in A$. So $\ker(\hat{p}) = \ker(\hat{q})$, where the domain of the point evaluation functions is A. Suppose $g \in A$ is such that $g(p) \ne 0$; then $|g| \in A$ and $|g|(q) = |g|(p) \ne 0$; so $A = \ker(\hat{p}) \oplus \mathbb{K}|g|$ and $\hat{q} = \hat{p}$. So $q = p$, for otherwise there would be some $h \in A$ with $h(p) \ne h(q)$, contradicting $\hat{q} = \hat{p}$. Therefore $S = \{p\}$. With $\alpha = f(p)/g(p)$, we have $\alpha g \in A$ and $\|f - \alpha g\|_S = 0$, so that, by Machado's Lemma, $\mathrm{dist}_X(f, A) = 0$. Since A is closed, $f \in A$.

Q 11.3.1 Every space, locally compact or not, is a subspace of a compactification.

Q 11.3.2 Suppose $x \in S$. Then $\exists V \in \mathrm{nbd}_X(x)$ such that $\mathrm{Cl}_X(V)$ is compact.
- If S is open, then, since $\{x\}$ is compact, 11.3.5 ensures that V can be chosen such that $x \in V \subseteq Cl_S V \subseteq \mathrm{Cl}_X(V) \subseteq S$.
- If S is closed, then $V \cap S \in \mathrm{nbd}_S(x)$ and $\mathrm{Cl}_S(V \cap S) = \mathrm{Cl}_X(V \cap S)$ is a closed subset of the compact space $\mathrm{Cl}_X(V)$, so is compact by 11.1.12.

Q 11.3.3 If $y \in X \backslash \{x\}$, then there exist $U \in \mathrm{nbd}_X(x)$ and $V \in \mathrm{nbd}_X(y)$ with $U \cap V = \varnothing$. Then $X \backslash U$ is closed in X, therefore compact. $V \in \mathrm{nbd}_{X \backslash \{x\}}(y)$ and $\mathrm{Cl}_{X \backslash \{x\}}(V) \subseteq X \backslash U$ is compact by 11.1.12.

Q 11.3.4 • Use the Product Theorem to associate with each irrational real number z a sequence $s(z)$ of rational numbers which converges in the usual topology to z. For each $n \in \mathbb{N}$ and $z \in \mathbb{R} \backslash \mathbb{Q}$, let $t_n(z)$ denote the nth tail of $s(z)$. The set $\{\{q\} \mid q \in \mathbb{Q}\} \cup \{\{z\} \cup s_n(z) \mid z \in \mathbb{R} \backslash \mathbb{Q}, n \in \mathbb{N}\}$ is a base for a Hausdorff topology on $\mathbb{R}$, called a RATIONAL SEQUENCE TOPOLOGY. A straightforward check establishes that all these basic open sets are both closed and compact, so that $\mathbb{R}$ with a rational sequence topology is a locally compact Hausdorff space; moreover $\mathbb{Q}$ is dense in $\mathbb{R}$ with this topology. But $\mathbb{R} \backslash \mathbb{Q}$ is an uncountable closed discrete subspace of this space So $\mathbb{R}$ with a rational sequence topology is not normal by Q 8.3.5.
• The non-normal $\mathbb{R}$ with the rational sequence topology is locally compact Hausdorff; a one-point compactification is Hausdorff by 11.3.3 and so normal by 11.2.2.

Q 11.3.5 Suppose X is a normed linear space. If X is locally compact, then 0 has a neighbourhood whose closure is compact. So $\exists \epsilon \in \mathbb{R}^+$ such that $\epsilon b[0\,;1)$ has compact closure; it follows easily that $b[0\,;1]$ is compact, so that $\dim(X) < \infty$ by 11.2.6. Conversely, if X is finite dimensional, then $b[0\,;1]$ is compact by 11.1.21 and, for each $x \in X$, $x + b[0\,;1)$ is a neighbourhood of x whose closure is compact.

12. Completeness

Q 12.1.1 Suppose (X, d) is a metric space and S is a discrete subspace of X. Suppose $z \in X$ and $\mathrm{dist}(z, S) = 0$. Let $w \in S$ be such that $d(w, z) < 1/2$; then we have $d(z, s) \ge d(s, w) - d(w, z) > 1/2$ for all $s \in S \backslash \{w\}$; so $d(z, w) = 0$ and $z = w \in S$.

Q 12.1.2 $\mathbb{R} \backslash \{0\}$.

Q 12.1.3 $(0, 1]$ is the image of $[1, \infty)$ under $x \mapsto 1/x$.

Q 12.2.1 In fact, $\mathcal{C}_0(\mathbb{R})$ is a completion of $\mathcal{C}_c(\mathbb{R})$. Suppose $\epsilon \in \mathbb{R}^+$ and $f \in \mathcal{C}_0(\mathbb{R})$. There exists a compact subset K of $\mathbb{R}$ such that $|f(x)| < \epsilon$ for all $x \in \mathbb{R} \backslash K$. Define $g \colon \mathbb{R} \to \mathbb{R}$ by $g(x) = f(x)$ if $x \in K$ and $g(x) = 0$ otherwise. Then $g \in \mathcal{C}_c(\mathbb{R})$ and $\|f - g\| \le \epsilon$. Since ϵ is arbitrary, $\mathrm{dist}(f, \mathcal{C}_c(\mathbb{R})) = 0$. Since f is arbitrary, $\mathcal{C}_c(\mathbb{R})$ is dense in $\mathcal{C}_0(\mathbb{R})$. The function given by $f(x) = 1/|x|$ if $x \in \mathbb{R} \backslash [-1, 1]$ and $f(x) = 1$ otherwise is in $\mathcal{C}_0(\mathbb{R}) \backslash \mathcal{C}_c(\mathbb{R})$, so the sets are distinct and $\mathcal{C}_c(\mathbb{R})$ is not complete.

Q 12.2.2 Let $\epsilon \in \mathbb{R}^+$. For each $n \in \mathbb{N}$, define $S_n = \{x \in X \mid f(x) - f_n(x) < \epsilon\}$; then (S_n) is increasing with respect to inclusion because (f_n) is increasing and $f_n(x) \to f(x)$ for each $x \in X$. That each S_n is open is shown as follows. Suppose $x \in S_n$ and set $\alpha = f(x) - f_n(x)$. Then $\alpha \in [0, \epsilon)$. Since both f_n and f are continuous, there exists $\delta \in \mathbb{R}^+$ such that, for each $y \in X$ with $d(x, y) < \delta$, we have both $|f_n(x) - f_n(y)| < (\epsilon - \alpha)/2$ and $|f(x) - f(y)| < (\epsilon - \alpha)/2$. Then $|f(y) - f_n(y)| \leq |f(y) - f(x)| + |f(x) - f_n(x)| + |f_n(x) - f_n(y)| < \epsilon$, so that $y \in S_n$. Now, since (f_n) converges pointwise to f, the sequence (S_n) of open sets forms a cover for X. The compactness of X ensures that this cover has a finite subcover. But (S_n) is increasing. So $\exists k \in \mathbb{N}$ such that $X = S_k$. Therefore, for every $x \in X$ and for every $n \in \mathbb{N}$ with $n \geq k$, we have $|f(x) - f_n(x)| < \epsilon$. So (f_n) converges uniformly to f.

Q 12.2.3 Suppose T is bounded and $(x, y) \in \overline{\Gamma(T)}$. Let $\epsilon \in \mathbb{R}^+$. The neighbourhood $\flat_X[x \,; \epsilon] \times \flat_Y[y \,; \epsilon]$ of (x, y) in $X \times Y$ contains some member (a, Ta) of $\Gamma(T)$. Then $\|y - Tx\| \leq \|y - Ta\| + \|T\| \|a - x\| < \epsilon(1 + \|T\|)$. Since ϵ is arbitrary, $y = Tx$ and $(x, y) \in \Gamma(T)$. For the converse, endow $X \times Y$ with the complete norm $(x, y) \mapsto \max\{\|x\|, \|y\|\}$ which determines the product topology. Suppose $\Gamma(T)$ is closed in $X \times Y$; then $\Gamma(T)$ is complete by 12.1.8. The projection $(x, Tx) \mapsto x$ of $\Gamma(T)$ onto X is a bijective bounded linear map. By 12.2.9, it has bounded inverse, so that T is bounded.

Q 12.2.4 For each $n \in \mathbb{N}$, the set $F_n = \bigcap\{T^{-1}(\flat_{\text{codom}(T)}[0 \,; n]) \mid T \in \mathcal{B}\}$ is closed because all members of $\mathcal{B}$ are continuous. By hypothesis, $\bigcup(F_n) = X$. By Baire's Theorem and Q 7.2.7, $\exists n \in \mathbb{N}$ such that F_n includes a ball $\flat[a \,; r]$ of X. Therefore $Ta + T(\flat[0 \,; r]) \subseteq \flat_{\text{codom}(T)}[0 \,; n]$ for all $T \in \mathcal{B}$. So, for all $x \in \flat_X[0 \,; 1)$ and $T \in \mathcal{B}$, we have $\|Tx\| \leq (n + \sup\{\|Ta\| \mid T \in \mathcal{B}\})/r$.

Q 12.3.1 Suppose Y is a completion of X and ϕ is an isometric isomorphism from X onto a dense subspace of Y. Suppose $a, b \in Y$. Continuity of the inner product on X ensures that, if (p_n) and (q_n) are sequences in X for which $\phi(p_n) \to a$ and $\phi(q_n) \to b$, then $\lim\langle p_n, q_n \rangle$ is well defined and independent of the particular sequences (p_n) and (q_n) with the stated properties. Define $\langle a, b \rangle = \lim\langle p_n, q_n \rangle$ and check that it is an inner product which determines the metric.

Q 12.3.2 • Suppose (x_n) and (y_n) are sequences in M and N respectively such that $(x_n + y_n)$ converges to $z \in H$. Then, since M and N are orthogonal, we have $\|x_n - x_m\|^2 + \|y_n - y_m\|^2 = \|(x_n + y_n) - (x_m + y_m)\|^2$, so that (x_n) and (y_n) are Cauchy sequences. Since M and N are closed in H, they are complete; so there exist $x \in M$ and $y \in N$ such that $x_n \to x$ and $y_n \to y$. Then $z = x + y \in M + N$.

• Let $\{e_n \mid n \in \mathbb{N}\}$ be the orthonormal basis for ℓ_2 given in 12.3.8. Consider the sets $A = \{x \in \ell_2 \mid x_{2n} = 0 \; \forall n \in \mathbb{N}\}$ and $B = \{x \in \ell_2 \mid x_{2n} = x_{2n-1}/2^n \; \forall n \in \mathbb{N}\}$. It is easy to check that A and B are closed subspaces of ℓ_2. For each $n \in \mathbb{N}$, let $a_n = \sum_{i=1}^{n} -e_{2i-1}$ and $b_n = \sum_{i=1}^{n}(e_{2i-1} + e_{2i}/2^i)$; then $a_n \in A$ and $b_n \in B$. Let $c = \sum_{i=1}^{\infty} e_{2i}/2^i$. Then $c \in \ell_2 \backslash (A + B)$; but $a_n + b_n \to c$ in ℓ_2, so that $c \in \overline{A + B}$.

Q 12.3.3 Suppose H is a Hilbert space whose algebraic dimension is infinite. For each finite subset F of H, $\langle F \rangle$ is a proper closed subspace of H; it includes no ball of H, so is nowhere dense. Since H is complete, Baire's Theorem and Q 7.2.7 ensure that no countable union of such subspaces equals H.

Bibliography

First to possess his books; for without them
He's but a sot, as I am, nor hath not
One spirit to command   *The Tempest, III,ii.*

This book was written for the most part without reference to earlier presentations of the same material. Some ideas, examples and exercises were, however, suggested by the following works. In particular, the presentation of Set Theory in Chapter 1 owes much to the excellent discussions in [3].

[1] Bernard Aupetit. *A Primer on Spectral Theory*. Springer-Verlag, 1991.

[2] Garrett Birkhoff and Saunders Mac Lane. *A Survey of Modern Algebra*. Macmillan, third edition, 1965.

[3] A. Fraenkel and Y. Bar-Hillel. *Foundations of Set Theory*. North Holland, second edition, 1973.

[4] Paul R. Halmos. *A Hilbert Space Problem Book*. Springer-Verlag, second edition, 1982.

[5] Robin E. Harte. *Invertibility and Singularity*. Marcel Dekker, 1988.

[6] Harro G. Heuser. *Functional Analysis*. Wiley, 1982.

[7] G.J.O. Jameson. *Topology and Normed Spaces*. Chapman and Hall, 1974.

[8] John L. Kelley. *General Topology*. Springer-Verlag, second edition, 1975.

[9] Gerard J. Murphy. *C^*-algebras and Operator Theory*. Academic Press, 1990.

[10] Gert K. Pedersen. *Analysis Now*. Springer-Verlag, 1989.

[11] Walter Rudin. *Real and Complex Analysis*. McGraw-Hill, second edition, 1974.

[12] George F. Simmons. *Introduction to Topology and Modern Analysis*. McGraw-Hill, 1963.

[13] Lynn Arthur Steen and J. Arthur Seebach. *Counterexamples in Topology*. Springer-Verlag, second edition, 1978.

Index

*O, there is a nobleman in town, one Paris, that would fain
lay knife aboard; but she, good soul, had as lief
see a toad, a very toad, as see him. I anger her
sometimes and tell her that Paris is the properer
man; but, I'll warrant you, when I say so, she looks
as pale as any clout in the versal world. Doth not
rosemary and Romeo begin both with a letter?*

Romeo and Juliet, II,iv.

A SERIES OF TEXTBOOKS FOR UNDERGRADUATES IN MATHEMATICAL SCIENCES

SPRINGER

UNDERGRADUATE

MATHEMATICS

SERIES

SUMS is a series for undergraduates in the mathematical sciences. From core foundational material to final year topics, **SUMS** books take a fresh and modern approach and are ideal for self-study for a one- or two-semester course.

SUMS books provide the student with:

⊕ a user-friendly approach with a clear and logical format;

⊕ plenty of examples, problems and fully-worked solutions;

⊕ up-to-date material which can be used either to support a course of lectures or for self-study.

All books are, of course, available from all good booksellers (who can order them even if they are not in stock), but if you have difficulties you can contact the publishers direct, by telephoning +44 (0) 1483 418822 (in the UK & Europe), +1/212/460/1500 (in the USA), or by emailing orders@svl.co.uk

www.springer.de   www.springer-ny.com

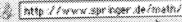